Principles of Highway Engineering and Traffic Analysis

Second Edition

Principles of Highway Engineering and Traffic Analysis

Second Edition

Fred L. Mannering
University of Washington

Walter P. Kilareski
The Pennsylvania State University

John Wiley & Sons, Inc.
New York / Chichester / Weinheim / Brisbane / Singapore / Toronto

COVER PHOTO	Bob Anderson/Masterfile
ACQUISITIONS EDITOR	Sharon Smith
MARKETING MANAGER	Harper Mooy
PRODUCTION EDITOR	Ken Santor
COVER DESIGNER	Madelyn Lesure
ILLUSTRATION COORDINATOR	Gene Aiello

This book was set in Times Roman by Bi-Comp, Inc., and printed and bound by Malloy Lithographing, Inc. The cover was printed by Phoenix Color Corporation.

Recognizing the importance of preserving what has been written, it is a policy of John Wiley & Sons, Inc. to have books of enduring value published in the United States printed on acid-free paper, and we exert our best efforts to that end.

The paper on this book was manufactured by a mill whose forest management programs include sustained yield harvesting of its timberlands. Sustained yield harvesting principles ensure that the number of trees cut each year does not exceed the amount of new growth.

Library of Congress Cataloging in Publication Data:
Mannering, Fred L.
   Principles of highway engineering and traffic analysis / Fred L.
Mannering, Walter P. Kilareski.—2nd ed.
     p. cm.
   Includes bibliographical references and index.
   ISBN 0-471-13085-0 (pbk. : alk. paper)
   1. Highway engineering. 2. Traffic engineering. I. Kilareski,
Walter P. II. Title.
TE147.M28 1998
625.7—dc21
                                                                      97-3077
                                                                         CIP

Printed in the United States of America

10 9 8 7 6 5 4

Preface

The first edition of *Principles of Highway Engineering and Traffic Analysis* sought to redefine how entry-level transportation engineering courses are taught. When the first edition was published, we saw the need for an entry-level transportation engineering book that focused on highway transportation and provided (1) the depth of coverage needed to serve as a basis for future transportation courses, and (2) the material needed to answer questions likely to appear on the exam for professional registration in civil engineering. The subsequent use of our book at some of the largest and most prestigious civil engineering schools in the United States suggests that our vision of a highly focused, well-written, entry-level book was shared by many other educators.

In this second edition of *Principles of Highway Engineering and Traffic Analysis* we continue the spirit of the first edition by again focusing exclusively on highway transportation, and by providing the depth of coverage necessary to solve the highway-related problems that are most likely to be encountered in engineering practice. The focus on highway transportation seems a natural one in light of the dominance of the highway mode in the United States.

Using the first edition as a basis, along with the comments of other instructors and students, we have carefully selected topics that are fundamental to highway engineering and traffic analysis. This material serves as a core upon which instructors can expand, if they wish, with material that they personally feel deserves additional attention. The core of material provided by the book, however, is important for effective teaching because instructors can be confident that students are learning the fundamentals needed to undertake upper-level transportation courses, enter transportation employment with a basic knowledge of highway engineering and traffic analysis, and answer transportation-related questions on the civil engineering professional registration exam.

This book also addresses the complaint of some students that transportation is not as mathematically challenging or rigorous as other civil engineering disciplines. This is not easily done because there is a dichotomy in transportation engineering with regard to mathematical rigor, with relatively simple mathematics used in practice-oriented material and complex mathematics used in research. Thus it is common for instructors either to insult students' mathematical knowledge or to exceed it. This book makes an effort to find the elusive middle ground of mathematical rigor that matches junior and senior engineering students' mathematical abilities.

This edition of *Principles of Highway Engineering and Traffic Analysis* has evolved from well over a decade of teaching introductory transportation engineering classes at the University of Washington and The Pennsylvania State University, feedback from users of the first edition, and experiences in teaching professional engineering exam review courses. The book's material and presentation style (which is characterized by the liberal use of example problems) are largely responsible for transforming a much-maligned introductory transportation engineering course into a course that consistently rates among the best civil engineering courses, as rated by students. We are hopeful that our favorable experiences with this material can now be shared with others.

The book begins with a short introductory chapter that stresses the significance of highway transportation in U.S. society. This chapter provides students with a basic overview of the problems facing the field of highway engineering and traffic analysis. The chapters that follow are arranged in sequences that focus on highway engineering (Chapters 2, 3, and 4) and traffic analysis (Chapters 5, 6, 7, and 8).

Chapter 2 introduces the basic elements of road vehicle performance. This chapter represents a major departure from the vehicle performance material presented in all other transportation engineering books, in that it is far more involved and detailed. The additional level of detail is justified on two grounds. First, because students own and drive automobiles, they have a basic interest that can be linked to their freshman and sophomore coursework in physics, statics, and dynamics. Traditionally, the absence of such a link has been a common criticism of introductory transportation engineering courses. Second, with continuing advances in vehicle technology, it is more important than ever that transportation engineering students understand the principles involved and, ultimately, the effect that rapidly changing vehicle technologies will have on highway engineering practice.

Chapter 3 presents current design practices for the geometric alignment of highways. This chapter provides a high level of detail on vertical-curve design and the basic elements of horizontal-curve design. Basic material on highway classifications, turning radii, and so on is not included due to the great potential for alienating introductory level students with too many design-oriented details. Such material rarely, if ever, appears on professional registration exams and can be readily learned in practice if necessary.

Chapter 4 overviews the current theory and practice of pavement design and provides a basic introduction to highway drainage. The pavement-related material covers both flexible and rigid pavements in a thorough and consistent manner. The material in this chapter also links well with soils, materials, and hydrology courses that are likely to be part of a student's curriculum.

Chapter 5 provides an introduction to the basic tools of traffic analysis including traffic flow, speed, and density models, as well as queuing theory. Considerable effort was expended to make the material in this chapter accessible to junior and senior engineering students.

Chapter 6 applies the traffic analysis tools introduced in Chapter 5 to signalized intersections. Using pretimed signals as the basis, this chapter presents both theoretical and practical elements associated with traffic signal timing.

Chapter 7 presents some of the current techniques used to assess highway level of service. Fundamentals are discussed along with the complexities involved in measuring highway level of service.

Chapter 8, the final chapter, provides an overview of traffic forecasting. This chapter concentrates on a theoretically and mathematically consistent approach to traffic forecasting that closely follows the approach most commonly used in practice. The material provides students with an important understanding of the current state of traffic forecasting, and some critical insight into the deficiencies of forecasting methods currently used.

Readers will note this book's extensive use of metric units. The U.S. Federal Highway Administration (FHWA) has ruled that all highway products that receive federal aid must be in metric units after 1999, and almost all state transportation departments have developed plans to convert from U.S. customary units to metric. This transition has required that national guidelines, such as those published by the American Association of State Highway and Transportation Officials (AASHTO), be converted to metric units. Unfortunately, the conversion of all such guidelines is not yet complete. As a result, we have used metric units in all areas except those where the profession has not yet completed the conversion (pavement design and highway level of service). In these areas we provide both U.S. customary and metric units. Also, because many existing highway designs in the United States are in U.S. customary units, we have provided a U.S. customary-units version of Chapter 3 (Geometric Design of Highways) in an Appendix to assist in understanding existing designs and the transition to metric designs.

Finally, users of the book will find the end-of-chapter problems to be extremely useful in supporting the material presented in the book. This second edition includes many more end-of-chapter problems (relative to the first edition), and all problems continue to be precise and challenging.

Fred L. Mannering
Walter P. Kilareski

Contents

Appendix A
Geometric Design of Highways: U.S. Customary Units   293

Appendix B
Unit Conversions   333

Index   337

Chapter 1

Introduction to Highway Engineering and Traffic Analysis

1.1 INTRODUCTION

Highway transportation is a critical underpinning of the industrial and technological complex of the United States. Virtually every aspect of the U.S. economy and way of life is tied directly or indirectly to highways. From the movement of freight and people to the impact on residential, commercial, and industrial locations, highways have had, and continue to have, a profound effect on American society. The manner in which highways have come to dominate the U.S. transportation system has been studied for years as a cultural, political, and economic phenomenon. Without doubt the demand for unrestricted mobility and unlimited access to the country's resources played an important role and helped to quickly move highway transportation to its dominant position from the middle of the twentieth century onward. The construction of the U.S. Interstate Highway System in the 1960s and 1970s (the largest civil engineering project ever undertaken in the history of mankind), which was an outgrowth of this cultural, political, and economic phenomenon, had a profound impact on the American way of life and serves as a monument to the nation's overwhelming commitment to highway transportation.

Given the significant impact that highway transportation has on American society, it is important for highway engineers to strive toward two goals: (1) providing a high level of service (i.e., seek to minimize travel times and delays), and (2) providing a high level of safety. These two goals are not only often contradictory (e.g., higher speeds minimize travel time but may also decrease safety), but must be achieved in the context of ever-changing constraints. Such constraints can be broadly classified as economic (the cost of highway-related projects), political (the community-related impacts of projects), and environmental (the impact of projects on the environment measured in terms of air, water, and noise impacts, and quality of life). As a further complicating concern, engineers must also address the likely short- and long-term impacts of highway-related projects on vehicle traffic, which is an outgrowth of traveler behavior.

Although attempting to provide higher and higher levels of highway service and safety seems an impossible task at times, engineers have an ethical responsibility to undertake this task. In so doing, they face challenges that are both technical and behavioral in nature. In the following sections of this chapter we discuss the important dimensions of these challenges.

1.2 TECHNOLOGICAL CHALLENGES

As with all fields of engineering, technological developments offer the promise of solving complex problems and achieving lofty goals. The quest for technological development or, equivalently, the technological challenges that face highway transportation include challenges relating to infrastructure, vehicle technologies, and traffic-control technologies.

1.2.1 Infrastructure

The United States made an extraordinary capital investment in highways during the 1960s and 1970s by constructing the Interstate Highway System and upgrading and constructing many other highways. The economic and political climate that permitted such an ambitious construction program was quite unusual, and it is unlikely that such conditions will ever exist again. For example, in today's economic and political environment, if a project of the magnitude of the interstate system was proposed, it would be readily dismissed as infeasible due to the costs associated with land acquisition and construction and the community and environmental impacts. Indeed, the United States was extremely fortunate to have constructed the interstate system at a time when technology, economics, and politics favored such endeavors.

It is important, however, to realize that the costs associated with a highway system go well beyond the initial construction costs. That is, although highways can be viewed as durable, long-lasting investments, they still require maintenance and rehabilitation at fairly regular intervals. Most of the U.S. Interstate Highway System was designed with pavements that were expected to last 25 to 30 years before major rehabilitation was necessary. Thus an unfortunate consequence of the extensive interstate-construction programs of the 1960s and 1970s is that massive rehabilitation is needed in the late 1990s and beyond. Although there are many compelling reasons to defer rehabilitation (including associated construction costs and the impact of reconstruction on traffic), such deferral can result in unacceptable losses in levels of service and safety.

The challenge confronting engineers in the infrastructure rehabilitation and maintenance area is to develop new techniques and technologies to economically combat aging highway infrastructure. Such developments are needed in the areas of reconstruction practices and productivity, environmental impact assessment, materials research, traffic control, and vehicle routing and work-zone safety during reconstruction.

1.2.2   Vehicle Technologies

Until the 1970s, vehicle technologies evolved rather slowly and often in response to gradual trends in the vehicle market as opposed to an underlying trend toward technological development. Beginning in the 1970s, four factors began a cycle of unparalleled advances in vehicle technology that continues to this day: (1) government regulations concerning air quality and vehicle occupant safety, (2) energy shortages and fuel-price increases, (3) government regulations mandating fuel efficiency increases, and (4) intense competition from foreign vehicle manufacturers. The aggregate effect of these factors has resulted in consumers who demand new technology. Vehicle manufacturers have found it necessary to reallocate resources and restructure manufacturing and inventory control processes to meet this demand. In recent years, consumer demand and competition among vehicle manufacturers has resulted in the widespread implementation of new technologies (e.g., safety-related items such as air bags and antilock brakes). There is little doubt that this trend will continue.

Although the development of new vehicle technologies is usually in the domain of disciplines such as mechanical and/or electrical engineering, the influence of such technologies on highway design and traffic analysis is an important concern for the highway engineer. Highway engineers must be able to account for new technologies in the design and rehabilitation of highways, in their ongoing effort to provide the highest possible level of service and safety.

1.2.3   Traffic Control Technologies

Intersection traffic signals are a familiar traffic control technology. At signalized intersections, the goals of providing the highest possible level of service and safety are brought into sharp focus. Procedures for setting traffic signals (allocating available green times to conflicting traffic movements) have made significant advances over the years. Today, many signals respond to prevailing traffic flows, groups of signals are sequenced to allow for a smooth through-flow of traffic, and, in some cases, computers control entire networks of signals. Still, a simple drive down virtually any highway in the country with a poorly timed signalized intersection underscores the need for further advancement in traffic control technologies.

In recent years, an influx of technological development has been stimulated by the U.S. government and defense contractors. Numerous safety, navigational, and congestion-mitigating technologies are now reaching the market under the broad heading of Intelligent Transportation Systems (ITS). Such technological efforts, in theory, offer the potential to significantly reduce traffic congestion and improve safety on highways by providing an unprecedented level of traffic control. There are, however, many obstacles associated with ITS implementation, including system reliability and human–machine interface concerns. It remains to be seen whether the ITS effort will be cost effective.

Highway engineers must participate in the development and implementation of the many new traffic control technologies that will undoubtedly appear in the coming years. To do this, a firm understanding of the principles of highway engineering and traffic analysis is needed.

1.3 BEHAVIORAL CHALLENGES

Highway traffic–the volume of which directly influences highway levels of service and safety–is a behavioral phenomenon that is an outgrowth of people's travel-related choices. In recent years, these choices have resulted in traffic volumes that have produced crippling congestion in many urban areas. Despite such congestion, the trend in traffic growth continues upward. The expected supply-demand effect (i.e., a limited, congested highway infrastructure acting to restrain the growth of private-vehicle usage) has simply not been a significant factor in urban traffic congestion. It has been found that people are willing to tolerate high amounts of congestion just to continue to use private vehicles, most of which have only a single occupant. The continued growth in traffic congestion is a serious obstacle to the goal of providing higher levels of service.

Addressing the issue of future traffic growth is a perplexing problem. The current economic and political environment does not favor large-scale construction programs aimed at increasing highway capacity. Technological innovations in traffic control have the potential to offer some relief, but it is questionable whether such innovations can keep pace with the rapid growth in traffic volumes. This points toward the more fundamental issue of the behavioral processes and preferences that generate highway traffic.

1.3.1 Dominance of Single-Occupant Private Vehicles

Of the available urban transportation modes (e.g., bus, commuter train, subway, and private vehicle), private vehicles in general, and single-occupant private vehicles in particular, offer a level of travel mobility that is unequaled. The private vehicle is such an overwhelmingly dominant choice that travelers are willing to pay substantial capital and operating costs, confront high levels of congestion, and struggle with parking-related problems, just to have the flexibility of travel departure time and destination choices uniquely provided by private vehicles. The dominance of the private vehicle mode is supported by historical data that show continuing increased use. For example, from 1960 to the present day, the percentage of trips taken in private vehicles has risen from about 69% to 90% (public transit, walking, and other modes make up the balance). Over this same period, the average private-vehicle occupancy rate (average number of persons in the vehicle) has dropped from 1.22 to 1.12, showing that the single-occupant vehicle has been and continues to be the preferred mode of travel.

Dealing with the issues of growth in private-vehicle use and low vehicle occupancy presents engineers with a classic dilemma in striving toward the goal of providing higher levels of service. On one hand, programs that encourage travelers to take alternate modes of transportation (e.g., bus-fare incentives, and increases in private-vehicle parking fees) or to increase their vehicle occupancy (e.g., high-occupancy vehicle lanes, and employer-based ridesharing programs) have the potential to relieve traffic congestion on highways and provide remaining highway users with a higher level of service. On the other hand, such programs have the adverse effect of directing travelers toward inferior modes that inherently provide lower levels of mobility and, consequently, service (i.e., no other mode offers the departure-time and destination-choice flexibility provided by the private, single-occupant vehicle). With this dilemma in mind, it is clear that controversial congestion-related compromises must be reached.

1.3.2 General Demographic Trends

Travelers' commuting patterns are inextricably interwined with their socioeconomic characteristics, such as age, income, household size, education, and job type, as well as the distribution of residential, commercial, and industrial developments within a given region. Many American metropolitan areas have experienced population declines in central cities coupled with rapid growth in suburban areas. In many respects, the population shift from central cities to the suburbs was made possible by the increased mobility provided by major highway projects undertaken during the 1960s and 1970s. This mobility enabled Americans to improve their quality of life by gaining access to affordable housing and land, while still being able to work in the central city. Conventional wisdom suggested that as overall metropolitan traffic congestion grew, thus making the suburb-to-city commuting pattern much less attractive, commuters would seek to avoid traffic congestion by reverting back to public transport modes and/or once again choosing to reside in the central city. Another trend, however, has emerged. Employment centers have developed in the suburbs and now provide a viable alternative to the suburb-to-city commute. The result is a continuing tendency toward low-density, private-vehicle-based development as Americans seek to retain the high quality of life associated with such development. Ongoing demographic trends present engineers with an ever-moving target that further complicates the problem of providing high levels of service and safety.

In addition to shifts in residential and employment-based development, there are long-term population/behavioral concerns that must also be considered. For example, due to the distribution of births (i.e., the baby boom following the Second World War) and advances in medical technology that prolong life, the average age of the U.S. population continues to increase. Because older people tend to have slower reaction times (i.e., they take longer to respond to driving situations that require action), engineers must confront the possibility of changing highway design guidelines and practices to accommodate slower reaction times and the potentially higher variance of reaction times among highway users.

1.4 HIGHWAY SAFETY

Highway safety involves both technical and behavioral components. Since the 1960s, considerable investment has been made to improve highway safety by setting new highway design guidelines and developing countermeasures (some technical and some behavioral) aimed at reducing the frequency and severity of highway accidents. This effort sought to reverse an upward trend in U.S. highway fatalities and injuries that saw highway fatalities exceed 50,000 per year as recently as the 1970s. Fortunately, efforts to improve highway design (e.g., more stringent design guidelines, and breakaway signs), vehicle occupant protection (e.g., safety belts, padded dashboards, collapsible steering columns, driver- and passenger-side airbags, and improved bumper design), and vehicle accident avoidance (e.g., antilock braking systems), combined with accident countermeasures (e.g., campaigns to reduce drunk driving), have reduced U.S. highway fatalities to less than 40,000 per year, but it is unclear as to whether this downward trend will continue. Still, this high loss of life is a steep price to pay for the mobility that highways provide. Many of these deaths can be attributed to behavioral concerns that are beyond the control of the engineering profession (e.g., more than 40% of highway fatalities involve alcohol consumption). Nevertheless, it is the engineer's responsibility to continually seek ways to reduce the frequency and severity of highway accidents within existing economic, political, and behavioral constraints.

1.5 MEETING THE CHALLENGE

As the preceding sections show, the highway engineering and traffic analysis problem is exceedingly complex and virtually impossible to solve. In spite of its insolvable nature, or perhaps because of it, this problem provides a challenge that is simply unequaled by any other engineering-related problem. Meeting this challenge requires the full use of engineers' mathematical and technological expertise as well as the ability to develop imaginative solutions to subsets of the problem. The intent of this book is to provide a fundamental background that will enable engineering students to better understand and to begin to fully appreciate the many challenging aspects of highway engineering and traffic analysis.

Chapter 2

Road Vehicle Performance

2.1 INTRODUCTION

The performance of road vehicles forms the basis for highway design guidelines and traffic analysis. For example, in highway design, determination of the length of freeway acceleration and deceleration lanes, maximum highway grades, stopping-sight distances, passing-sight distances, and numerous accident prevention devices all rely on a basic understanding of vehicle performance. Similarly, vehicle performance is a major consideration in the selection and design of traffic control devices, determination of speed limits, and timing and control of traffic signal systems.

Studying vehicle performance serves two important functions. First, it provides insight into highway design and traffic operations and the compromises that are necessary to accommodate the wide variety of vehicles (from high-powered sports cars to heavily ladened trucks) that use highways. Second, it forms a basis on which to assess the impact of advancing vehicle technologies on existing highway design guidelines. This second function is particularly important in light of the ongoing unprecedented advances in vehicle technology. Such advances will necessitate more frequent updating of highway design guidelines, as well as engineers who have a better understanding of the fundamental principles underlying vehicle performance.

The objective of this chapter is to introduce the basic principles of road vehicle performance. In so doing, primary attention will be given to the straight-line performance of vehicles (acceleration, deceleration, top speed, and the ability to ascend grades). Cornering performance of vehicles will be overviewed in Chapter 3, but detailed presentations of this material are better suited to more specialized sources (Brewer and Rice 1983; Campbell 1978; Wong 1978).

2.2 TRACTIVE EFFORT AND RESISTANCE

Tractive effort and resistance are the two primary opposing forces that determine the straight-line performance of road vehicles. Tractive effort is simply the force

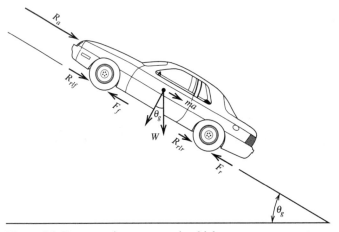

Figure 2.1 Forces acting on a road vehicle.

available at the roadway surface to perform work; it is expressed in newtons. Resistance (also expressed in newtons) is defined as the force impeding vehicle motion. The three major sources of vehicle resistance are (1) aerodynamic resistance, (2) rolling resistance (which originates from the roadway surface/tire interface), and (3) grade or gravitational resistance. To illustrate these forces, consider the vehicle force diagram shown in Fig. 2.1. In this figure, R_a is the aerodynamic resistance, R_{rlf} is the rolling resistance of the front tires, R_{rlr} is the rolling resistance of the rear tires, F_f is the available tractive effort of the front tires, F_r is the available tractive effort of the rear tires, W is the total vehicle weight in newtons, θ_g is the angle of the grade in degrees, m is the vehicle mass in kilograms, and a is the rate of acceleration in m/s² (meters per second squared).

Summing the forces along the vehicle's longitudinal axis provides the basic equation of vehicle motion:

$$F_f + F_r = ma + R_a + R_{rlf} + R_{rlr} + R_g \tag{2.1}$$

where R_g is the grade resistance and is equal to $W \sin \theta_g$. For exposition purposes it is convenient to let F be the sum of available tractive effort delivered by the front and rear tires $(F_f + F_r)$ and similarly to let R_{rl} be the sum of rolling resistance $(R_{rlf} + R_{rlr})$. This notation allows Eq. 2.1 to be written as

$$F = ma + R_a + R_{rl} + R_g \tag{2.2}$$

Sections 2.3 to 2.8 present a thorough discussion of the components and implications of Eq. 2.2.

2.3   AERODYNAMIC RESISTANCE

Aerodynamic resistance is a resistive force that can have significant impacts on vehicle performance. At high speeds, where this component of resistance can become overwhelming, proper vehicle aerodynamic design is essential. Attention to aerodynamic efficiency in design has long been the rule in race and sports cars; more recently, concerns over fuel efficiency and overall vehicle performance have resulted in more efficient aerodynamic designs in common passenger cars.

Aerodynamic resistance originates from a number of sources. The primary source (typically accounting for over 85% of total aerodynamic resistance) is the turbulent flow of air around the vehicle body. This turbulence is a function of the shape of the vehicle, particularly the rear portion, which has been shown to be a major source of air turbulence. To a much lesser extent (on the order of 12% of total aerodynamic resistance), the friction of the air passing over the body of the vehicle contributes to resistance. Finally, approximately 3% of the total aerodynamic resistance can be attributed to airflow through vehicle components such as radiators and air vents.

Based on these sources, the equation for determining aerodynamic resistance is

$$R_a = \frac{\rho}{2} C_D A_f V^2 \tag{2.3}$$

where ρ is the air density in kilograms per cubic meter (kg/m^3), C_D is the coefficient of drag and is unitless, A_f is the frontal area of the vehicle (projected area of the vehicle in the direction of travel) in square meters (m^2), and V is the speed of the vehicle in meters per second (m/s). To be truly accurate, for aerodynamic resistance computations, V is actually the speed of the vehicle relative to the prevailing wind speed. To simplify the exposition of concepts soon to be presented, the wind speed is assumed to be equal to zero for all problems and derivations in this book.

Air density is a function of both elevation and temperature, as indicated in Table 2.1. Equation 2.3 indicates that as the air becomes more dense, total aerodynamic resistance increases. The drag coefficient (C_D) is a term that implicitly accounts for all three of the aerodynamic resistance sources previously discussed. The coefficient of drag is measured from empirical data either from wind tunnel experiments or actual field tests in which a vehicle is allowed to decelerate

Table 2.1 Typical Values of Air Density Under Specified Atmospheric Conditions

Altitude (m)	Temperature (°C)	Pressure (kPa)	Air Density (kg/m^3)
0	15	101.4	1.2256
1500	9.7	84.4	1.0567
3000	−4.5	70.1	0.9096

**Table 2.2 Ranges of Drag
Coefficients for Typical
Road Vehicles**

Vehicle Type	Drag Coefficient (C_D)
Automobile	0.25–0.55
Bus	0.5–0.7
Tractor-Trailer	0.6–1.3
Motorcycle	0.27–1.8

from a known speed with other sources of resistance (rolling and grade resistances) accounted for. Table 2.2 gives some approximation of the range of drag coefficients for different types of road vehicles. Table 2.3 presents drag coefficients for selected automobiles covering three decades from the 1960s to 1990s. The general trend toward lower drag coefficients over this period of time reflects the continuing efforts of the automotive industry to improve overall vehicle efficiency by minimizing resistance forces.

Figure 2.2 illustrates the effect of automobile operating conditions on drag coefficients. As indicated, even minor factors, such as opening windows, can have a significant effect on a vehicle's drag coefficient and thus total aerodynamic resistance. Projected frontal areas (approximated as the height of the vehicle multiplied by its width) typically range from 1.0 m² to 2.5 m² for passenger cars and are also a major factor in determining aerodynamic resistance.

Because aerodynamic resistance is proportional to the square of a vehicle's speed, it is clear that such resistance will increase rapidly at higher speeds. The magnitude of this increase can be underscored by considering an expression for the power (P_{R_a}) required to overcome aerodynamic resistance. With power being the product of force and speed, the multiplication of Eq. 2.3 with speed gives

$$P_{R_a} = \frac{\rho}{2} C_D A_f V^3 \tag{2.4}$$

**Table 2.3 Drag Coefficients of
Selected Automobiles**

Vehicle	Drag Coefficient (C_D)
1967 Chevrolet Corvette	0.50
1967 Volkswagen Beetle	0.46
1977 Triumph TR7	0.40
1977 Jaguar XJS	0.36
1987 Acura Integra	0.34
1987 Ford Taurus	0.32
1997 Lexus LS400	0.29
1997 Infiniti Q45	0.29

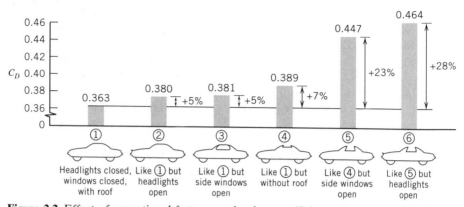

Figure 2.2 Effect of operational factors on the drag coefficient of an automobile. (Reproduced by permission from L. J. Janssen and W. H. Hucho, "The Effect of Various Parameters on the Aerodynamic Drag of Passenger Cars," *Advances in Road Vehicle Aerodynamics 1973*, BMRA Fluid Engineering, Cranfield, England.)

where P_{R_a} is in watts (Nm/s). Thus the amount of power required to overcome aerodynamic resistance increases with the cube of speed, indicating, for example, that eight times as much power is required to overcome aerodynamic resistance if vehicle speed is doubled.

2.4 ROLLING RESISTANCE

Rolling resistance refers to the resistance generated from a vehicle's internal mechanical friction, and pneumatic tires and their interaction with the roadway surface. The primary source of this resistance is the deformation of the tire as it passes over the roadway surface. The force needed to overcome this deformation accounts for approximately 90% of the total rolling resistance. Depending on the vehicle's weight and the material composition of the roadway surface, the penetration of the tire into the surface and the corresponding surface compression can also be a significant source of rolling resistance. However, for typical vehicle weights and pavement types, penetration/compression constitutes only around 4% of the total rolling resistance. Finally, frictional motion due to the slippage of the tire on the roadway surface, and, to a lesser extent, air circulation around the tire and wheel (i.e., the fanning effect), are sources accounting for roughly 6% of the total rolling resistance (Taborek 1957).

Considering the sources of rolling resistance, three factors are worthy of note. First, the rigidity of the tire and the roadway surface will influence the degree of tire penetration, surface compression, and tire deformation. Hard, smooth, and dry roadway surfaces provide the lowest rolling resistance. Second, tire conditions, including inflation pressure and temperature, can also have a substantial impact on rolling resistance. High tire inflation on hard paved surfaces decreases rolling resistance as a result of reduced friction but increases rolling resistance on soft unpaved surfaces due to additional surface penetration. Also, higher tire temperatures make the tire body more flexible and thus less resistance is encountered during tire deformation. The third factor is the vehicle's operating speed, which affects tire deformation. Increasing speed results in additional tire flexing and vibration and thus a higher rolling resistance.

Due to the wide range of factors that determine rolling resistance, a simplifying approximation is used. Studies have shown that overall rolling resistance can be approximated as the product of a friction term (coefficient of rolling resistance) and the weight of the vehicle acting normal to the roadway surface. The coefficient of rolling resistance (f_{rl}) for road vehicles operating on paved surfaces is approximated as

$$f_{rl} = 0.01 \left(1 + \frac{V}{44.73}\right) \tag{2.5}$$

where V is the vehicle's speed in m/s. By inspection of Fig. 2.1, the rolling resistance (in newtons) will simply be the coefficient of rolling resistance multiplied by $W \cos \theta_g$, the vehicle weight acting normal to the surface. For most highway applications θ_g is quite small, so it can be assumed that $\cos \theta_g = 1$, giving the equation for rolling resistance (R_{rl}) as

$$R_{rl} = f_{rl} W \tag{2.6}$$

From this, the amount of power required to overcome rolling resistance is

$$P_{R_{rl}} = f_{rl} W V \tag{2.7}$$

EXAMPLE 2.1

An 11.0-kN car is driven at sea level ($\rho = 1.2256$ kg/m³) on a paved surface and has $C_D = 0.38$ and 2.0 m² of frontal area. It is known that at maximum speed, 38 kW are being expended to overcome rolling and aerodynamic resistance. Determine the car's maximum speed.

SOLUTION

It is known that at maximum speed (V_m),

$$\text{available power} = R_a V_m + R_{rl} V_m$$

or

$$\text{available power} = \frac{\rho}{2} C_D A_f V_m^3 + f_{rl} W V_m$$

Substituting, we obtain

$$38,000 = \frac{1.2256}{2}(0.38)(2.0)V_m^3 + 0.01\left(1 + \frac{V_m}{44.73}\right)(11,000)V_m$$

or

$$38,000 = 0.4657V_m^3 + 2.459V_m^2 + 110V_m$$

Solving for V_m gives

$$V_m = \underline{\underline{39.95 \text{ m/s}}}, \quad \text{or} \quad \underline{\underline{143.82 \text{ km/h}}}$$

2.5   GRADE RESISTANCE

Grade resistance is simply the gravitational force acting on a vehicle. As suggested in Fig. 2.1, the expression for grade resistance (R_g) is

$$R_g = W \sin \theta_g \tag{2.8}$$

As was the case in the development of the rolling resistance formula (Eq. 2.6), highway grades are usually very small so that $\sin \theta_g \cong \tan \theta_g$. Rewriting Eq. 2.8 we have

$$R_g \cong W \tan \theta_g = WG \tag{2.9}$$

where G is the grade defined as the vertical rise per some specified horizontal distance (i.e., opposite side of the force triangle, Fig. 2.1, divided by the adjacent). Grades are generally specified in percent for ease of understanding. Thus a roadway that rises 5 m vertically per 100 m horizontal ($G = 0.05$ and $\theta_g = 2.86^0$) is said to have a 5% grade.

EXAMPLE 2.2

An 8.9-kN car is traveling at an elevation of 1500 m ($\rho = 1.0567$ kg/m^3) on a paved surface. If the car is traveling at 110 km/h and has $C_D = 0.40$ and $A_f = 2.0$ m^2 and the available tractive effort is 1135 N, what is the maximum grade that this car could ascend and still maintain the 110 km/h speed?

SOLUTION

To maintain the speed, the available tractive effort will be exactly equal to the summation of resistances. Thus no tractive effort will remain for vehicle acceleration (i.e., $ma = 0$). Therefore, Eq. 2.2 can be written as

$$F = R_a + R_{rl} + R_g$$

For grade resistance,

$$R_g = WG$$
$$= 8900G$$

For aerodynamic resistance (using Eq. 2.3),

$$R_a = \frac{\rho}{2} C_D A_f V^2$$

$$= \frac{1.0567}{2} (0.4)(2)(110 \times 1000/3600)^2$$

$$= 394.63 \text{ N}$$

For rolling resistance (using Eq. 2.6),

$$R_{rl} = f_{rl} W$$

$$= 0.01 \left(1 + \frac{110 \times 1000/3600}{44.73} \right) \times 8900$$

$$= 149.8 \text{ N}$$

Therefore,

$$F = 1135 = 394.63 + 149.8 + 8900G$$

$$G = \underline{\underline{0.0664}} \text{ or a } 6.64\% \text{ grade}$$

2.6 AVAILABLE TRACTIVE EFFORT

With the resistance terms in the basic equation of vehicle motion (Eq. 2.2) discussed, attention can now be directed toward available tractive effort (F) as used in Example 2.2. The tractive effort available to overcome resistance and/ or to accelerate the vehicle is determined either by the force generated by the vehicle's engine or by some maximum value that will be a function of the vehicle's weight distribution and the characteristics of the roadway surface/tire interface. The basic concepts underlying these two determinants of available tractive effort are presented here.

2.6.1 Maximum Tractive Effort

No matter how much force a vehicle's engine makes available at the roadway surface, there is a point beyond which additional force merely results in the spinning of tires and does not overcome resistance or accelerate the vehicle. To understand what determines this point of maximum tractive effort (i.e., the limiting value beyond which tire spinning begins) a force and moment-generating diagram is provided in Fig. 2.3. In this figure, L is the wheelbase, h is the height of the center of gravity above the roadway surface, l_f is the distance from the front axle to the center of gravity, l_r is the distance from the rear axle to the center of gravity, W_f is the weight of the vehicle on the front axle, W_r is the weight of the vehicle on the rear axle, and other terms are as defined for Fig. 2.1.

To determine the maximum tractive effort that the roadway surface/tire contact can support, it is necessary to examine the normal loads on the axles. The normal load on the rear axle (W_r) is given by summing the moments about point A (see Fig. 2.3):

$$W_r = \frac{R_a h + W l_f \cos \theta_g + mah \pm Wh \sin \theta_g}{L} \qquad \textbf{(2.10)}$$

In this equation the grade moment ($Wh \sin \theta_g$) is positive for vehicles on an upward slope and negative on a downward slope. Rearranging terms (assuming $\cos \theta_g = 1$ for the small grades encountered in highway applications) and substituting into Eq. 2.2 gives

$$W_r = \frac{l_f}{L} W + \frac{h}{L} (F - R_{rl}) \qquad \textbf{(2.11)}$$

From basic physics, the maximum tractive effort as determined by the roadway surface/tire interaction will be the normal force multiplied by the coefficient of

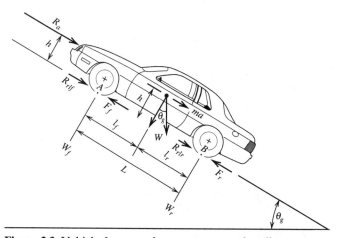

Figure 2.3 Vehicle forces and moment-generating distances.

road adhesion, μ. Thus for a rear-wheel-drive car,

$$F_{max} = \mu W_r \tag{2.12}$$

Substituting Eq. 2.11 into Eq. 2.12,

$$F_{max} = \mu \left[\frac{l_f}{L} W + \frac{h}{L} (F_{max} - R_{rl}) \right] \tag{2.13}$$

$$F_{max} = \frac{\mu W (l_f - f_{rl} h)/L}{1 - \mu h/L} \tag{2.14}$$

Similarly, by summing moments about point B (see Fig. 2.3), it can be shown that for a front-wheel-drive vehicle,

$$F_{max} = \frac{\mu W (l_r + f_{rl} h)/L}{1 + \mu h/L} \tag{2.15}$$

Note that in Eqs. 2.14 and 2.15, because of canceling units, h, l_f, l_r, and L can be in any unit of length (e.g., centimeters, meters, etc.). However, all of these terms must be in the same chosen unit of measure.

<hr>

EXAMPLE 2.3

An 11.0-kN car is designed with a 305-cm wheel base. The center of gravity is located 55 cm above the pavement and 100 cm behind the front axle. If the coefficient of road adhesion is 0.6, what is the maximum tractive effort that can be developed if the car is (a) a front-wheel-drive vehicle, and (b) a rear-wheel-drive vehicle?

SOLUTION

For the front-wheel-drive case, Eq. 2.15 can be used:

$$F_{max} = \frac{\mu W (l_r + f_{rl} h)/L}{1 + \mu h/L}$$

From Eq. 2.5, $f_{rl} = 0.01$ because $V = 0$ m/s, so

$$F_{max} = \frac{\{0.6 \times 11{,}000 \times [205 + 0.01(55)]\}/305}{1 + (0.6 \times 55)/305}$$

$$= \underline{\underline{4013.70 \text{ N}}}$$

For the rear-wheel-drive case, Eq. 2.14 can be used:

$$F_{max} = \frac{\{0.6 \times 11{,}000 \times [100 - 0.01(55)]\}/305}{1 - (0.6 \times 55)/305}$$

$$= \underline{\underline{2413.13 \text{ N}}}$$

2.6.2 Engine-Generated Tractive Effort

The amount of tractive effort generated by the vehicle's engine is a function of a variety of engine and driveline design factors. For engine design, critical factors in determining output include the shape of the combustion chamber, the quantity of air drawn into the combustion chamber during the induction phase, the type of fuel used, and fuel intake design. Although a complete description of engine design is beyond the scope of this book, an understanding of how engine output is measured and used is important to the study of vehicle performance. The two most commonly used measures of engine output are torque and power. Torque is the work generated by the engine (the twisting moment) and is expressed in newton-meters (Nm). Power is the rate of engine work, usually expressed in kilowatts (kW) for engines, and is related to the engine's torque by the following equation:

$$P_e = \frac{2\pi M_e n_e}{1000} \tag{2.16}$$

where P_e is engine power in kW, M_e is engine torque in Nm, n_e is engine speed in revolutions per second (i.e., the speed of the crankshaft). Figure 2.4 presents a torque–power diagram for a typical gasoline-powered engine.

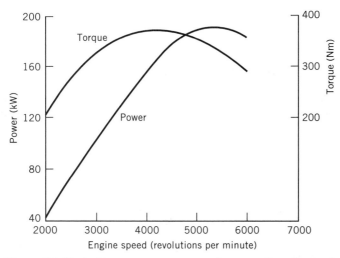

Figure 2.4 Typical torque–power curves for a gasoline-powered automobile engine.

> ### EXAMPLE 2.4

It is known that a small experimental engine has a torque curve of the form $M_e = an_e - bn_e^2$ where M_e is engine torque in Nm, n_e is engine speed in revolutions per second, and a and b are unknown parameters. If the engine develops a maximum torque of 125 Nm at 3200 rev/min (revolutions per minute) what is the engine's maximum power?

SOLUTION

At maximum torque, $n_e = 53.33$ rev/s (3200/60), and

$$\frac{dM_e}{dn_e} = 0 = a - 2bn_e$$

$$a = 2(53.33)b = 106.67b$$

Also, at maximum torque,

$$M_e = an_e - bn_e^2$$

$$125 = a(53.33) - b(53.33)^2$$

Using these two equations to solve for the two unknowns (a and b), we find that $b = 0.044$ and $a = 4.692$. Using Eq. 2.16 and $M_e = an_e - bn_e^2$,

$$P_e = \frac{2\pi(an_e - bn_e^2)n_e}{1000}$$

$$= \frac{2\pi(4.692n_e - 0.044n_e^2)n_e}{1000}$$

The first derivative of the power equation is used to solve for the engine speed at maximum power:

$$\frac{dP_e}{dn_e} = 0 = (0.00628)(9.384n_e - 0.132n_e^2)$$

$$n_e = 71.09 \text{ rev/s}$$

so the engine's maximum power is

$$P_e = \frac{2\pi(4.692n_e - 0.044n_e^2)n_e}{1000}$$

$$= \frac{2\pi[4.692(71.09) - 0.044(71.09)^2]71.09}{1000}$$

$$= \underline{\underline{49.64 \text{ kW}}}$$

Given the output measures of a vehicle's engine, focus can be directed toward the relationship between engine-generated torque and the tractive effort ultimately delivered to the driving wheels. Unfortunately, the tractive effort needed for acceptable vehicle performance (i.e., to provide adequate acceleration characteristics) is greater at lower vehicle speeds and, because maximum engine torque is developed at fairly high engine speeds (crankshaft revolutions), the use of gasoline-powered engines requires some form of gear reduction as illustrated in Fig. 2.5. This gear reduction provides the mechanical advantage necessary for acceptable vehicle acceleration.

With gear reductions, two factors determine the amount of tractive effort reaching the driving wheels. First, the mechanical efficiency of the driveline (i.e., the gear reduction devices including the transmission and differential) must be considered. Typically, 5% to 25% of the tractive effort generated by the engine is lost in gear reduction devices, which corresponds to a mechanical efficiency of the driveline (η_d) of 0.75 to 0.95. Second, the overall gear reduction ratio (ε_0), which includes the gear reductions of the transmission and differential, plays a key role in the determination of tractive effort. By definition, the overall gear reduction ratio refers to the relationship between the revolutions of the engine's crankshaft and the revolutions of the road wheels. For example, an overall gear reduction ratio of 4 to 1 ($\varepsilon_0 = 4$) means that the engine's crankshaft turns four revolutions for every one turn of the road wheel.

With these terms defined, the engine-generated tractive effort reaching the

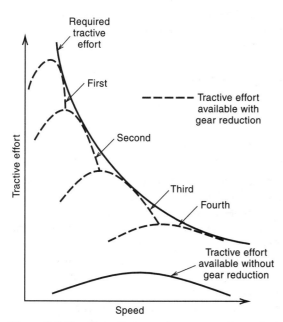

Figure 2.5 Tractive effort requirements and tractive effort generated by a typical gasoline-powered vehicle.

driving wheels (F_e) is given as

$$F_e = \frac{M_e \varepsilon_0 \eta_d}{r} \tag{2.17}$$

where r is the radius of the road wheel in meters and all other terms are as previously defined. It follows that the relationship between vehicle speed and engine speed (in crankshaft revolutions per second, n_e) is

$$V = \frac{2\pi r n_e (1 - i)}{\varepsilon_0} \tag{2.18}$$

where V is the vehicle speed (m/s) and i is the slippage of the driveline generally taken as 2% ($i = 0.02$) to 5% ($i = 0.05$) for passenger cars.

To summarize this section, the available tractive effort (F in Eq. 2.2) at any given speed is the lesser of the maximum tractive effort (F_{max}) and the engine tractive effort (F_e).

2.7 VEHICLE ACCELERATION

As defined in the previous section, available tractive effort (F) can be used to determine a number of vehicle performance characteristics including vehicle acceleration and top speed. For determining vehicle acceleration, Eq. 2.2 can be applied with an additional term added to account for the inertia of the vehicle's rotating parts that must be overcome during acceleration. This term is referred to as the mass factor (γ_m) and is introduced in Eq. 2.2 as

$$F - \sum R = \gamma_m ma \tag{2.19}$$

where the mass factor is approximated as

$$\gamma_m = 1.04 + 0.0025\varepsilon_0^2 \tag{2.20}$$

Two measures of vehicle acceleration are worthy of note: the time to accelerate, and the distance to accelerate. For both, the force available to accelerate is $F_{net} = F - \sum R$. The basic relationship between the force available to accelerate, F_{net}; the available tractive effort, F (the lesser of F_{max} and F_e); and the summation of resistances is illustrated in Fig. 2.6. In this figure, F_{net} will be the vertical distance between the lesser of the F_{max} and the F_e curves and the total resistance curve. So, referring to Fig. 2.6, at speed V', F_{net} will be $F_{max} - \sum R$, and at speed V'', F_{net} will be $F_e - \sum R$. It follows that when $F_{net} = 0$, the vehicle cannot accelerate and is at its maximum speed for specified conditions (i.e., grade, air density, engine torque, and so on). Such was the case for the vehicle described

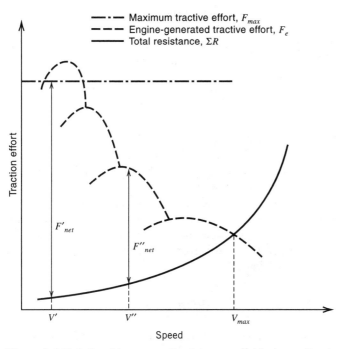

Figure 2.6 Relationship among the forces available to accelerate, available tractive effort, and total vehicle resistance.

in Example 2.2. When F_{net} is greater than zero (i.e., the vehicle is traveling at a speed less than its maximum speed), Eq. 2.19 can be written in differential form:

$$F_{net} = \gamma_m m \frac{dV}{dt} \quad \text{or} \quad dt = \frac{\gamma_m m dV}{F_{net}}$$

Because F_{net} is itself a function of vehicle speed [$F_{net} = f(V)$], integration gives the time to accelerate as

$$t = \gamma_m m \int_{V_1}^{V_2} \frac{dV}{f(V)} \tag{2.21}$$

where V_1 is the initial vehicle speed and V_2 is the final vehicle speed. Similarly, it can be shown that the distance to accelerate is

$$D_a = \gamma_m m \int_{V_1}^{V_2} \frac{V dV}{f(V)} \tag{2.22}$$

To solve Eqs. 2.21 and 2.22, numerical integration is necessary because the functional forms of these equations do not lend themselves to closed-form solutions. Such numerical integration is straightforward but requires a computer. Thus we will not provide an example of solving these equations.

EXAMPLE 2.5

A car is traveling at 16 km/h with $C_D = 0.30$, $A_f = 2.0$ m², $W = 13.3$ kN, and $\rho = 1.0567$ kg/m³. Its engine is producing 130 Nm of torque, and is in a gear that gives an overall gear reduction ratio of 4.5 to 1. The wheel radius is 36 cm, and the mechanical efficiency of the driveline is 80%. The road has a paved surface with a coefficient of road adhesion equal to 0.20. The driver decides to accelerate to avoid an accident. If the car has a wheelbase of 305 cm and a center of gravity 50 cm above the roadway surface and 127 cm behind the front axle, what would the acceleration be if the car was front-wheel drive, and what would the acceleration be if it was rear-wheel drive?

SOLUTION

We begin by computing the resistances, tractive effort generated by the engine, and mass factor because all of these factors will be the same for both front- and rear-wheel drive.

The air resistance is (from Eq. 2.3)

$$R_a = \frac{\rho}{2} C_D A_f V^2$$

$$= \frac{1.0567}{2} (0.3)(2)(16 \times 1000/3600)^2$$

$$= 6.26 \text{ N}$$

The rolling resistance is (from Eq. 2.6)

$$R_{rl} = f_{rl} W$$

$$= 0.01 \left(1 + \frac{16 \times 1000/3600}{44.73} \right) \times 13{,}300$$

$$= 146.22 \text{ N}$$

The engine-generated tractive effort is (from Eq. 2.17)

$$F_e = \frac{M_e \varepsilon_0 \eta_d}{r}$$

$$= \frac{130(4.5)(0.8)}{0.36}$$

$$= 1300 \text{ N}$$

The mass factor is (from Eq. 2.20)

$$\gamma_m = 1.04 + 0.0025\varepsilon_0^2$$

$$= 1.04 + 0.0025(4.5)^2$$

$$= 1.091$$

Recall that, to determine acceleration, we need the resistances (already computed) and the available tractive effort, F, which is the lesser F_e or F_{max}. For the case of the front-wheel-drive car, Eq. 2.15 can be applied to determine F_{max}:

$$F_{max} = \frac{\mu W(l_r + f_{rl}h)/L}{1 + \mu h/L}$$

$$= \frac{\{0.2 \times 13{,}300 \times [178 + 0.011(50)]\}/305}{1 + (0.2 \times 50)/305}$$

$$= \underline{1507.76 \text{ N}}$$

Thus for a front-wheel-drive car, $F = 1300$ N (the lesser of 1300 and 1507.76), and the acceleration is (from Eq. 2.19)

$$F - \sum R = \gamma_m ma$$

$$a = \frac{F - \sum R}{\gamma_m m} = \frac{1300 - 152.48}{1.091(13{,}300/9.807)} = \underline{0.776 \text{ m/s}^2}$$

For the case of the rear-wheel-drive car, Eq. 2.14 can be applied to determine F_{max}:

$$F_{max} = \frac{\{0.2 \times 13{,}300 \times [127 - 0.011(50)]\}/305}{1 - (0.2 \times 50)/305}$$

$$= \underline{1140.19 \text{ N}}$$

Thus for a rear-wheel-drive car, $F = 1140.19$ N (the lesser of 1300 and 1140.19), and the acceleration is (from Eq. 2.19)

$$a = \frac{F - \sum R}{\gamma_m m} = \frac{1140.19 - 152.48}{1.091(13{,}300/9.807)} = \underline{0.668 \text{ m/s}^2}$$

2.8 FUEL EFFICIENCY

Given the factors discussed in the preceding sections of this chapter, it becomes clear which elements determine a vehicle's fuel efficiency. One of the most critical determinants relates to engine design (i.e., how the engine-generated tractive effort is produced). Engine designs that increase the quantity of air entering the combustion chamber, decrease internal engine friction, and improve fuel delivery to the combustion chamber all lead to improved fuel efficiency. Improvements to mechanical components (the vehicle's driveline), including decreasing slippage and improving the mechanical efficiency of the driveline, also produce improvements in overall fuel efficiency.

In terms of resistance-reducing options, decreasing overall vehicle weight (W) will lower grade and rolling resistances, thus reducing fuel consumption (all other factors held constant). Similarly, aerodynamic improvements such as lower drag coefficients (C_D) and reduced frontal areas (A_f) can produce significant fuel savings. Finally, improved tire designs with lower rolling resistance can improve overall fuel efficiency.

2.9 PRINCIPLES OF BRAKING

In terms of highway design and traffic analysis, the braking characteristics of road vehicles is arguably the single most important aspect of vehicle performance. Knowledge of the braking behavior of road vehicles is critical in the determination of stopping-sight distance, roadway surface design, and accident avoidance systems. Moreover, ongoing advances in braking technology make it essential for transportation engineers to have a basic comprehension of the underlying principles involved.

2.9.1 Braking Forces

To begin the discussion of braking principles, consider the force and moment-generating distance diagram in Fig. 2.7, where F_{bf} and F_{br} are the front and rear braking forces, respectively, and other terms are as previously defined. During vehicle braking there is a load transfer from the rear to the front axle. To illustrate this, expressions for the normal loads on the front and rear axles can be written by summing the moments about roadway surface/tire contact points A and B (as was done in deriving Eq. 2.12, with $\cos \theta_g$ assumed to be equal to 1 because of the small θ_g encountered in highway applications):

$$W_f = \frac{1}{L}[Wl_r + h(ma - R_a \pm W \sin \theta_g)] \tag{2.23}$$

and

$$W_r = \frac{1}{L}[Wl_f - h(ma - R_a \pm W \sin \theta_g)] \tag{2.24}$$

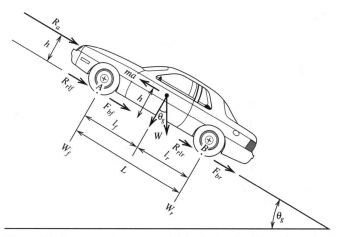

Figure 2.7 Forces acting on a vehicle during braking with driveline resistance ignored.

where, in this case, the contribution of grade resistance ($W \sin \theta_g$) is negative for uphill grades and positive for downhill grades.

Also, from the summation of forces along the vehicle's longitudinal axis,

$$F_b + f_{rl}W = ma - R_a \pm W \sin \theta_g \tag{2.25}$$

where $F_b = F_{bf} + F_{br}$. Substituting Eq. 2.25 into Eqs. 2.23 and 2.24 gives

$$W_f = \frac{1}{L}[Wl_r + h(F_b + f_{rl}W)] \tag{2.26}$$

and

$$W_r = \frac{1}{L}[Wl_f - h(F_b + f_{rl}W)] \tag{2.27}$$

Because the maximum vehicle braking force ($F_{b\,max}$) is equal to the coefficient of road adhesion, μ, multiplied by the weights normal to the roadway surface, front maximum vehicle braking force is,

$$F_{bf\,max} = \mu W_f$$

$$= \frac{\mu W}{L}[l_r + h(\mu + f_{rl})] \tag{2.28}$$

and rear maximum vehicle braking force is,

$$F_{br\,max} = \mu W_r$$

$$= \frac{\mu W}{L}[l_f - h(\mu + f_{rl})] \tag{2.29}$$

Table 2.4 Typical Values of Coefficients of Road Adhesion for Vehicle Braking

	Coefficient of Road Adhesion	
Pavement	Maximum	Slide
Good, dry	1.00	0.80
Good, wet	0.90	0.60
Poor, dry	0.80	0.55
Poor, wet	0.60	0.30
Packed snow or ice	0.25	0.10

Source: S. G. Shadle, L. H. Emery, and H. K. Brewer, "Vehicle Braking, Stability, and Control," *SAE Transactions,* vol. 92, paper 830562, 1983.

To develop maximum braking forces, the tires should be at the point of an impending slide. If the tires begin to slide (i.e., the brakes lock), a significant reduction in road adhesion will result. An indication of the extent of the reduction in road adhesion as the result of tire slide, under various pavement and weather conditions, is presented in Table 2.4. It is clear from this table that the braking forces will decline dramatically when the wheels are locked (resulting in tire slide). Avoiding these locked conditions is the function of antilock braking systems in cars. Such systems will be discussed later in this chapter.

2.9.2 Braking Force Ratio and Efficiency

On a given roadway surface, the maximum attainable vehicle deceleration (using the vehicle's braking system) is equal to μg, where μ is the coefficient of adhesion and g is the gravitational constant (9.807 m/s^2). To approach this maximum vehicle deceleration, vehicle braking systems must correctly distribute braking forces between the vehicle's front and rear brakes. This is typically done by the allocation of hydraulic pressures within the braking system. This front/rear proportioning of braking forces (within the vehicle's braking system) will be optimal (achieving a deceleration rate equal to μg) when it is in the exact same proportion as the ratio of the maximum braking forces on the front and rear axles ($F_{bf\,max}/F_{br\,max}$). Thus maximum braking forces (with the tires at the point of impending slide) will be developed when the brake force ratio (front force over rear force) is

$$BFR_{f/r\,max} = \frac{l_r + h(\mu + f_{rl})}{l_f - h(\mu + f_{rl})} \qquad (2.30)$$

where $BFR_{f/r\,max}$ is the brake force ratio, allocated by the vehicle's braking system, that results in maximum (optimal) braking forces, and other terms are as previously defined. It follows that the percentage of braking force that the braking system should allocate to the front axle (PBF_f) for maximum braking is

$$PBF_f = 100 - \frac{100}{1 + BFR_{f/r\,max}} \qquad (2.31)$$

and the percentage of braking force that the braking system should allocate to the rear axle (PBF_r) for maximum braking is

$$PBF_r = \frac{100}{1 + BFR_{f/r\,max}} \qquad (2.32)$$

EXAMPLE 2.6

A car has a wheelbase of 250 cm and a center of gravity that is 100 cm behind the front axle at a height of 60 cm. If the car is traveling at 130 km/h on a road with poor pavement that is wet, determine the percentage of braking forces that should be allocated to the front and rear brakes (by the vehicle's braking system) to ensure that maximum braking forces are developed.

SOLUTION

The coefficient of rolling resistance is

$$f_{rl} = 0.01 \left(1 + \frac{130 \times 1000/3600}{44.73}\right)$$

$$= 0.0181$$

and $\mu = 0.6$ from Table 2.4 (maximum because we want the tires to be at the point of impending slide). Applying Eq. 2.30 gives

$$BFR_{f/r\,max} = \frac{l_r + h(\mu + f_{rl})}{l_f - h(\mu + f_{rl})}$$

$$= \frac{150 + 60(0.6 + 0.0181)}{100 - 60(0.6 + 0.0181)}$$

$$= 2.97$$

Using Eq. 2.31, the percentage of the force allocated to the front brakes should be

$$PBF_f = 100 - \frac{100}{1 + BFR_{f/r\,max}}$$

$$= 100 - \frac{100}{1 + 2.97}$$

$$= 74.81\%$$

Using Eq. 2.32 (or simply $100 - PBF_f$), the percentage of the force allocated to the rear brakes should be

$$PBF_r = \frac{100}{1 + BFR_{f/r\,max}}$$

$$= \frac{100}{1 + 2.97}$$

$$= \underline{\underline{25.19\%}}$$

It is clear from Eq. 2.30 that the design of a vehicle's braking system is not an easy task because the optimal brake-force proportioning changes with both vehicle and road conditions. For example, the addition of vehicle cargo and/or passengers will not only change the weight of the vehicle (which affects f_{rl} in Eq. 2.30), but will also change the distribution of the weight, shifting the height of the center of gravity and its location along the vehicle's longitudinal axis. This will change the optimal brake-force proportioning (i.e., $BFR_{f/r\,max}$). Similarly, changes in road conditions will produce different coefficients of adhesion, again changing optimal brake-force proportioning.

Figures 2.8 and 2.9 illustrate the effects of vehicle loadings on the optimal distribution of braking forces on a road with a coefficient of adhesion equal to

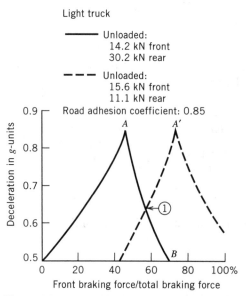

Figure 2.8 Effect of brake proportioning on the braking performance of a light truck. (Reproduced by permission of the Society of Automotive Engineers from D. J. Bickerstaff and G. Hartley, "Light Truck Tire Traction Properties and Their Effects on Braking Performance," *SAE Transactions,* vol. 83, paper 741137, 1974.)

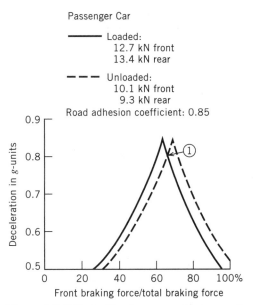

Passenger Car

⎯⎯⎯ Loaded:
         12.7 kN front
         13.4 kN rear

− − − Unloaded:
         10.1 kN front
         9.3 kN rear

Road adhesion coefficient: 0.85

Figure 2.9 Effect of brake proportioning on the braking performance of a passenger car. (Reproduced by permission of the Society of Automotive Engineers from D. J. Bickerstaff and G. Hartley, "Light Truck Tire Traction Properties and Their Effects on Braking Performance," *SAE Transactions,* vol. 83, paper 741137, 1974.)

0.85. Figure 2.8 shows the effect of loading conditions on the braking performance of a light truck. This figure's curves define the points of impending wheel lockup. For example, with 20% of the braking force allocated to the front brakes by the vehicle's braking system, a 0.6 g (5.88 m/s^2) deceleration will be achieved before the brakes lock and the coefficient of adhesion drops to slide values. Point A defines the optimal brake-force proportioning under loaded conditions as roughly 42% of the total braking force allocated to the front brakes (as would be computed by Eq. 2.31). At this point, a maximum deceleration of 0.85 g's (8.33 m/s^2) will be achieved as limited by the coefficient of road adhesion ($\mu = 0.85$). Note that when only 20% of the braking force is allocated to the front brakes, the deceleration (5.88 m/s^2) is substantially less than the maximum (8.33 m/s^2). Point A' indicates that optimal brake-force proportioning under unloaded conditions has about 72% of the braking force allocated to the front brakes. This disparity in optimal brake-force allocation between loaded and unloaded conditions (points A and A') presents a dilemma that necessitates a compromised braking system design (e.g., the compromised brake proportioning represented by point 1). This compromise gives suboptimal braking-force allocations under both loaded and unloaded conditions. Figure 2.9 shows that the disparity of optimal braking forces between loaded and unloaded conditions is less pronounced for passenger cars but still necessitates brake system compromises.

As indicated in Eq. 2.30, optimal brake-force distributions are not only a function of vehicle loading conditions but also the road's coefficient of adhesion. Table 2.5 presents optimal values for passenger cars with coefficients of road

Table 2.5 Examples of Brake Proportioning for Automobiles

Brake Parameters	Car 1 (An Early 1970s-Vintage Car)	Car 2 (An Early 1970s-Vintage Car)	Car 3 (An Early 1980s-Vintage Car)	Car 4 (An Early 1980s-Vintage Car)
Wheelbase, L (cm)	228.6	226.7	225.0	266.5
Unloaded				
weight W (kN)	8.90	13.34	9.72	13.68
front weight (%)	60	54	60	62.4
center of gravity, h (cm)	51	51	48	53
Loaded				
weight W (kN)	11.56	16.90	11.83	16.17
front weight (%)	51	47	52.4	54.2
center of gravity, h (cm)	51	51	46	53
Actual front brake				
proportioning (%)	60	60	79.7	80.6
Ideal front brake				
proportioning at $\mu = 0.6$				
unloaded, front (%)	73	65	72.9	74.4
loaded, front (%)	64	58	64.6	66.2
Ideal front brake				
proportioning at $\mu = 0.8$				
unloaded, front (%)	78	69	77.2	78.4
loaded, front (%)	69	62	68.7	70.2

Source: S. G. Shadle, L. H. Emery, and H. K. Brewer, "Vehicle Braking, Stability, and Control," *SAE Transactions,* vol. 92, paper 830562, 1983.

adhesion equal to 0.60 and 0.80. In examining this table, it is important to note that studies have indicated that if wheel lockup is to occur, it is preferable to have the front wheels lock first because rear wheels locking first can result in uncontrollable vehicle spin. Front-wheel lockup will result in the loss of steering control but the vehicle will at least continue to brake in a straight line. With this in mind, and looking historically at the evolution of automotive brake-force proportioning, Table 2.5 shows that typical automobiles of the early 1970s had poorly proportioned brakes relative to their early 1980s' counterparts in terms of actual versus ideal front-brake proportioning, and that underbraking on the front increased the likelihood of a dangerous rear-wheel-first lockup.

Because true optimal brake-force proportioning is seldom achieved in standard non-antilock braking systems, it is useful to define a braking efficiency term that reflects the degree to which the braking system is operating below optimal. Simply stated, braking efficiency is defined as the ratio of the maximum rate of deceleration, expressed in g's (g_{max}), achievable prior to any wheel lockup, to the coefficient of road adhesion:

$$\eta_b = \frac{g_{max}}{\mu} \tag{2.33}$$

As an example of the application of this equation, suppose we want to find the braking efficiency of the loaded truck in Fig. 2.8 when 40% of the available braking force is allocated to the front brakes. The figure shows that the deceleration rate is 0.75 g's, so the application of Eq. 2.33 gives a braking efficiency of 0.8823, or 88.23% ($\eta_b = 0.75/0.85 = 0.8823$).

2.9.3   Antilock Braking Systems

Many modern cars have braking systems that are designed to prevent the wheels from locking during braking applications (antilock braking systems). In theory, antilock braking systems serve two purposes. First, they prevent the coefficient of road adhesion from dropping to slide values (see Table 2.4). Second, they have the potential to raise the braking efficiency to 100%. In practice, designing an antilock braking system that avoids slide coefficients of adhesion and achieves 100% braking efficiency ($\eta_b = 1.0$) is a difficult task. This is because most antilock braking systems technologies detect which wheels have locked and release them momentarily before reapplying the brake on the locking wheel. The wheel-lock detection speed, speed of brake-force reallocation, and braking system design (the amount of braking forces that can be accommodated by the vehicle's front and rear brake disks and calipers) all impact the overall effectiveness of the antilock braking system. Early antilock braking systems often fell short of achieving 100% braking efficiency and, in many cases, an expert driver using a non-antilock-brake car could modulate the brakes to achieve shorter stopping distances than could be achieved with cars equipped with antilock brakes. However, advances in antilock braking system technology bring us closer and closer to 100% braking efficiency.

2.9.4   Theoretical Stopping Distance

With a basic understanding of brake-force proportioning and the resulting brake efficiency, attention can now be directed toward developing expressions for minimum stopping distances. By inspection of Fig. 2.7, it can be seen that the relationship between the stopping distance, braking force, vehicle mass, and vehicle speed is

$$a\,dS = \left[\frac{F_b + \sum R}{\gamma_b m}\right] dS \tag{2.34}$$

$$= V dV$$

where γ_b is the mass factor accounting for moments of inertia during braking and is given the value of 1.04 for automobiles (Wong 1978). Integrating to determine stopping distance gives

$$S = \int_{V_1}^{V_2} \gamma_b m \frac{V dV}{F_b + \sum R} \tag{2.35}$$

Substituting in the resistances (see Fig. 2.7), we obtain

$$S = \gamma_b m \int_{V_1}^{V_2} \frac{V dV}{F_b + R_a + f_{rl} W \pm W \sin \theta_g} \tag{2.36}$$

where V_1 is the initial vehicle speed, V_2 is the final vehicle speed, $f_{rl} W$ is the rolling resistance, and the grade resistance ($W \sin \theta_g$) is positive for uphill slopes and negative for downhill slopes. To simplify notation, let

$$K_a = \frac{\rho}{2} C_D A_f \tag{2.37}$$

so that Eq. 2.3 is

$$R_a = K_a V^2 \tag{2.38}$$

Continuing, assume that the effect of speed on the coefficient of rolling resistance, f_{rl}, is constant and can be approximated using the average of initial (V_1) and final (V_2) speeds in Eq. 2.5: that is, $V = (V_1 + V_2)/2$. With this assumption (which will introduce only a very small amount of error), and letting $m = W/g$, and $F_b = \mu W$, integration of Eq. 2.36 gives

$$S = \frac{\gamma_b W}{2g K_a} \ln \left(\frac{\mu W + K_a V_1^2 + f_{rl} W \pm W \sin \theta_g}{\mu W + K_a V_2^2 + f_{rl} W \pm W \sin \theta_g} \right) \tag{2.39}$$

If the vehicle is assumed to stop ($V_2 = 0$),

$$S = \frac{\gamma_b W}{2g K_a} \ln \left(1 + \frac{K_a V_1^2}{\mu W + f_{rl} W \pm W \sin \theta_g} \right) \tag{2.40}$$

With braking efficiency considered, the actual braking force is

$$F_b = \eta_b \mu W \tag{2.41}$$

Therefore, by substitution into Eq. 2.40, the theoretical stopping distance is

$$S = \frac{\gamma_b W}{2g K_a} \ln \left(1 + \frac{K_a V_1^2}{\eta_b \mu W + f_{rl} W \pm W \sin \theta_g} \right) \tag{2.42}$$

Similarly, Eq. 2.39 can be written to include braking efficiency. Finally, if aerodynamic resistance is ignored (due to its comparatively small contribution to braking), deceleration is constant during braking and integration of Eq. 2.35 gives

$$S = \frac{\gamma_b (V_1^2 - V_2^2)}{2g(\eta_b \mu + f_{rl} \pm \sin \theta_g)} \tag{2.43}$$

EXAMPLE 2.7

A new experimental 11-kN car is traveling at 145 km/h down a 10% grade with $C_D = 0.25$, $A_f = 2.0$ m^2, and $\rho = 1.2256$ kg/m^3. The car has an advanced antilock braking system that gives a braking efficiency of 100%, and the coefficient of road adhesion is 0.7. Determine the theoretical minimum stopping distance for the case when aerodynamic resistance is considered, and for the case when aerodynamic resistance is ignored.

SOLUTION

With aerodynamic resistance considered, Eq. 2.42 can be applied with $\gamma_b = 1.04$, $\theta_g = 5.71°$, and

$$f_{rl} = 0.01 \left[1 + \frac{\left(\dfrac{145 \times 1000/3600 + 0}{2} \right)}{44.73} \right] = 0.0145$$

$$K_a = \frac{1.2256}{2}(0.25)(2) = 0.3064$$

Then

$$S = \frac{1.04(11{,}000)}{2(9.807)(0.3064)}$$

$$\ln \left[1 + \frac{0.3064(145 \times 1000/3600)^2}{(1.0)(0.7)(11{,}000) + (0.0145)(11{,}000) - 11{,}000 \sin(5.71°)} \right]$$

$$= \underline{\underline{134.97 \text{ m}}}$$

With aerodynamic resistance excluded, Eq. 2.43 is used:

$$S = \frac{1.04(1.45 \times 1000/3600)^2}{2(9.807)[0.7 + 0.0145 - \sin(5.71°)]}$$

$$= \underline{\underline{139.87 \text{ m}}}$$

EXAMPLE 2.8

A car is traveling at 130 km/h and has 80% braking efficiency. The brakes are applied to miss an object that is 45 m from the point of brake application, and the coefficient of road adhesion is 0.85. Ignoring aerodynamic resistance and assuming theoretical minimum stopping distance, estimate how fast the car will be going when it strikes the object if (a) the surface is level, and (b) the surface is on a 5% upgrade.

SOLUTION

In both cases rolling resistance will be approximated as

$$f_{rl} = 0.01 \left[1 + \frac{\left(\dfrac{130 \times 1000/3600 + V_2}{2} \right)}{44.73} \right] = 0.014 + 0.0001118 V_2$$

Applying Eq. 2.43 for the level grade with $\gamma_b = 1.04$, and $\theta_g = 0°$,

$$S = \frac{\gamma_b (V_1^2 - V_2^2)}{2g(\eta_b \mu + f_{rl} \pm \sin \theta_g)}$$

$$45 = \frac{1.04[(130 \times 1000/3600)^2 - V_2^2]}{2(9.807)[0.8(0.85) + (0.014 + 0.0001118 V_2) \pm 0]}$$

$$V_2 = \underline{\underline{26.70 \text{ m/s}}} \quad \text{or} \quad \underline{\underline{96.11 \text{ km/h}}}$$

On a 5% grade with $\theta_g = 2.86°$,

$$45 = \frac{1.04[(130 \times 1000/3600)^2 - V_2^2]}{2(9.807)[0.8(0.85) + (0.014 + 0.0001118 V_2) + 0.05]}$$

$$V_2 = \underline{\underline{25.89 \text{ m/s}}} \quad \text{or} \quad \underline{\underline{93.20 \text{ km/h}}}$$

2.9.5   Practical Stopping Distance

As mentioned earlier, one of the most critical concerns in the design of a highway is to provide adequate driver sight distance to permit a safe stop. The theoretical assessment of vehicle stopping distance presented in the previous section provided the principles of braking for an individual vehicle under specified roadway surface conditions. However, highway engineers face a more complex problem because they must design for a variety of driver skill levels (which can affect whether or not the brakes lock and reduce the coefficient of adhesion to slide values), vehicle types (with varying aerodynamics, weight distributions, and brake efficiencies), and weather conditions (which change the roadway's coefficient of adhesion). As a result of the wide variability inherent in the determination of braking distance, an equation is required that provides an estimate of typical observed braking distances, and is more simplistic and usable than Eq. 2.42. To begin deriving such an equation, note that if aerodynamic resistance, rolling resistance, moments of inertia, and braking efficiency are ignored, the summation of longitudinal vehicle forces in Fig. 2.7 can be written as

$$F_b \pm W \sin \theta_g = ma \tag{2.44}$$

where the grade resistance ($W \sin \theta_g$) is positive for uphill grades and negative for downhill grades. As before, because highway grades are typically small, it is assumed that $\sin \theta_g = \tan \theta_g = G$, where G is the percent grade divided by 100 as previously defined.

Now define a friction term, f, such that $F_b = fW$. Solving Eq. 2.44 for acceleration, a, and substituting $F_b = fW$ and $m = W/g$ yields

$$a = - \left[\frac{g}{W}(fW \pm WG) \right] \quad \text{or} \quad a = -[g(f \pm G)] \tag{2.45}$$

with the minus sign denoting deceleration. Now use the basic physics equation

$$V_2^2 = V_1^2 + 2ad \tag{2.46}$$

where d is the deceleration distance (practical stopping distance) in meters, V_1 is the initial vehicle speed (m/s), and V_2 is the final vehicle speed (m/s). Substituting Eq. 2.45 into Eq. 2.46 gives

$$d = \frac{V_1^2 - V_2^2}{2g(f \pm G)} \tag{2.47}$$

If $V_2 = 0$ (i.e., the vehicle comes to a complete stop), the practical-stopping-distance equation is

$$d = \frac{V_1^2}{2g(f \pm G)} \tag{2.48}$$

It is important to note the similarity between Eq. 2.48 and Eq. 2.43 (the theoretical stopping distance ignoring aerodynamic resistance). If the braking mass factor is excluded and $\sin \theta_g$ is assumed to be equal to G, the friction term in Eq. 2.48 (f) is equal to $\eta_b \mu + f_{rl}$.

The values of the friction term, f, used in Eq. 2.48 are from actual vehicle stopping experiments. With known initial vehicle speed and grade, test vehicles are brought to a stop, the distance from the brake application to the stop is measured, and Eq. 2.48 is applied to solve for f. The values of f used in highway design are conservative estimates that are based on the assumption of some of the worst driver skills (i.e., a tendency to lock the wheels), roadway and tire conditions, and vehicle braking efficiencies likely to be encountered. Table 2.6 presents f values that are the current guideline in highway design.

It is important to note that the f values determined by the experimental method just described implicitly include the effects of aerodynamic resistance, braking efficiency, coefficient of road adhesion (with locked wheels), and inertia during braking (the braking mass factor). Thus the f's are a function of vehicle technology at the time that the experiments were conducted. It is important to recognize that as vehicle braking technology and other vehicle characteristics change, f's will also change and so will the accepted highway design guidelines. The relationship

Table 2.6 Coefficients of Friction Used in Practical Stopping-Distance Computations

Initial Vehicle Speed (km/h)	Coefficient of Friction (f)	Braking Distance on Level Road (m)
30	0.40	8.8
40	0.38	16.6
50	0.35	28.1
60	0.33	42.9
70	0.31	62.2
80	0.30	83.9
90	0.30	106.2
100	0.29	135.6
110	0.28	170.0
120	0.28	202.3

between changing vehicle characteristics and changing highway design guidelines is one that must always be kept in the design engineer's mind.

EXAMPLE 2.9

A car ($C_D = 0.4$, $A_f = 2.2$ m², $W = 13.3$ kN, and $\rho = 1.2256$ kg/m³) is being used in a test on level pavement to arrive at f values for practical-stopping-distance equations. The coefficient of road adhesion is 0.75 and the vehicle's braking efficiency is 0.85. If, during the test, the tires are at the point of impending skid and theoretical minimum stopping distance is achieved (considering aerodynamic resistance), what value of f will result from bringing the vehicle to a stop from 100 km/h?

SOLUTION

First, to calculate the theoretical minimum stopping distance, Eq. 2.42 is applied with $\gamma_b = 1.04$, $\theta_g = 0°$, and

$$f_{rl} = 0.01 \left[1 + \dfrac{\left(\dfrac{100 \times 1000/3600 + 0}{2} \right)}{44.73} \right] = 0.0131$$

$$K_a = \frac{1.2256}{2}(0.40)(2.2) = 0.5393$$

Thus, from Eq. 2.42,

$$S = \frac{1.04(13{,}300)}{2(9.807)(0.5393)} \ln\left[1 + \frac{0.5393(100 \times 1000/3600)^2}{(0.85)(0.75)(13{,}300) + (0.0131)(13{,}300) \pm 0}\right]$$

$$= \underline{\underline{61.42 \text{ m}}}$$

In solving for f, Eq. 2.48 is used:

$$d = \frac{V_1^2}{2gf}$$

$$f = \frac{V_1^2}{2gd}$$

$$= \frac{(100 \times 1000/3600)^2}{2(9.807)61.43}$$

$$= \underline{\underline{0.640}}$$

EXAMPLE 2.10

Consider the test conditions in Example 2.9. Assume that a new car ($C_D = 0.25$, $A_f = 2.0 \text{ m}^2$, $W = 9.8 \text{ kN}$) has an antilock braking system that gives a braking efficiency of 100% and is tested under the same conditions. How inaccurate will the stopping distance predicted by the practical-stopping-distance equation be (with $f = 0.640$)? How inaccurate will the practical-stopping-distance equation be if the new car has the same 85% braking efficiency as the original test car?

SOLUTION

For the case with 100% braking efficiency ($\eta_b = 1.00$) with $\gamma_b = 1.04$, $\theta_g = 0°$, $f_{rl} = 0.0131$ (from Example 2.9), and

$$K_a = \frac{1.2256}{2}(0.25)(2.0) = 0.3064$$

Eq. 2.42 gives

$$S = \frac{1.04(9800)}{2(9.807)(0.3064)} \ln\left[1 + \frac{0.3064(100 \times 1000/3600)^2}{(1.0)(0.75)(9800) + (0.0131)(9800) \pm 0}\right]$$

$$= \underline{\underline{52.78 \text{ m}}}$$

Now, applying Eq. 2.48,

$$d = \frac{V_1^2}{2gf} = \frac{(100 \times 1000/3600)^2}{2(9.807)(0.640)} = \underline{\underline{61.47 \text{ m}}}$$

Therefore, the error is 8.69 m. By a similar procedure, the case with 85% braking efficiency gives

$$S = \frac{1.04(9800)}{2(9.807)(0.3064)} \ln \left[1 + \frac{0.3064(100 \times 1000/3600)^2}{(0.85)(0.75)(9800) + (0.0131)(9800) \pm 0} \right]$$

$$= \underline{\underline{61.75 \text{ m}}}$$

which results in an error of 0.28 m.

2.9.6 Distance Traveled During Perception/Reaction

Until now the focus has been directed toward the distance required to stop a vehicle from the point of brake application. However, in providing a driver sufficient stopping-sight distance, it is also necessary to consider the distance traveled during the time the driver is perceiving and reacting to the need to stop. Thus the stopping-sight-distance requirement, using practical stopping distance, d, is

$$d_s = d_p + d \tag{2.49}$$

where d_s is the total stopping distance (including perception/reaction) and d_p is the distance traveled during perception/reaction and is equal to

$$d_p = V_1 t_p \tag{2.50}$$

where t_p is the time (in seconds) required to perceive and react to the need to stop.

The perception/reaction time of a driver is a function of a number of factors, including the driver's age, physical condition, and emotional state, as well as the complexity of the situation and the strength of the stimuli requiring a stopping action. For highway design, a conservative perception/reaction time has been determined to be 2.5 seconds (AASHTO 1994). For comparison, average drivers have perception/reaction times of approximately 1.0 to 1.5 seconds.

EXAMPLE 2.11

Two drivers have reaction times of 2.5 seconds. One is obeying a 90-km/h speed limit and the other is traveling illegally at 120 km/h. How much distance will each of the drivers cover while perceiving/reacting to the need to stop?

SOLUTION

For the driver traveling at 90 km/h,

$$d_p = V_1 t_p = (90 \times 1000/3600)(2.5) = \underline{\underline{62.5 \text{ m}}}$$

For the driver traveling at 120 km/h,

$$d_p = V_1 t_p = (120 \times 1000/3600)(2.5) = \underline{\underline{83.33 \text{ m}}}$$

Therefore, driving at 120 km/h increases the distance covered during perception/reaction by a substantial 20.83 m.

NOMENCLATURE FOR CHAPTER 2

a	acceleration (deceleration if negative)
A_f	frontal area of vehicle
$BFR_{f/r\,max}$	brake-force ratio (front over rear) for maximum braking force
C_D	coefficient of aerodynamic drag
D_a	distance to accelerate
d	practical stopping distance
d_p	distance traveled during driver perception/reaction
d_s	total stopping distance (vehicle stopping distance plus perception/reaction distance)
F	total available tractive effort
F_b	total braking force
F_{bf}	front-axle braking force
$F_{bf\,max}$	maximum front-axle braking force
F_{br}	rear-axle braking force
$F_{br\,max}$	maximum rear-axle braking force
F_e	engine-generated tractive effort
F_f	available tractive effort at the front axle
F_{max}	maximum tractive effort
F_r	available tractive effort at the rear axle
f	friction term for practical stopping distance

f_{rl}	coefficient of rolling resistance
G	grade in m/m (percent grade divided by 100; $G = 0.05$ is a 5% grade)
g	gravitational constant (9.807 m/s²)
g_{max}	maximum deceleration achieved before wheel lockup
h	height of vehicle's center of gravity above the roadway surface
i	driveline slippage
K_a	elements of aerodynamic resistance that are not a function of speed
L	vehicle wheelbase
l_f	distance from vehicle's center of gravity to front axle
l_r	distance from vehicle's center of gravity to rear axle
M_e	engine torque
m	mass
n_e	engine speed in crankshaft revolutions per second
P_e	engine-generated power
P_{R_a}	power required to overcome aerodynamic resistance
$P_{R_{rl}}$	power required to overcome rolling resistance
PBF_f	optimal percent of braking force on the front axle
PBF_r	optimal percent of braking force on the rear axle
R_a	aerodynamic resistance
R_g	grade resistance
R_{rl}	rolling resistance
r	vehicle wheel radius
S	minimum theoretical stopping distance
t_p	driver perception/reaction time
V	vehicle speed
W	total vehicle weight
W_f	vehicle weight acting normal to the roadway surface on the front axle
W_r	vehicle weight acting normal to the roadway surface on the rear axle
γ_b	braking mass factor

γ_m	acceleration mass factor
ε_0	gear reduction ratio
η_b	braking efficiency
η_d	driveline efficiency
θ_g	angle of grade
μ	coefficient of road adhesion
ρ	air density

REFERENCES

ASSHTO (American Association of State Highway and Transportation Officials). "A Policy on Geometric Design of Highways and Streets." Washington, DC, 1994.

Brewer, H. K., and R. S. Rice. "Tires: Stability and Control." *SAE Transactions,* vol. 92, paper 830561, 1983.

Campbell, C. *The Sports Car: Its Design and Performance.* Cambridge, MA: Robert Bently, Inc., 1978.

Taborek, J. J. "Mechanics of Vehicles." *Machine Design,* 1957.

Wong, J. H. *Theory of Ground Vehicles.* New York: Wiley, 1978.

PROBLEMS

2.1. A new sports car has a drag coefficient of 0.29, a frontal area of 1.9 m², and is traveling at 160 km/h. How much power is required to overcome aerodynamic resistance if $\rho = 1.2256$ kg/m³?

2.2. A vehicle manufacturer is considering an engine for a new sedan ($C_D = 0.30$, $A_f = 2.0$ m²). The car will be tested at a maximum speed of 160 km/h on a paved surface at sea level ($\rho = 1.2256$ kg/m³). The car currently weighs 9.3 kN, but the designer selected an underpowered engine because aerodynamic and rolling resistance were ignored. If 9 N of additional vehicle weight is added for each kW of additional power produced by the engine (needed to overcome the neglected resistance), what will be the final weight of the car if it is to achieve the 160 km/h speed?

2.3. For Example 2.3, how far back from the front axle would the center of gravity have to be to ensure that the maximum tractive effort developed for front- and rear-wheel drive options is equal (assume all other variables are unchanged)?

2.4. A rear-wheel-drive 13.3-kN drag race car has a 510-cm wheelbase and a center of gravity 50 cm above the pavement and 355 cm behind the front axle. The owners wish to achieve an initial acceleration from rest of 4.6 m/s^2 on a level surface. Assuming $\gamma_m = 1.00$, what is the minimum coefficient of road adhesion needed to achieve this acceleration?

2.5. Suppose the drag race car in Problem 2.4 has a center of gravity 178 cm above the pavement and is run on pavement with a coefficient of adhesion equal to 0.95. Assuming $\gamma_m = 1.00$, how far back from the front axle would the center of gravity have to be to develop a maximum acceleration from rest of 1.2 g's (i.e., 11.77 m/s^2)?

2.6. A car is traveling on a paved road with $C_D = 0.35$, $A_f = 2$ m^2, $W = 13.3$ kN, and $\rho = 1.2256$ kg/m^3. Its engine is running at 3000 rpm and is producing 340 Nm of torque. The car's gear reduction ratio is 3.5 to 1, driveline efficiency is 90%, driveline slippage is 3.5%, and the road-wheel radius is 38 cm. What will the car's maximum acceleration rate be under these conditions on a level road? (Assume that the available tractive effort is the engine-generated tractive effort.)

2.7. A car ($C_D = 0.35$, $A_f = 2.3$ m^2, $W = 11$ kN, and $\rho = 1.2256$ kg/m^3) has 35.5-cm radius wheels, a driveline efficiency of 90%, is in a gear that gives an overall gear reduction ratio of 3.2 to 1, and has driveline slippage of 3.5%. The engine develops maximum torque of 270 Nm at 3500 rpm. What is the maximum grade this vehicle could ascend on a paved surface while its engine is developing maximum torque?

2.8. An 11-kN car has a maximum speed of 242 km/h (on a paved surface, level terrain, at sea level). The car has 35.5-cm radius wheels, is in a gear that gives an overall gear reduction ratio of 3.0 to 1, and has a driveline efficiency of 90%. At the car's top speed, the engine produces 270 Nm of torque. If the car's frontal area is 2.3 m^2, what is the aerodynamic drag coefficient?

2.9. A rear-wheel-drive car weighs 11 kN, has a 203-cm wheelbase, a center of gravity 50 cm above the paved roadway surface and 76 cm behind the front axle, a driveline efficiency of 75%, 35.5-cm radius wheels, and is in a gear with an overall gear reduction ratio of 11 to 1. The car's torque/engine speed curve is

$$M_e = 6n_e - 0.045n_e^2$$

If the car is on a roadway with a coefficient of adhesion of 0.75, what will be its maximum acceleration from rest?

2.10. Consider the car in Problem 2.9. It is known that the car achieves maximum speed at an overall gear reduction ratio of 2 to 1 with a driveline slippage of 3.5%. How fast would the car be going if it could achieve its maximum speed when its engine is producing maximum power?

2.11. Consider the situation described in Example 2.5. If the vehicle is redesigned with 33-cm radius wheels (assume that the mass factor is unchanged) and a center of gravity located at the same height but at the midpoint of the wheelbase, determine the maximum acceleration for front- and rear-wheel-drive alternatives.

2.12. An engineer designs a rear-wheel-drive car that weighs 8.9 kN without an engine. It has a wheelbase of 255 cm, driveline efficiency of 80%, 35.5-cm radius wheels, and the center of gravity without the engine is 55 cm above the roadway surface and 140 cm behind the front axle. An engine that weighs 13 N for each Nm of torque developed is to be placed in the front portion of the car. Calculations show that for every 89 N of engine weight added, the center of gravity stays at the same height but moves 2.5 cm closer to the front axle. Assume that the car is to start from rest in a gear with an overall gear reduction ratio of 10 to 1, on level pavement with a coefficient of adhesion of 0.8, and select an engine size (weight and associated torque) that will result in the highest possible available tractive effort.

2.13. If the car in Example 2.7 had $C_D = 0.45$, and $A_f = 2.3$ m², what would have been the difference in minimum theoretical stopping distances with and without aerodynamic resistance considered (all other factors the same as in Example 2.7)?

2.14. A car ($C_D = 0.40$, $A_f = 2.4$ m², $W = 15.6$ kN, and $\rho = 1.2256$ kg/m³) is driven on a surface with a coefficient of road adhesion of 0.5, and with the coefficient of rolling friction approximated as 0.015 for all speeds. Assuming minimum theoretical stopping distances, if the vehicle comes to a stop 76 m after brake application on a level surface and has a braking efficiency of 0.78, what is the initial speed if (a) aerodynamic resistance is considered, and (b) if aerodynamic resistance is ignored?

2.15. A level test road has a coefficient of road adhesion of 0.75, and a car being tested has a coefficient of rolling friction that is approximated as 0.018 for all speeds. The vehicle is tested unloaded and achieves the theoretical minimum stop in 61 m (from brake application). The initial speed was 100 km/h. Ignoring aerodynamic resistance, what is the unloaded braking efficiency?

2.16. A small truck is to be driven down a 4%-grade at 120 km/h. The coefficient of road adhesion is 0.95, and it is known that the braking efficiency is 80% when the truck is empty, and that it decreases by one percentage point for every 445 N of cargo added. Ignoring aerodynamic resistance, if the driver wants the truck to be able to achieve a minimum theoretical stopping distance of 90 m from the point of brake application, what is the maximum amount of cargo (in newtons) that can be carried?

2.17. Consider the conditions in Example 2.8. The car has $C_D = 0.50$, $A_f = 2.3$ m², $W = 15.6$ kN, and $\rho = 1.2256$ kg/m³, and a coefficient of rolling resistance that is approximated as 0.018 for all speeds. If aerodynamic resistance is considered in stopping, estimate how fast the car will be going when it strikes the object on a level grade and on a 5% up grade (all other conditions, speed, etc., as in Example 2.8).

2.18. An engineering student is driving on a level roadway and sees a construction sign 180 m ahead in the middle of the roadway. The student strikes the sign at a speed of 56 km/h. If the student was traveling at 90 km/h when the sign was first spotted, what was the student's associated perception/reaction time? (Use the practical-stopping-distance formula.)

2.19. Two cars are traveling on a level road at 100 km/h. The coefficient of road adhesion is 0.8. The driver of car 1 has a 2.5-s perception/reaction time and the driver of car 2 has a 2.0-s perception/reaction time. Both cars are traveling side by side and the drivers are able to stop their respective cars in the same distance from first seeing a roadway obstacle (i.e., perception/reaction plus vehicle stopping distance). If the braking efficiency of car 2 is 0.75, determine the braking efficiency of car 1. (Assume minimum theoretical stopping distance and ignore aerodynamic resistance.)

2.20. An engineering student claims that a country road can be safely negotiated at 110 km/h in rainy weather. Because of the winding nature of the road, one stretch of level pavement has a sight distance of only 180 m. Using the practical-stopping-sight-distance formula, comment on the student's claim.

2.21. A driver is traveling at 90 km/h on a wet road. An object is spotted on the road 140 m ahead and the driver is able to come to a stop just before hitting the object. Assuming standard reaction time and using the practical-stopping-distance equation, determine the grade of the road.

2.22. A test of a driver's reaction time is being conducted on a special testing track with wet pavement (i.e., f-values in Table 2.6 apply) and a driving speed of 90 km/h. When the driver is sober, a stop can be made just in time to avoid hitting an object that is first visible 160 m ahead. After a few drinks, and under the same conditions, the driver fails to stop in time and strikes the object at a speed of 55 km/h. Determine the driver's reaction time before and after drinking. (Use the practical-stopping-distance equation.)

Chapter 3

Geometric Design of Highways

3.1 INTRODUCTION

With the understanding of vehicle performance provided in Chapter 2, we can now turn to roadway design. The design of a roadway necessitates the determination of specific design elements which include the number of lanes, lane width, median type (if any) and width, length of acceleration and deceleration lanes for on and off ramps, need for truck climbing lanes for roadways with steep grades, curve radii requires for vehicle turning, and the roadway alignment required to provide adequate stopping- and passing-sight distances. Many of these design elements are influenced by the performance characteristics of vehicles. For example, vehicle acceleration and deceleration characteristics have a direct impact on the design of acceleration and deceleration lanes (i.e., the length needed to provide a safe and orderly flow of traffic), and the roadway alignment needed to provide adequate passing- and stopping-sight distances. Furthermore, vehicle performance characteristics determine the need for truck climbing lanes on steep grades (i.e., where the poor performance of large trucks necessitates a separate lane), as well as the number of roadway lanes required because the observed spacing between vehicles in traffic is directly related to vehicle performance characteristics (this will be discussed further in Chapter 5). In addition, the physical dimensions of vehicles affect a number of design elements such as the radii required for low-speed turning, height of highway overpasses, lane widths, and so on.

When one considers the diversity of vehicles, in terms of performance and physical dimensions, and the interaction of these characteristics with the many elements comprising the design of a roadway, it is clear that proper highway design is a complex procedure that requires numerous compromises. Moreover, it is important for design guidelines to evolve over time in response to changes in vehicle performance and dimensions, and to evidence collected regarding the effectiveness of existing highway design practices (e.g., the relationship between accident rates and various roadway design characteristics). Current guidelines of highway design are presented in detail in "A Policy on Geometric Design of Highways and Streets 1994," published by the American Association of State Highway and Transportation Officials (AASHTO 1994). The 1990 version of this publication has the same design guidelines, but they are presented in U.S. customary units (AASHTO 1990).

Because of the sheer number of geometric elements involved in the design of a roadway, presenting a detailed discussion of each design element is beyond the scope of this book, and the reader is referred to AASHTO 1994 for a complete discussion of current design practices. Instead, this book focuses exclusively on the key elements of highway alignment, which are arguably the most important components of geometric design. As will be shown, the alignment topic is particularly well suited for demonstrating the effect of vehicle performance (specifically braking performance) and vehicle dimensions (e.g., height of the driver's eye, headlight height, and vehicle height) on the design of highways. By concentrating on the specifics of the highway alignment problem, the reader will develop an understanding of the procedures and compromises inherent in the design of all highway geometric elements.

3.2 PRINCIPLES OF HIGHWAY ALIGNMENT

The alignment of a highway is a three-dimensional problem measured in x, y, and z dimensions. This is illustrated, from a driver's perspective, in Fig. 3.1. However, in highway design practice, three-dimensional design computations are cumbersome, and, what is perhaps more important, the implementation and construction of a design based on three-dimensional coordinates has historically

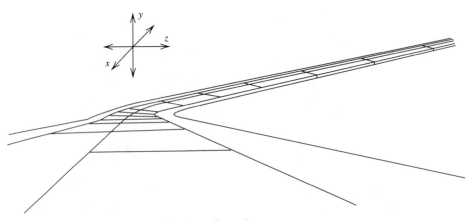

Figure 3.1 Highway alignment in three dimensions.

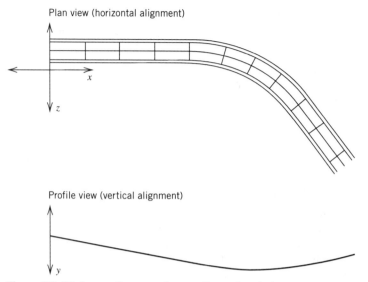

Figure 3.2 Highway alignment in two-dimensional views.

been prohibitively difficult. As a consequence, the three-dimensional highway alignment problem is reduced to two two-dimensional alignment problems, as illustrated in Fig. 3.2. One of the alignment problems in this figure corresponds roughly to x and z coordinates and is referred to as horizontal alignment. The other corresponds to highway length (measured along some constant elevation) and y coordinates (i.e., elevation). Referring to Fig. 3.2, note that the horizontal alignment of a highway is referred to as the *plan view*, which is roughly equivalent to the perspective of an aerial photo of the highway. The vertical alignment is represented in a *profile view*, which gives the elevation of all points measured along the length of the highway (again, with length measured along a constant elevation reference).

Aside from considering the alignment problem as two two-dimensional problems, one further simplification is made. Instead of using x and z coordinates, highway positioning and length are defined as the distance along the highway (usually measured along the centerline of the highway, on a horizontal, constant elevation plane) from a specified point. This distance is measured in terms of stations, with each station consisting of 1 km of highway alignment distance. The notation for stationing distance is such that a point on a highway 1258.5 m from a specified point of origin is said to be at station 1 + 258.500 (i.e., 1 station and 258.500 m) with the point of origin being at station 0 + 000.000. This stationing concept, when combined with the highway's alignment direction given in the plan view (horizontal alignment) and the elevation corresponding to stations given in the profile view (vertical alignment), gives a unique identification of all highway points in a manner that is virtually equivalent to using true x, y, and z coordinates.

3.3   VERTICAL ALIGNMENT

Vertical alignment specifies the elevations of points along a roadway. The elevations of these roadway points are usually determined by the need to provide proper drainage (from rainfall runoff) and an acceptable level of driver safety. A primary concern in vertical alignment is establishing the transition of roadway elevations between two grades. This transition is achieved by means of a vertical curve.

Vertical curves can be broadly classified into crest vertical curves and sag vertical curves as illustrated in Fig. 3.3. In this figure, G_1 is the initial roadway grade (also referred to as the initial tangent grade, viewing Fig. 3.3 from left to right), G_2 is the final roadway (tangent) grade, A is the absolute value of the difference in grades (initial minus final, usually expressed in percent), L is curve length, PVC is the point of the vertical curve (the initial point of the curve), PVI is the point of vertical intersection (intersection of initial and final grades), and PVT is the point of vertical tangent, which is the final point of the vertical curve (i.e., the point where the curve returns to the final grade or, equivalent, the final tangent). In practice, the vast majority of vertical curves are arranged such that half of the curve length is positioned before the PVI and half after (as illustrated in Fig. 3.3). Curves that satisfy this criterion are said to be *equal tangent vertical curves.*

In terms of referencing points on a vertical curve, it is important to note that the profile views presented in Fig. 3.3 correspond to all highway points even if a horizontal curve occurs concurrently with a vertical curve (as in Fig. 3.1 and Fig. 3.2). Thus each roadway point is uniquely defined by stationing (which is measured along a horizontal plane) and elevation. This will be made clearer through forthcoming examples.

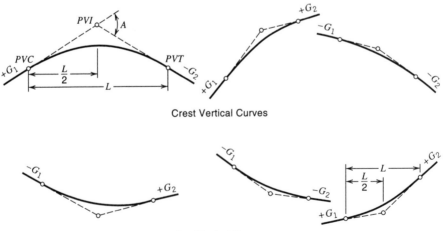

Crest Vertical Curves

Sag Vertical Curves

Figure 3.3 Types of vertical curves. (Reproduced by permission from American Association of State Highway and Transportation Officials, "A Policy on Geometric Design of Highways and Streets," Washington, DC, 1994.).

3.3.1 Vertical Curve Fundamentals

In connecting roadway grades (tangents) with an appropriate vertical curve, a mathematical relationship defining roadway elevations at all points (or, equivalently, stations) along the vertical curve is needed. A parabolic function has been found suitable in this regard because, among other things, it provides a constant rate of change of slope and implies equal curve tangents. The general form of the parabolic equation, as applied to vertical curves, is

$$y = ax^2 + bx + c \qquad (3.1)$$

where y is the roadway elevation x meters from the beginning of the vertical curve (i.e., from PVC). By definition, c is the elevation of the PVC, because $x = 0$ corresponds to the PVC. In defining a and b, note that the first derivative of Eq. 3.1 gives the slope and is

$$\frac{dy}{dx} = 2ax + b \qquad (3.2)$$

At the PVC, $x = 0$, so, using Eq. 3.2,

$$b = \frac{dy}{dx} = G_1 \qquad (3.3)$$

where G_1 is the initial slope (in m/m) as previously defined. Also note that the second derivative of Eq. 3.1 is the rate of change of slope and is

$$\frac{d^2y}{dx^2} = 2a \qquad (3.4)$$

However, the average rate of change of slope, by observation of Fig. 3.3, can also be written as

$$\frac{d^2y}{dx^2} = \frac{G_2 - G_1}{L} \qquad (3.5)$$

Equating Eqs. 3.4 and 3.5 gives

$$a = \frac{G_2 - G_1}{2L} \qquad (3.6)$$

where L is the curve length in meters, G_1 is the initial grade in m/m, and G_2 is the final grade in m/m. These equations define all of the terms in the parabolic vertical curve equation (Eq. 3.1). The following example gives a typical application of this equation.

EXAMPLE 3.1

A 200-m equal tangent sag vertical curve has the *PVC* at station 3 + 700.000 and elevation 321 m. The initial grade is −3.5% and the final grade is 0.5%. Determine the elevation and stationing of the *PVI*, *PVT*, and lowest point on the curve.

SOLUTION

Because the curve is equal tangent, the *PVI* will be 100 m (measured in a horizontal, profile plane) from the *PVC*, and the *PVT* will be 200 m from the *PVC*. Therefore, the stationing of the *PVI* and *PVT* are 3 + 800.000 and 3 + 900.000, respectively. For the elevations of the *PVI* and *PVT*, it is known that a −3.5% grade can be equivalently written as −0.035 m/m. Because the *PVI* is 100 m from the *PVC*, which is known to be at elevation 321 m, the elevation of the *PVI* is

$$321 - 0.035 \text{ m/m} \times (100 \text{ m}) = 317.5 \text{ m}$$

Similarly, with the *PVI* at elevation 317.5 m, the elevation of the *PVT* is

$$317.5 + 0.005 \text{ m/m} \times (100 \text{ m}) = 318.0 \text{ m}$$

It is clear from the values of the initial and final grades that the lowest point on the vertical curve will occur when the first derivative of the parabolic function (Eq. 3.1) is zero because the initial and final grades are opposite in sign. When initial and final grades are not opposite in sign, the low (or high) point on the curve will not be where the first derivative is zero because the slope along the curve will never be zero. For example, a sag curve with an initial grade of −2.0 percent and a final grade of −1.0 percent will have its lowest elevation at the *PVT*, and the first derivative of Eq. 3.1 will not be zero at any point along the curve. However, in our example problem the derivative will be equal to zero at some point, so the low point will occur when

$$\frac{dy}{dx} = 2ax + b = 0$$

From Eq. 3.3 we have

$$b = G_1 = -0.035$$

and from Eq. 3.6,

$$a = \frac{0.005 - (-0.035)}{2(200)} = 0.0001$$

Substituting for a and b gives

$$\frac{dy}{dx} = 2(0.0001)x + (-0.035) = 0$$

$$x = 175 \text{ m}$$

This gives the stationing of the low point at $3 + 875.000$ (i.e., 175 m from the PVC). For the elevation of the lowest point on the vertical curve, the values of a, b, c (elevation of the PVC), and x are substituted into Eq. 3.1, giving

$$y = 0.0001(175)^2 + (-0.035)(175) + 321$$

$$= \underline{\underline{317.94 \text{ m}}}$$

Another interesting vertical curve problem that is sometimes encountered is one in which the curve must be designed so that the elevation of a specific location is met. An example might be to have the roadway connect with another (at the same elevation), or to have the roadway at some specified elevation pass under another roadway. This type of problem is referred to as a curve-through-a-point problem and is demonstrated by the following example.

EXAMPLE 3.2

An equal tangent vertical curve is to be constructed between grades of -2.0% (initial) and 1.0% (final). The PVI is at station $11 + 000.000$ and at elevation 420 m. Due to a street crossing the roadway, the elevation of the roadway at station $11 + 071.000$ must be at 421.5 m. Design the curve.

SOLUTION

The design problem is one of determining the length of the curve required to ensure that station $11 + 071.000$ is at elevation 421.5 m. To begin, we use Eq. 3.1:

$$y = ax^2 + bx + c$$

From Eq. 3.3,

$$b = G_1 = -0.02$$

and, from Eq. 3.6,

$$a = \frac{G_2 - G_1}{2L}$$

Substituting $G_1 = -0.02$ and $G_2 = 0.01$, we have

$$a = \frac{G_2 - G_1}{2L} = \frac{0.01 - (-0.02)}{2L} = \frac{0.015}{L}$$

Now note that c (the elevation of the *PVC*) in Eq. 3.1 will be equal to the elevation of the *PVI* plus $G_1 \times 0.5L$. (This is just using the slope of the initial grade to determine the elevation difference between the *PVI* and *PVC*.) With G_1 in m/m and the curve length L in meters, we have

$$c = 420 + 0.02(0.5L) = 420 + 0.01L$$

Finally, the value of x to be used in Eq. 3.1 will be $0.5L + 71$ because the point of interest (station $11 + 071.000$) is 71 meters from the *PVI* (which is station $11 + 000.000$). Substituting $b = -0.02$; the expressions for a, c, and x; and $y = 421.5$ m (the given elevation) into Eq. 3.1 gives

$$421.5 = (0.015/L)(0.5L + 71)^2 + (-0.02)(0.5L + 71) + (420 + 0.01L)$$

$$0 = 0.00375\,L^2 - 1.855L + 75.615$$

Solving this quadratic equation gives $L = 44.816$ m (which is not feasible because we know that the point of interest is 71 m beyond the *PVI*, so the curve length must be longer than 44.816 m) or $L = 449.851$ m (which is the only feasible solution). This means that the curve must be 449.851 m long. Using this value of L,

elevation of $PVC = c = 420 + 0.01L = 420 + 4.49851 = 424.50$ m

stationing of $PVC = 11 + 000.000 - (449.851)/2 = 10 + 775.075$

elevation of $PVT =$ elevation of $PVI + (0.5L)G_2 = 420 + [0.5(449.851)](0.01)$

$$= 422.25 \text{ m}$$

stationing of $PVT = 11 + 000 + (449.851)/2 = 11 + 224.926$

and

$$x = 0.5L + 71 = 224.926 + 71 = 295.926 \text{ m from the } PVC$$

To check the elevation of the curve at station $11 + 071.000$, we apply Eq. 3.1 with $x = 295.926$:

$$y = ax^2 + bx + c$$

$$y = \left[\frac{0.03}{2(449.851)}\right](295.926)^2 + (-0.02)(295.926) + 424.5$$

$$y = \underline{\underline{421.5 \text{ m}}}$$

Therefore, all calculations are correct.

Some additional properties of vertical curves can now be formalized. For example, offsets, which are vertical distances from the initial tangent to the curve as illustrated in Fig. 3.4, are extremely important in vertical curve design and construction. In Fig. 3.4, Y is the offset at any distance, x, from the PVC; Y_m is the midcurve offset; and Y_f is the offset at the end of the vertical curve. From the properties of an equal tangent parabola, it can be readily shown that

$$Y = \frac{A}{200L}x^2 \tag{3.7}$$

where Y is the offset in meters, A is the absolute value of the difference in grades ($|G_1 - G_2|$ expressed in percent), L is the length of the vertical curve in meters, and x is the distance from the PVC in meters. Note that in this equation, 200 is used in the denominator instead of 2 because A is expressed in percent instead of m/m (this division by 100 also applies to Eq. 3.8 and 3.9 below). It follows from Fig. 3.4 that

$$Y_m = \frac{AL}{800} \tag{3.8}$$

and

$$Y_f = \frac{AL}{200} \tag{3.9}$$

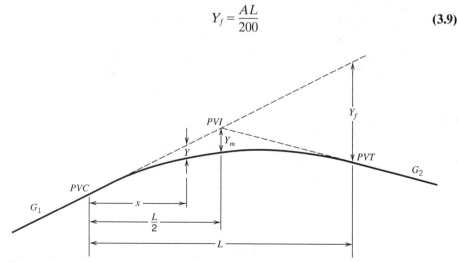

Figure 3.4 Offsets for equal tangent vertical curves.

Another useful vertical curve property is one that simplifies the computation of the high and low points of crest and sag vertical curves, respectively (given that the high or low point does not occur at the curve ends, *PVC* or *PVT*). Recall that in Example 3.1, the first derivative was used to determine the location of the low point. The alternative to this is to use a *K*-value defined as (with *L* in meters and *A* in percent)

$$K = \frac{L}{A} \tag{3.10}$$

The *K*-value can be used directly to compute the high/low points for crest/sag vertical curves by

$$x_{hl} = K \times |G_1| \tag{3.11}$$

where x_{hl} is the distance from the *PVC* to the high/low point in meters, and G_1 is the initial grade in percent. In words, the *K*-value is the horizontal distance, in meters, required to affect a 1% change in the slope of the vertical curve. Aside from high/low point computations, it will be shown in Sections 3.3.3 and 3.3.4 that the *K*-value has many important applications in the design of vertical curves.

EXAMPLE 3.3

A vertical curve crosses a 1-m diameter pipe at right angles. The pipe is located at station 11 + 025.000 and its centerline is at elevation 1091.60 m. The *PVI* of the vertical curve is at station 11 + 000.000 and elevation 1095.20 m. The vertical curve is equal tangent, 150 m long, and connects an initial grade of +1.20% and a final grade of −1.08%. Using offsets, determine the depth, below the surface of the curve, to the top of the pipe, and determine the station of the highest point on the curve.

SOLUTION

The *PVC* is at station 10 + 925.000 (11 + 000.000 minus 0 + 075.000, which is half of the curve length), so the pipe is 100 m (11 + 025.000 minus 10 + 925.000) from the beginning of the curve (*PVC*). The elevation of the *PVC* will be the elevation of the *PVI* minus the drop in grade over one-half the curve length:

$$1095.2 - (75 \text{ m} \times 0.012 \text{ m/m}) = 1094.30 \text{ m}$$

Using this, the elevation of the initial tangent above the pipe is

$$1094.3 + (100 \text{ m} \times 0.012 \text{ m/m}) = 1095.50 \text{ m}$$

Using Eq. 3.7 to determine the offset above the pipe at $x = 100$ m (the distance of the pipe from the *PVC*), we have

$$Y = \frac{A}{200L} x^2$$

$$Y = \frac{|1.2 - (-1.08)|}{200(150)} (100)^2 = 0.76 \text{ m}$$

Thus the elevation of the curve above the pipe is 1094.74 m (1095.50 − 0.76). The elevation of the top of the pipe is 1092.1 m (elevation of the centerline plus one half of the pipe's diameter), so the pipe is 2.64 m below the surface of the curve (1094.74 − 1092.1).

To determine the location of the highest point on the curve, we find K from Eq. 3.10:

$$K = \frac{150}{|1.2 - (-1.08)|} = 65.79$$

The distance from the *PVC* to the highest point is (from Eq. 3.11)

$$x = K \times |G_1| = 65.79 \times 1.2 = 78.948 \text{ m}$$

This gives the station of the highest point at 11 + 003.948 (10 + 925.00 plus 0 + 078.948). Note that this example could also be solved by applying Eq. 3.1, setting Eq. 3.2 equal to zero (for determining the location of the highest point on the curve), and following the procedure used in Example 3.1.

3.3.2 Minimum and Desirable Stopping-Sight Distances

Construction of a vertical curve is generally a costly operation requiring the movement of significant amounts of earthen material. Thus one of the primary challenges facing highway designers is to minimize construction costs (usually by making the vertical curve as short as possible) while still providing an adequate level of safety. An adequate level of safety is usually defined as a level of safety that provides drivers with sufficient sight distance to allow them to safely stop their vehicles to avoid collisions with objects obstructing their forward motion. The provision of adequate roadway drainage is sometimes an important concern as well, but is not discussed in terms of vertical curves in this book (see AASHTO 1994). Referring back to the vehicle braking performance concepts discussed in Chapter 2, we can compute the necessary stopping-sight distance (SSD) simply as the summation of vehicle stopping-sight distance (Eq. 2.48) and the distance traveled during perception/reaction time (Eq. 2.50). That is,

$$\text{SSD} = \frac{V_1^2}{2g(f \pm G)} + V_1 t_p \tag{3.12}$$

where SSD is the stopping-sight distance, V_1 is the initial vehicle speed in m/s, g is the gravitational constant in m/s^2, f is the coefficient of braking friction, G is the grade, and t_p is the perception/reaction time.

Given Eq. 3.12, the question now becomes one of selecting appropriate values for the computation of SSD. Values of the coefficient of braking friction, f, are selected to be representative of poor driver skills, low braking efficiencies, and wet pavements, thus providing conservative or near-worst-case stopping-sight distances (see Table 2.6 for such examples of f). Also, a standard driver perception/reaction time of 2.5 seconds is typically used and is a conservative estimate, as individual reaction times are generally less than this. In terms of the assumed initial vehicle speeds used in Eq. 3.12, two values are worthy of note. One is the design speed of the highway, which is defined as the maximum safe speed that a highway can be negotiated assuming near-worst-case conditions (wet weather conditions). The second value is the average vehicle running speed, which is obtained from actual vehicle speed observations under low traffic volume conditions, and which usually ranges from about 83 to 100% of the highway's design speed, depending on the actual design speed of the highway. At lower highway design speeds, the average running speed is equal to or only slightly less than the highway's design speed. At higher design speeds, the average running speed is significantly less than the highway's design speed. The average running speed reflects drivers' general tendency to drive below the design speed of the highway even under low-volume, uncongested traffic conditions. The application of Eq. 3.12 (assuming $G = 0$) produces the stopping-sight distances presented in Table 3.1 for average running speeds (the first number in the "assumed speed for condition" column) and design speeds (the second number in the "assumed speed for condition" column). Because average running speeds are equal to or lower than the corresponding design speeds (depending on the design speed value), the stopping-sight distance required will be generally lower for average running speeds. Stopping-sight distances derived from average running speeds are referred to as *minimum* SSDs, whereas stopping-sight distances derived from design speeds are referred to as *desirable* SSDs. Whenever possible, desirable SSDs should be used. Many states require special exceptions for designers to use stopping-sight distances based on anything less than the design speed of the highway.

Table 3.1 Stopping-Sight Distance (Minimum and Desirable for Wet Pavements)

Design Speed (km/h)	Assumed Speed for Condition (km/h)	Brake Reaction Time (s)	Brake Reaction Distance (m)	Coefficient of Friction f	Braking Distance on Level (m)	Stopping-Sight Distance for Design (m)
30	30–30	2.5	20.8–20.8	0.40	8.8–8.8	29.6–29.6
40	40–40	2.5	27.8–27.8	0.38	16.6–16.6	44.4–44.4
50	47–50	2.5	32.6–34.7	0.35	24.8–28.1	57.4–62.8
60	55–60	2.5	38.2–41.7	0.33	36.1–42.9	74.3–84.6
70	63–70	2.5	43.7–48.6	0.31	50.4–62.2	94.1–110.8
80	70–80	2.5	48.6–55.5	0.30	64.2–83.9	112.8–139.4
90	77–90	2.5	53.5–62.5	0.30	77.7–106.2	131.2–168.7
100	85–100	2.5	59.0–69.4	0.29	98.0–135.6	157.0–205.0
110	91–110	2.5	63.2–76.4	0.28	116.3–170.0	179.5–246.4
120	98–120	2.5	68.0–83.3	0.28	134.9–202.3	202.9–285.6

Source: American Association of State Highway and Transportation Officials, "A Policy on Geometric Design of Highways and Streets," Washington, DC, 1994.

3.3.3 Stopping-Sight Distance and Crest Vertical Curve Design

In providing sufficient SSD on a vertical curve, the length of curve (L in Fig. 3.3) is the critical concern. Longer lengths of curve provide more SSD, all else being equal, but are more costly to construct. Shorter curve lengths are relatively inexpensive to construct but may not provide adequate SSD. What is needed, then, is an expression for minimum curve length given a required SSD. In developing such an expression, crest and sag vertical curves are considered separately.

For the case of a crest vertical curve, consider the diagram presented in Fig. 3.5. In this figure, S is the sight distance, L is the length of the curve, H_1 is the driver's eye height, H_2 is the height of a roadway object, and other terms are as previously defined. Using the properties of a parabola for an equal tangent curve, it can be shown that the minimum length of curve, L_m, for a required sight distance, S, is

$$L_m = 2S - \frac{200(\sqrt{H_1} + \sqrt{H_2})^2}{A} \qquad \text{for } S > L \qquad \textbf{(3.13)}$$

and

$$L_m = \frac{AS^2}{200(\sqrt{H_1} + \sqrt{H_2})^2} \qquad \text{for } S < L \qquad \textbf{(3.14)}$$

For the sight distance required to provide adequate SSD, current AASHTO design guidelines (1994) use a driver eye height, H_1, of 1.07 m (1070 mm) and a roadway object height, H_2 (the height of an object to be avoided by stopping before a collision), of 0.15 m (150 mm). In applying Eqs. 3.13 and 3.14 to determine the minimum length of curve required to provide adequate SSD, we set the sight distance, S, equal to the stopping-sight distance, SSD. Substituting AASHTO guidelines for H_1 and H_2 and $S = $ SSD into Eqs. 3.13 and 3.14 gives

$$L_m = 2\,\text{SSD} - \frac{404}{A} \qquad \text{for SSD} > L \qquad \textbf{(3.15)}$$

$$L_m = \frac{A\,\text{SSD}^2}{404} \qquad \text{for SSD} < L \qquad \textbf{(3.16)}$$

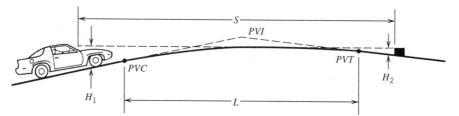

Figure 3.5 Stopping-sight-distance considerations for crest vertical curves.

EXAMPLE 3.4

A highway is being designed to AASHTO guidelines with a 120-km/h design speed. At one section, an equal tangent vertical curve must be designed to connect grades of +1.0% and −2.0%. Determine the length of curve required assuming provisions are to be made for minimum SSD and desirable SSD.

SOLUTION

The first concern is to determine the SSD to use. If we ignore the effects of grades (i.e., $G = 0$), the SSDs can be read directly from Table 3.1. In this case, the assumed average running speed (which is used to determine minimum SSD) is 98 km/h and the corresponding SSD is 202.9 m. If we assume that $L >$ SSD (an assumption that is typically made), Eq. 3.16 gives

$$L_m = \frac{A \text{ SSD}^2}{404} = \frac{3(202.9)^2}{404} = \underline{\underline{305.706 \text{ m}}}$$

Since $305.706 > 202.9$, the assumption that $L >$ SSD was correct.

For desirable SSD, the assumed speed is the design speed (120 km/h) and the corresponding SSD is 285.6 m (from Table 3.1). Again assuming that $L >$ SSD, Eq. 3.16 gives

$$L_m = \frac{3(285.6)^2}{404} = \underline{\underline{605.698 \text{ m}}}$$

Since $605.698 > 285.6$, the assumption that $L >$ SSD was again correct. Note that the provision of desirable SSD requires a significantly longer curve than providing SSD for the average running speed (minimum SSD).

The assumption that $G = 0$, made at the beginning of Example 3.4, is not strictly correct. If $G \neq 0$, we cannot use the SSD values in Table 3.1 and instead must apply Eq. 3.12 with the appropriate G value. In this problem, if we use the initial grade in Eq. 3.12 (+1.0%), we will underestimate the stopping-sight distance because the vertical curve has a slope as steeply positive as this only at the *PVC*. If we use the final grade in Eq. 3.12 (−2.0%), we will overestimate the stopping-sight distance because the vertical curve has a slope as steeply negative as this only at the *PVT*. If we knew where the vehicle began to brake, we could use the first derivative of the parabolic curve function (from Eq. 3.2) to give G in Eq. 3.12 and set up the equation to solve for SSD exactly. In practice, policies vary as to how this grade problem is handled. Fortunately, because sight distance tends to be greater on downgrades (which require longer stopping distances) than on upgrades, a self-correction for the effect of grades is generally provided. As a consequence, some design agencies ignore the effect of grades completely, while others assume G is equal to zero for grades less than 3% and use simple adjustments to the SSD, depending on the initial and final grades, if grades exceed 3% (see AASHTO 1994). For the remainder of this book, we will ignore the effect of grades (i.e., assume that $G = 0$ in Eq. 3.12). However, it must be pointed out that the use of SSD-grade corrections is very easy and straightforward and all of the equations presented herein still apply.

Using Eqs. 3.15 and 3.16 can be simplified if the initial assumption that $L >$ SSD is made, in which case Eq. 3.16 is always used. The advantage of this assumption is that the relationship between A and L_m is linear, and Eq. 3.10 can be used to give

$$L_m = KA \tag{3.17}$$

where

$$K = \frac{\text{SSD}^2}{404} \tag{3.18}$$

Again, K is the horizontal distance, in meters, required to effect a 1% change in the slope (as in Eq. 3.10). With known SSD for a given design speed (assuming $G = 0$), K-values can be computed for crest vertical curves as shown in Table 3.2. Thus the minimum curve length can be obtained (as shown in Eq. 3.17) simply by multiplying A by the K-value read from Table 3.2.

Some discussion about the assumption that $L >$ SSD is warranted. Basically this assumption is made because there are two complications that could arise when SSD $> L$. First, if SSD $> L$ the relationship between A and L_m is not

Table 3.2 Design Controls for Crest Vertical Curves Based on Minimum and Desirable Stopping-Sight Distance

Design Speed (km/h)	Assumed Speed for Condition (km/h)	Coefficient of Friction f	Stopping-Sight Distance for Design (m)	Rate of Vertical Curvature, K (Length [m] per % of A)	
				Computed	Rounded for Design
30	30–30	0.40	29.6–29.6	2.17–2.17	3–3
40	40–40	0.38	44.4–44.4	4.88–4.88	5–5
50	47–50	0.35	57.4–62.8	8.16–9.76	9–10
60	55–60	0.33	74.3–84.6	13.66–17.72	14–18
70	63–70	0.31	94.1–110.8	21.92–30.39	22–31
80	70–80	0.30	112.8–139.4	31.49–48.10	32–49
90	77–90	0.30	131.2–168.7	42.61–70.44	43–71
100	85–100	0.29	157.0–205.0	61.01–104.02	62–105
110	91–110	0.28	179.5–246.4	79.75–150.28	80–151
120	98–120	0.28	202.9–285.6	101.90–201.90	102–202

Source: American Association of State Highway and Transportation Officials, "A Policy on Geometric Design of Highways and Streets," Washington, DC, 1994.

linear; thus K-values cannot be used in the $L = KA$ formula (Eq. 3.10). Second, at low values of A, it is possible to get negative minimum curve lengths (see Eq. 3.15). As a result of these complications, the assumption that $L > $ SSD is almost always made in practice, and thus Eqs. 3.17 and 3.18 and the K-values presented in Table 3.2 are valid. It is important to note that the assumption that $L > $ SSD (upon which Eqs. 3.17 and 3.18 are made) is usually a safe one because, in many cases, L is greater than SSD. When it is not (i.e., SSD $> L$), the use of the $L > $ SSD formula (Eq. 3.16 instead of Eq. 3.15) gives longer curve lengths and thus the error is on the conservative, safe side.

A final point relates to the smallest allowable length of curve. Very short vertical curves can be difficult to construct and may not be worth the effort. As a result, it is common practice to set minimum curve length limits that range from 30 to 100 m depending on individual jurisdictional guidelines. A common alternative to these limits is to set the minimum curve length limit (in meters) at 0.6 times the design speed, in km/h (AASHTO 1994).

EXAMPLE 3.5

Solve Example 3.4 using the K-values in Table 3.2.

SOLUTION

From Example 3.4, $A = 3$. For minimum SSD, $K = 102$ (from Table 3.2) at a 120-km/h design speed (assumed running speed is 98 km/h). Therefore, application of Eq. 3.17 gives

$$L_m = KA = 102(3) = \underline{\underline{306 \text{ m}}}$$

which is just slightly different than the 305.76 m obtained in Example 3.4. This difference is due to round-off error. In this example, the rounded K of 102 was used as opposed to the computed K of 101.90. It is important to note that such small differences are not significant in design.

For desirable SSD, $K = 202$ (from Table 3.2) at a 120-km/h design speed, giving

$$L_m = 202(3) = \underline{\underline{606 \text{ m}}}$$

which is close to the 605.698 obtained in Example 3.4 (again, the difference can be attributed to round-off error in the K-value).

EXAMPLE 3.6

If the grades in Example 3.4 intersect at station 10 + 000.000, determine the stationing of the PVC, PVT, and curve high points for both minimum and desirable SSD curve lengths.

SOLUTION

Using the curve lengths from Example 3.5, $L = 306$ m for minimum SSD. Because the curve is equal tangent (as are virtually all curves used in practice), one-half of the curve will occur before the PVI and one-half after, so that

PVC is at $10 + 000.000 - L/2 = 10 + 000.000 - 0 + 153.000 = 9 + 847.000$

PVT is at $10 + 000.000 + L/2 = 10 + 000.000 + 0 + 153.000 = 10 + 153.000$

For the stationing of the high point, Eq. 3.11 is used:

$$x_{hl} = K \times |G_1| = 102(1) = 102 \text{ m}$$

or at station $9 + 847.000 + 0 + 102.000 = 9 + 949.000$

Similarly, for desirable SSD, the PVC can be shown to be at station $9 + 697.000$, PVT at station $10 + 303.000$, and the high point at station $9 + 899.000$.

3.3.4 Stopping-Sight Distance and Sag Vertical Curve Design

Sag vertical curve design differs from crest vertical curve design in the sense that sight distance is governed by nighttime conditions because, in daylight, sight distance on a sag vertical curve is unrestricted. Thus the critical concern for sag vertical curve design is the length of roadway illuminated by the vehicle headlights, which is a function of the height of the headlight above the roadway, H, and the inclined angle of the headlight beam relative to the horizontal plane of the car, β. The sag vertical curve sight-distance design problem is illustrated in Fig. 3.6. By using the properties of a parabola for an equal tangent vertical curve (as was done for the crest vertical curve case), it can be shown that the minimum length of curve, L_m, for a required sight distance, S, is

$$L_m = 2S - \frac{200(H + S \tan \beta)}{A} \qquad \text{for } S > L \qquad \textbf{(3.19)}$$

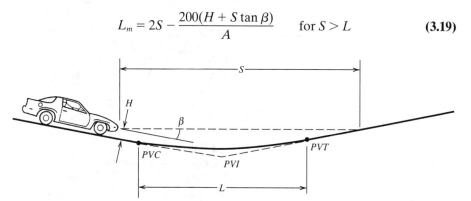

Figure 3.6 Stopping-sight-distance considerations for sag vertical curves.

$$L_m = \frac{AS^2}{200(H + S \tan \beta)} \qquad \text{for } S < L \qquad \textbf{(3.20)}$$

For the sight distance required to provide adequate SSD, current AASHTO design guidelines (1994) use a headlight height of 0.6 m (600 mm) and an upward angle of 1 degree. Substituting these design guidelines and S = SSD (as was done in the crest vertical curve case) into Eqs. 3.19 and 3.20 gives

$$L_m = 2\,\text{SSD} - \frac{120 + 3.5\,\text{SSD}}{A} \qquad \text{for SSD} > L \qquad \textbf{(3.21)}$$

$$L_m = \frac{A\,\text{SSD}^2}{120 + 3.5\,\text{SSD}} \qquad \text{for SSD} < L \qquad \textbf{(3.22)}$$

As was the case for crest vertical curves, K-values can be computed by assuming $L > \text{SSD}$, which gives us the linear relationship between L_m and A as shown in Eq. 3.22. Thus for sag vertical curves (with $L_m = KA$),

$$K = \frac{\text{SSD}^2}{120 + 3.5\,\text{SSD}} \qquad \text{for SSD} < L \qquad \textbf{(3.23)}$$

The K-values corresponding to minimum and desirable SSDs are presented in Table 3.3. As was the case for crest vertical curves, some caution should be exercised in using this table because the assumption that $G = 0$ (for determining

Table 3.3 Design Controls for Sag Vertical Curves Based on Minimum and Desirable Stopping-Sight Distance

Assumed Design Speed (km/h)	Speed for Condition (km/h)	Coefficient of Friction f	Stopping-Sight Distance for Design (m)	Rate of Vertical Curvature, K (Length [m] per % of A)	
				Computed	Rounded for Design
30	30–30	0.40	29.6–29.6	3.88–3.88	4–4
40	40–40	0.38	44.4–44.4	7.11–7.11	8–8
50	47–50	0.35	57.4–62.8	10.20–11.54	11–12
60	55–60	0.33	74.3–84.6	14.45–17.12	15–18
70	63–70	0.31	94.1–110.8	19.62–24.08	20–25
80	70–80	0.30	112.8–139.4	24.62–31.86	25–32
90	77–90	0.30	131.2–168.7	29.62–39.95	30–40
100	85–100	0.29	157.0–205.0	36.71–50.06	37–51
110	91–110	0.28	179.5–246.4	42.95–61.68	43–62
120	98–120	0.28	202.9–285.6	49.47–72.72	50–73

Source: American Association of State Highway and Transportation Officials, "A Policy on Geometric Design of Highways and Streets," Washington, DC, 1994.

SSD) is used. Also, assuming that $L >$ SSD is a safe, conservative assumption (as was the case for crest vertical curves), and the smallest allowable curve lengths for sag curves are the same as those for crest curves (see discussion in Section 3.3.3).

EXAMPLE 3.7

A highway has a design speed of 100 km/h, but an engineering mistake has resulted in the need to connect an already constructed tunnel and bridge with sag and crest vertical curves. The profile view of the tunnel and bridge is given in Fig. 3.7. Devise a vertical alignment to connect the tunnel and bridge by determining the highest possible common design speed (for desirable SSD) for the sag and crest (equal tagent) vertical curves needed. Compute the stationing and elevations of *PVC*, *PVI*, and *PVT* curve points.

SOLUTION

From left to right (see Fig. 3.7), a sag vertical curve (with subscripting s) and a crest vertical curve (with subscripting c) are needed to connect the tunnel and bridge. From given information, it is known that $G_{1s} = 0\%$ (i.e., the initial slope of the sag vertical curve) and $G_{2c} = 0\%$ (i.e., the final slope of the crest vertical curve). To obtain the highest possible design speed, we will want to use all of the horizontal distance available. This means we will want to connect the curve such that the *PVT* of the sag curve (PVT_s) will be the *PVC* of the crest curve (PVC_c). If this is the case, $G_{2s} = G_{1c}$, and because $G_{1s} = G_{2c} = 0$, then $A_s = A_c = A$, the common algebraic difference in the grades.

Because 450 m separate the tunnel and bridge,

$$L_s + L_c = 450$$

Also, the summation of the end-of-curve offset for the sag curve and the beginning-of-curve offset (relative to the final grade) for the crest curve must equal

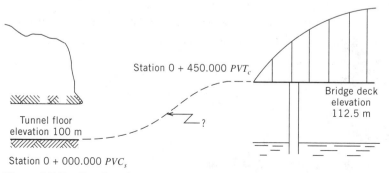

Station 0 + 450.000 PVT_c

Bridge deck elevation 112.5 m

Tunnel floor elevation 100 m

Station 0 + 000.000 PVC_s

Figure 3.7 Profile view (vertical-alignment diagram) for Example 3.7.

12.5 m. Using the equation for the final offset, Eq. 3.9, we have

$$\frac{AL_s}{200} + \frac{AL_c}{200} = 12.5$$

Rearranging,

$$\frac{A}{200}(L_s + L_c) = 12.5$$

Because $L_s + L_c = 450$,

$$\frac{A}{200}(450) = 12.5$$

Solving for A gives $A = 5.556\%$. The problem now becomes one of finding K-values that allow $L_s + L_c = 450$. Because $L = KA$ (Eq. 3.17), we can write

$$K_s A + K_c A = 450$$

Substituting $A = 5.556$,

$$K_s + K_c = 81$$

To find the highest possible design speed, Tables 3.2 and 3.3 are used to arrive at desirable K-values to solve $K_s + K_c = 81$. By looking at Tables 3.2 and 3.3, we see that the highest possible design speed is 80 km/h, at which speed $K_c = 49$ and $K_s = 32$ (i.e., the summation of K's is 81).

To arrive at the stationing of curve points, we first determine curve lengths as

$$L_s = K_s A = 32(5.556) = 177.778 \text{ m}$$

$$L_c = K_c A = 49(5.556) = 272.222 \text{ m}$$

Because the station of PVC_s is 0 + 000.000 (given), it is clear that $PVI_s = 0 + 088.889$, $PVT_s = PVC_c = 0 + 177.778$, $PVI_c = 0 + 313.889$, and $PVT_c = 0 + 450.000$. For elevations, $PVC_s = PVI_s = 100$ m, and $PVI_c = PVT_c = 112.5$ m. Finally, the elevation of PVT_s and PVC_c can be computed as

$$100 + \frac{AL_s}{200} = 100 + \frac{5.556(177.778)}{200} = 104.94 \text{ m}$$

EXAMPLE 3.8

Consider the conditions described in Example 3.7. Suppose a design speed of only 50 km/h (desirable) is needed. Determine the lengths of curves required to connect the bridge and tunnel while keeping the connecting grade as small as possible.

SOLUTION

It is known that the 450 m separating the bridge and tunnel are more than enough to connect a 50-km/h alignment because Example 3.7 showed that 80 km/h is possible. Therefore, to connect the bridge and tunnel and keep the connecting grade as small as possible, we will place a constant grade section between the sag and crest curves as shown in Fig. 3.8. The elevation change will be the final offsets of the sag and crest curves plus the change in elevation resulting from the constant grade section connecting the two curves. Let G_{con} be the grade of the constant grade section. This means that $G_{2s} = G_{1c} = G_{con}$, and, because $G_{1s} = G_{2c} = 0$ (as in Example 3.7), $G_{con} = A_s = A_c = A$. The equation that will solve the vertical alignment for this problem will be

$$\frac{AL_s}{200} + \frac{AL_c}{200} + \frac{A(450 - L_s - L_c)}{100} = 12.5$$

where the third term in this equation accounts for the elevation difference attributable to the constant grade section connecting the sag and crest curves (the 100 in the denominator of this term converts A from percent to m/m). Using $L = KA$ we have

$$\frac{A^2K_s}{200} + \frac{A^2K_c}{200} + \frac{A(450 - K_sA - K_cA)}{100} = 12.5$$

From Table 3.2, $K_c = 10$, and from Table 3.3, $K_s = 12$ (50 km/h desirable). Putting these values in the preceding equation gives

$$0.11A^2 + 4.50A - 0.22A^2 = 12.5$$

$$-0.11A^2 + 4.50A - 12.5 = 0$$

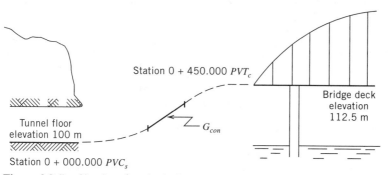

Station 0 + 450.000 PVT_c

Bridge deck
elevation
112.5 m

Tunnel floor
elevation 100 m

G_{con}

Station 0 + 000.000 PVC_s

Figure 3.8 Profile view (vertical-alignment diagram) for Example 3.8.

Solving this gives either $A = 2.997$ or $A = 37.92$; $A = 2.997$ is chosen because $A = 37.92$ results in curve lengths that exceed 450 m. For this value of A, the curve lengths are

$$L_s = K_s A = 12(2.997) = 35.964 \text{ m}$$

$$L_c = K_c A = 10(2.997) = 29.970 \text{ m}$$

and the length of the constant grade section will be 384.066 m. This means that about 11.510 m of the elevation difference will occur in the constant grade section, with the remainder of the elevation difference attributable to final curve offsets.

Another variation of this type of problem is the case in which the initial and final grades are not equal to zero. This makes the problem a bit more complex as demonstrated in the following example.

EXAMPLE 3.9

Two sections of highway are separated by 700 m as shown in Fig. 3.9. Determine the curve lengths required for a 70-km/h (desirable) vertical alignment to connect these two highway segments while keeping the connecting grade as small as possible.

SOLUTION

Let Y_{fc} and Y_{fs} be the final offsets of the crest and sag curves, respectively. Let G_{con} be the slope of a constant grade section connecting the crest and sag curves. (We will assume that the horizontal distance is sufficient to connect the highway with a 70-km/h alignment. If this assumption is incorrect, the following equations will produce an obviously erroneous answer and a lower design speed will have

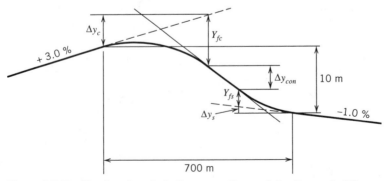

Figure 3.9 Profile view (vertical-alignment diagram) for Example 3.9.

to be chosen.) Finally, let Δy_{con} be the change in elevation over the constant grade section, and let Δy_c and Δy_s be the changes in elevation due to the extended curve tangents. The elevation equation is then (see Fig. 3.9)

$$Y_{fc} + Y_{fs} + \Delta y_{con} + \Delta y_s = 10 + \Delta y_c$$

Substituting offset equations and equations for elevation changes (with subscripting c for crest and s for sag),

$$\frac{A_c L_c}{200} + \frac{A_s L_s}{200} + \frac{G_{con}(700 - L_c - L_s)}{100} + \frac{1.0 \, L_s}{100} = 10 + \frac{3.0 \, L_c}{100}$$

Using $L = KA$, this equation becomes

$$\frac{A_c^2 K_c}{200} + \frac{A_s^2 K_s}{200} + \frac{G_{con}(700 - K_c A_c - K_s A_s)}{100} + \frac{1.0 \, K_s A_s}{100} = 10 + \frac{3.0 \, K_c A_c}{100}$$

From Tables 3.2 and 3.3, $K_c = 31$ and $K_s = 25$ (at 70 km/h desirable). Substituting and defining A's (and arranging the equation such that G_{con} will be positive, and assuming G_{con} will be greater than 1%) gives

$$\frac{(3 + G_{con})^2 31}{200} + \frac{(G_{con} - 1)^2 25}{200} + \frac{G_{con}[700 - 31(3 + G_{con}) - 25(G_{con} - 1)]}{100}$$

$$+ \frac{1.0[25(G_{con} - 1)]}{100} = 10 + \frac{3.0[31(3 + G_{con})]}{100}$$

or

$$-0.28 G_{con}^2 + 6.32 G_{con} - 11.52 = 0$$

which gives $G_{con} = 2.0$. (The other possible solution is 20.57, which is rejected because it results in curve lengths that exceed 700 m.) Using $L = KA$ gives $L_c = 155$ m (31×5) and $L_s = 25$ m (25×1). Elevations and the locations of curve points can be readily computed with this information.

3.3.5   Passing-Sight Distance and Crest Vertical Curve Design

In addition to stopping-sight distance, in some instances it may be desirable to provide adequate passing-sight distance, which can be an important issue in two-lane highway design (one lane in each direction). Passing-sight distance is a factor only in crest vertical curve design because, for sag curves, the sight distance is unobstructed looking up or down the grade, and, at night, the headlights of oncoming or opposing vehicles will be noticed. In determining the sight distance required to pass on a crest vertical curve, Eqs. 3.13 and 3.14 will apply, but while the driver's eye height, H_1, will remain 1.07 m (1070 mm), H_2 will now be set to 1.3 m (1300 mm), which is the assumed height of an opposing vehicle that could prevent a passing maneuver. Substituting these H-values into Eqs. 3.13 and 3.14 and letting the sight-distance, S, equal passing-sight distance, PSD,

$$L_m = 2\,\text{PSD} - \frac{946}{A} \qquad \text{for PSD} > L \qquad \textbf{(3.24)}$$

$$L_m = \frac{A\,\text{PSD}^2}{946} \qquad \text{for PSD} < L \qquad \textbf{(3.25)}$$

As was the case for stopping-sight distance, it is typically assumed that the length of curve is greater than the required sight distance (i.e., in this case, $L >$ PSD), so

$$K = \frac{\text{PSD}^2}{946} \qquad \textbf{(3.26)}$$

The passing-sight distance (PSD) used for design is assumed to consist of four distances: (1) the initial maneuver distance, which includes drivers' perception/reaction time and the time it takes to bring a vehicle from its trailing speed to the speed needed to pass; (2) the distance that the passing vehicle traverses while occupying the left lane; (3) the clearance length (a comfortable length between the vehicle being passed and the opposing or oncoming vehicle at the end of the pass, assumed to be 30 to 90 m); and (4) the distance traversed by an opposing vehicle during the passing maneuver. The determination of these distances is undertaken using assumptions regarding the time of the initial maneuver, average

Table 3.4 Design Controls for Crest Vertical Curves Based on Passing-Sight Distance

Design Speed (km/h)	Minimum Passing-Sight Distance for Design (m)	Rate of Vertical Curvature, K, Rounded for Design (Length [m] per % of A)
30	217	50
40	285	90
50	345	130
60	407	180
70	482	250
80	541	310
90	605	390
100	670	480
110	728	570
120	792	670

Source: American Association of State Highway and Transportation Officials, "A Policy on Geometric Design of Highways and Streets," Washington, DC, 1994.

vehicle acceleration, and speeds of passing, passed, and opposing vehicles. The summation of these four distances gives a single *required* passing-sight distance (i.e., there are not *minimums* and *desirables* as was the case with stopping-sight distance). The reader is referred to AASHTO 1994 for a complete description of the assumptions made in determining required passing-sight distances.

The minimum distances needed to pass (PSD) at various design speeds, along with the corresponding K-values as computed from Eq. 3.26, are presented in Table 3.4. Notice that the K-values in this table are much higher than those required for stopping-sight distance (as shown in Table 3.2). As a result, designing a crest curve to provide adequate passing-sight distance is often an expensive proposition (due to the length of curve required) and not very common.

EXAMPLE 3.10

An equal-tangent crest vertical curve is 480 m long and connects a +2.0% and a −1.5% grade. If the design speed of the roadway is 80 km/h, does this curve have adequate passing-sight distance?

SOLUTION

To determine the length of curve required to provide adequate passing-sight distance at a design speed of 80 km/h, we use $L = KA$ with $K = 310$ (as read from Table 3.4). This gives

$$L = 310(3.5) = 1085 \text{ m}$$

Because the curve is only 480 m long it is not long enough to provide adequate passing-sight distance.

3.4   HORIZONTAL ALIGNMENT

The critical aspect of horizontal alignment is the horizontal curve, with the focus on the design of the directional transition of the roadway in a horizontal plane. Stated differently, a horizontal curve provides a transition between two straight (or tangent) sections of roadway. A key concern in this directional transition is the ability of a vehicle to negotiate a horizontal curve. (Provision of adequate drainage is also important, but is not discussed in this book; see AASHTO 1994). As was the case with the straight-line vehicle performance characteristics discussed at length in Chapter 2, a highway engineer must design a horizontal alignment to accommodate a variety of vehicle cornering capabilities that range from nimble sports cars to ponderous trucks. A theoretical assessment of vehicle cornering at the same level of detail that was given to straight-line performance in Chapter 2 is beyond the scope of this book; see Campbell 1978, and Wong

1978. Instead, vehicle cornering performance is viewed only at the practical design-oriented level, with equations simplified in a manner similar to the stopping-distance equation, as discussed in Section 2.9.5.

3.4.1   Vehicle Cornering

Figure 3.10 illustrates the forces acting on a vehicle during cornering. In this figure, α is the angle of incline, W is the weight of the vehicle (in newtons), W_n and W_p are the weights normal and parallel to the roadway surface respectively, F_f is the side frictional force (centripetal, in newtons), F_c is the centripetal force (lateral acceleration × mass, in newtons), F_{cp} is the centripetal force acting parallel to the roadway surface, F_{cn} is the centripetal force acting normal to the roadway surface, and R_v is the radius defined to the vehicle's traveled path (in meters). Some basic horizontal curve relationships can be derived by noting that

$$W_p + F_f = F_{cp} \tag{3.27}$$

From basic physics this equation can be written, with $F_f = f_s(W_n + F_{cn})$, as

$$W \sin \alpha + f_s \left(W \cos \alpha + \frac{WV^2}{gR_v} \sin \alpha \right) = \frac{WV^2}{gR_v} \cos \alpha \tag{3.28}$$

where f_s is the coefficient of side friction (which is different from the coefficient of friction term, f, used in stopping-distance computations), g is the gravitational constant, and V is the vehicle speed (in meters per second). Dividing both sides of Eq. 3.28 by $W \cos \alpha$ gives

$$\tan \alpha + f_s = \frac{V^2}{gR_v} (1 - f_s \tan \alpha) \tag{3.29}$$

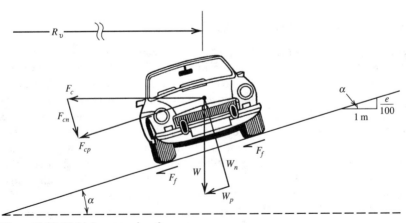

Figure 3.10 Vehicle cornering forces.

The term $\tan \alpha$ indicates the superelevation of the curve; it can be expressed in percent and is denoted e (i.e., $e = 100 \tan \alpha$). In words, the superelevation is the number of vertical meter rise per 100 meters of horizontal distance (see Fig. 3.10). The term $f_s \tan \alpha$ in Eq. 3.29 is conservatively set equal to zero for practical applications due to the small values that f_s and α typically assume (this is equivalent to ignoring the normal component of centripetal force). With $e = 100 \tan \alpha$, Eq. 3.29 can be arranged so that

$$R_v = \frac{V^2}{g\left(f_s + \dfrac{e}{100}\right)} \tag{3.30}$$

EXAMPLE 3.11

A roadway is being designed for a speed of 120 km/h. At one horizontal curve, it is known that the superelevation is 8.00% and the coefficient of side friction is 0.09. Determine the minimum radius of curve (measured to the traveled path) that will provide safe vehicle operation.

SOLUTION

The application of Eq. 3.30 (with 1000/3600 converting km/h to m/s) gives

$$R_v = \frac{V^2}{g\left(f_s + \dfrac{e}{100}\right)} = \frac{(120 \times 1000/3600)^2}{9.807\left(0.09 + \dfrac{8}{100}\right)} = \underline{\underline{666.457 \text{ m}}}$$

This value is the minimum radius, because radii larger than 666.457 m will generate centripetal forces lower than those capable of being safely supported by the superelevation and the side frictional force.

In the actual design of a horizontal curve, engineers must select appropriate values of e and f_s. The value selected for superelevation, e, is critical because high rates of superelevation can cause vehicle steering problems on the horizontal curve. Furthermore, in cold climates, ice on the roadway can reduce f_s such that vehicles traveling less than the design speed on an excessively superelevated curve could slide inward off the curve as a result of gravitational forces. AASHTO provides general guidelines for the selection of e and f_s for horizontal curve design, as shown in Table 3.5. The values presented in this table are grouped by five values of maximum e. The selection of any one of these five maximum e values depends on the type of road (e.g., higher maximum e's are permitted on freeways relative to arterials and local roads) and local design practice. Limiting values of f_s are simply a function of design speed. Table 3.5 also presents calculated radii (given V, e, and f_s) by applying Eq. 3.30.

Table 3.5 Minimum Radius Using Limiting Values of e and f_s

Design Speed (km/h)	Maximum e %	Limiting Values of f_S	Total $e/100 + f_S$	Calculated Radius, R_v (Meters)	Rounded Radius, R_v (Meters)
30	4.00	0.17	0.21	33.7	35
40	4.00	0.17	0.21	60.0	60
50	4.00	0.16	0.20	98.4	100
60	4.00	0.15	0.19	149.2	150
70	4.00	0.14	0.18	214.3	215
80	4.00	0.14	0.18	280.0	280
90	4.00	0.13	0.17	375.2	375
100	4.00	0.12	0.16	492.1	490
110	4.00	0.11	0.15	635.2	635
120	4.00	0.09	0.13	872.2	870
30	6.00	0.17	0.23	30.8	30
40	6.00	0.17	0.23	54.8	55
50	6.00	0.16	0.22	89.5	90
60	6.00	0.15	0.21	135.0	135
70	6.00	0.14	0.20	192.9	195
80	6.00	0.14	0.20	252.0	250
90	6.00	0.13	0.19	335.7	335
100	6.00	0.12	0.18	437.4	435
110	6.00	0.11	0.17	560.4	560
120	6.00	0.09	0.15	755.9	755
30	8.00	0.17	0.25	28.3	30
40	8.00	0.17	0.25	50.4	50
50	8.00	0.16	0.24	82.0	80
60	8.00	0.15	0.23	123.2	125
70	8.00	0.14	0.22	175.4	175
80	8.00	0.14	0.22	229.1	230
90	8.00	0.13	0.21	303.7	305
100	8.00	0.12	0.20	393.7	395
110	8.00	0.11	0.19	501.5	500
120	8.00	0.09	0.17	667.0	665
30	10.00	0.17	0.27	26.2	25
40	10.00	0.17	0.27	46.7	45
50	10.00	0.16	0.26	75.7	75
60	10.00	0.15	0.25	113.4	115
70	10.00	0.14	0.24	160.8	160
80	10.00	0.14	0.24	210.0	210
90	10.00	0.13	0.23	277.3	275
100	10.00	0.12	0.22	357.9	360
110	10.00	0.11	0.21	453.7	455
120	10.00	0.09	0.19	596.8	595

Table 3.5 (*Continued*)

Design Speed (km/h)	Maximum e %	Limiting Values of f_s	Total $e/100 + f_s$	Calculated Radius, R_v (Meters)	Rounded Radius, R_v (Meters)
30	12.00	0.17	0.29	24.4	25
40	12.00	0.17	0.29	43.4	45
50	12.00	0.16	0.28	70.3	70
60	12.00	0.15	0.27	105.0	105
70	12.00	0.14	0.26	148.4	150
80	12.00	0.14	0.26	193.8	195
90	12.00	0.13	0.25	255.1	255
100	12.00	0.12	0.24	328.1	330
110	12.00	0.11	0.23	414.2	415
120	12.00	0.09	0.21	539.9	540

Note: In recognition of safety considerations, use of $e_{max} = 4.00\%$ should be limited to urban conditions.

Source: American Association of State Highway and Transportation Officials, "A Policy on Geometric Design of Highways and Streets," Washington, DC, 1994.

3.4.2 Horizontal Curve Fundamentals

In connecting straight (tangent) sections of roadway with a horizontal curve, several options are available. The most obvious of these is the simple curve, which is just a standard curve with a single, constant radius. Other options include compound curves, which consist of two or more simple curves in succession, and spiral curves, which are curves with a continuously changing radius. To illustrate the basic principles involved in horizontal curve design, this book will focus on simple curves. For detailed information regarding compound and spiral curves, the reader is referred to standard route-surveying texts (Skelton 1949).

Figure 3.11 shows the basic elements of a simple horizontal curve. In this figure, R is the radius (usually measured to the centerline of the road), PC is the point of curve (the beginning point of the horizontal curve), T is the tangent length, PI is the point of tangent intersection, Δ is the central angle of the curve, PT is the point of tangent (the ending point of the horizontal curve), M is the middle ordinate, E is the external distance, and L is the length of curve.

A geometric and trigonometric analysis of Fig. 3.11 reveals the following relationships:

$$T = R \tan \frac{\Delta}{2} \tag{3.31}$$

$$E = R \left[\frac{1}{\cos(\Delta/2)} - 1 \right] \tag{3.32}$$

$$M = R \left(1 - \cos \frac{\Delta}{2} \right) \tag{3.33}$$

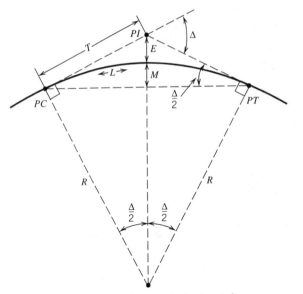

Figure 3.11 Elements of a simple horizontal curve.

$$L = \frac{\pi}{180} R\Delta \qquad (3.34)$$

It is important to note that horizontal curve stationing, curve length, and curve radius (R) are typically measured with respect to the centerline of the road (for a two-lane road). In contrast, the radius determined on the basis of vehicle forces $(R_v$ in Eq. 3.30) is measured from the innermost vehicle path, which is assumed to be the midpoint of the innermost vehicle lane. Thus, to be truly correct, a slight correction for lane width is required when equating the R_v of Eq. 3.30 with the R in Eqs. 3.31 to 3.34.

EXAMPLE 3.12

A horizontal curve is designed with a 725-m radius. The curve has a tangent length of 140 m and the *PI* is at station 3 + 103.000. Determine the stationing of the *PT*.

SOLUTION

Equation 3.31 is applied to determine the central angle, Δ:

$$T = R \tan \frac{\Delta}{2}$$

$$140 = 725 \tan \frac{\Delta}{2}$$

$$\Delta = 21.86°$$

So, from Eq. 3.34, the length of the curve is

$$L = \frac{\pi}{180} R\Delta$$

$$L = \frac{3.1416}{180} 725(21.86) = 276.609 \text{ m}$$

Given that the tangent is 140 m,

stationing $PC = 3 + 103.000 - 0 + 140.000 = 2 + 963.000$

Since horizontal curve stationing is measured along the alignment of the road,

stationing PT = stationing $PC + L$

$$= 2 + 963.000 + 0 + 276.609 = \underline{\underline{3 + 239.609}}$$

3.4.3 Stopping-Sight Distance and Horizontal Curve Design

As was the case for vertical curve design, adequate stopping-sight distance must be provided in the design of horizontal curves. Sight-distance restrictions on horizontal curves occur when obstructions are present as shown in Fig. 3.12. Such obstructions are frequently encountered in highway design due to the cost of right-of-way acquisition and/or the cost of moving earthen materials (e.g.,

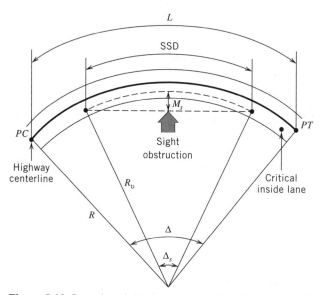

Figure 3.12 Stopping-sight-distance considerations for horizontal curves.

rock outcroppings). When such an obstruction exists, the stopping-sight distance is measured along the horizontal curve from the center of the traveled lane (the assumed location of the driver's eyes). As shown in Fig. 3.12, for a specified stopping distance, some distance, M_s (the middle ordinate of a curve that has an arc length equal to the stopping distance), must be visually cleared so that the line of sight is such that sufficient stopping-sight distance is available.

Equations for computing stopping-sight-distance (SSD) relationships for horizontal curves can be derived by first determining the central angle, Δ_s, for an arc equal to the required stopping-sight distance. (See Fig. 3.12 and note that this is not the central angle, Δ, of the horizontal curve whose arc is equal to L.) Assuming that the length of the horizontal curve exceeds the required SSD (as shown in Fig. 3.12), we have (as with Eq. 3.34)

$$SSD = \frac{\pi}{180} R_v \Delta_s \qquad (3.35)$$

where R_v is the radius to the vehicle's traveled path, which is also assumed to be the location of the driver's eyes for sight distance, and again is taken as the radius to the middle of the innermost lane; and Δ_s is the angle subtended by an arc equal to the SSD length. Rearranging terms gives

$$\Delta_s = \frac{180 \, SSD}{\pi R_v} \qquad (3.36)$$

Substituting this into the general equation for the middle ordinate of a simple horizontal curve (Eq. 3.33) gives

$$M_s = R_v \left(1 - \cos \frac{90 \, SSD}{\pi R_v} \right) \qquad (3.37)$$

where M_s is the middle ordinate necessary to provide adequate stopping-sight distance, as shown in Fig. 3.12. Solving Eq. 3.37 for SSD gives

$$SSD = \frac{\pi R_v}{90} \left[\cos^{-1} \left(\frac{R_v - M_s}{R_v} \right) \right] \qquad (3.38)$$

Note that Eqs. 3.35 to 3.38 can also be applied directly to determine sight-distance requirements for passing. If these equations are to be used for passing, distance values given in Table 3.4 would apply and SSD in the equations would be replaced by PSD.

EXAMPLE 3.13

A horizontal curve on a two-lane highway is designed with a 700-m radius, 3.6-m lanes, and a 100-km/h design speed. Determine the distance that must be cleared from the inside edge of the inside lane to provide sufficient sight distance for desirable and minimum SSD.

SOLUTION

Because the curve radius is usually taken to the centerline of the roadway, $R_v = R - 3.6/2 = 700 - 1.8 = 698.2$ m, which gives the radius to the middle of the inside lane (i.e., the critical driver location). From Table 3.1, the desirable SSD is 205 m, so applying Eq. 3.37 gives

$$M_s = R_v \left(1 - \cos \frac{90 \, \text{SSD}}{\pi R_v} \right)$$

$$= 698.2 \left[1 - \cos \frac{90(205)}{\pi(698.2)} \right] = \underline{\underline{7.513 \text{ m}}}$$

Therefore, 7.513 m must be cleared, as measured from the center of the inside lane, or 5.713 m as measured from the inside edge of the inside lane. Similarly, the minimum SSD is 157 m (Table 3.1), and the application of Eq. 3.37 gives $M_s = 4.408$ m. Thus, for minimum SSD, 2.608 m must be cleared from the inside edge of the inside lane.

EXAMPLE 3.14

A two-lane highway (two 3.6-m lanes) has a posted speed limit of 80 km/h, and, on one section, has both horizontal and vertical curves as shown in Fig. 3.13. A recent daytime accident (driver traveling eastbound and striking a stationary roadway object) resulted in a fatality and a lawsuit alleging that the 80-km/h posted speed limit was an unsafe speed for the curves in question and a major cause of the accident. Evaluate and comment on the roadway design.

SOLUTION

Begin with an assessment of the horizontal alignment. Two concerns must be considered: the adequacy of the curve radius and superelevation, and the adequacy of the sight distance on the eastbound (inside) lane. For the curve radius, note from Fig. 3.13 that

$$L = \text{station of } PC - \text{station of } PT$$

$$L = 4 + 600.000 - 4 + 160.000 = 440 \text{ m}$$

Rearranging Eq. 3.34, we get

$$R = \frac{180}{\pi \Delta} L = \frac{180}{\pi(80)} (440) = 315.126 \text{ m}$$

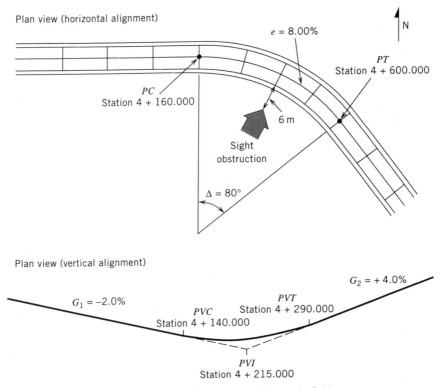

Figure 3.13 Horizontal and vertical alignment for Example 3.14.

Using the posted speed limit of 80 km/h with $e = 8.00\%$, we find that Eq. 3.30 can be rearranged (with the vehicle traveling in the middle of the inside lane, $R_v = R -$ half the lane width, or $R_v = 315.126 - 1.8 = 313.326$) to give

$$f_s = \frac{V^2}{gR_v} - \frac{e}{100} = \frac{(80 \times 1000/3600)^2}{9.807(313.326)} - \frac{8}{100} = 0.0807$$

From Table 3.5, the maximum f_s for 80 km/h is 0.14. Since 0.0807 does not exceed 0.14, the radius and superelevation are sufficient for the 80-km/h speed.

For sight distance, the available M_s is 6 m plus the 1.8-m distance to the center of the eastbound (inside) lane, or 7.8 m. Application of Eq. 3.38 gives

$$SSD = \frac{\pi R_v}{90} \left[\cos^{-1} \left(\frac{R_v - M_s}{R_v} \right) \right]$$

$$= \frac{\pi(313.326)}{90} \left[\cos^{-1} \left(\frac{313.326 - 7.8}{313.326} \right) \right]$$

$$= 140.119 \text{ m}$$

From Table 3.1, desirable SSD at 80 km/h is 139.4 m, so the 140.119 m of SSD provided is sufficient. Turning to the sag vertical curve, the length of curve is

$$L = \text{station of } PVT - \text{station of } PVC$$

$$L = 4 + 290.000 - 4 + 140.000 = 150 \text{ m}$$

Using $A = 6$ (from Fig. 3.13) and applying Eq. 3.10, we obtain

$$K = \frac{L}{A} = \frac{150}{6} = 25$$

For the 80-km/h speed, Table 3.3 indicates a necessary K-value of 32 desirable and 25 minimum. Thus the K-value of 25 reveals that the curve is adequate for minimum but underdesigned for desirable. However, because the accident occurred in daylight and sight distances on sag vertical curves are governed by nighttime conditions, this design did not contribute to the accident.

NOMENCLATURE FOR CHAPTER 3

A absolute value of the algebraic difference in grades (in percent)

D degree of curvature

e rate of superelevation

F_f frictional side force

F_c centripetal force

F_{cn} centripetal force normal to roadway surface

F_{cp} centripetal force parallel to roadway surface

f coefficient of stopping friction

f_s coefficient of side friction

G grade in m/m or percent depending on application

G_1 initial roadway grade in m/m or percent

G_2 final roadway grade in m/m or percent

g gravitational constant

H height of vehicle headlights

H_1 height of driver's eye

H_2 height of roadway object for stopping, height of oncoming car for passing

K	horizontal distance required to effect a 1% change in slope
L	length of curve
L_m	minimum length of curve
M	middle ordinate
M_s	middle ordinate for stopping-sight distance
PC	initial point of horizontal curve
PI	point of tangent intersection (horizontal curve)
PSD	passing-sight distance
PT	final point of horizontal curve
PVC	initial point of vertical curve
PVI	point of tangent intersection, vertical curve
PVT	final point of vertical curve
R	radius of curve measured to roadway centerline
R_v	radius of curve measured to center of vehicle
S	sight distance
SSD	stopping-sight distance
T	tangent length
t_p	driver perception/reaction time
V	vehicle speed
V_1	initial vehicle speed
W	vehicle weight
W_n	vehicle weight normal to roadway surface
W_p	vehicle weight parallel to roadway surface
x	distance from beginning of vertical curve
x_{hl}	distance from beginning of vertical curve to high or low point
Y	vertical curve offset
Y_f	end-of-curve offset, vertical curve
Y_m	midcurve offset, vertical curve
α	angle of superelevation
β	angle of upward headlight beam
Δ	central angle
Δ_s	central angle subtended by stopping-sight (SSD) arc

REFERENCES

AASHTO (American Association of State Highway and Transportation Officials). "A Policy on Geometric Design of Highways and Streets." Washington, DC, 1994.

Campbell, C. *The Sports Car: Its Design and Performance.* Cambridge, MA: Robert Bently, Inc., 1978.

Wong, J. H. *Theory of Ground Vehicles.* New York: Wiley, 1978.

Skelton, R. R. *Route Surveys.* New York: McGraw-Hill, 1949.

PROBLEMS

3.1. A 350-m-long sag vertical curve (equal tangent) has a *PVC* at station 3 + 040.000 and elevation 280 m. The initial grade is −3.5% and the final grade is +6.5%. Determine the elevation and stationing of the low point, *PVI*, and *PVT*.

3.2. A 160-m-long equal tangent crest vertical curve connects tangents that intersect at station 4 + 310.000 and elevation 411 m. The initial grade is +4.0% and the final grade is −2.5%. Determine the elevation and stationing of the high point, *PVC*, and *PVT*.

3.3. Consider Example 3.3. Solve this problem without using offsets (i.e., use Eq. 3.1).

3.4. Again consider Example 3.3. Does this curve provide adequate stopping-sight distance (desirable) for a speed of 100 km/h?

3.5. An equal tangent sag vertical curve is designed to provide desirable SSD. The *PVC* of the curve is at station 11 + 130.000 (elevation 322 m), the *PVI* at station 11 + 210.000 (elevation 320.8 m), and the low point is at station 11 + 190.000. Determine the design speed of the curve.

3.6. An equal tangent vertical curve was designed in 1996 (to 1994 AASHTO guidelines) for desirable SSD at a design speed of 120 km/h to connect grades $G_1 = +1.0\%$ and $G_2 = -2.0\%$. The curve is to be redesigned for a 120-km/h design speed in the year 2030. Vehicle braking technology has advanced such that coefficients of stopping friction, f's, have increased by 40% relative to the 1994 guidelines given in Table 3.1, but due to the higher percentage of older people in the driving population, design reaction times have increased by 20%. Also, because vehicles have become smaller the driver's eye height is assumed to be 0.90 m above the pavement, and roadway objects are assumed to be 0.10 m above the pavement. Compute the difference in design curve lengths for 1996 and 2030 designs.

3.7. A highway reconstruction project is being undertaken to reduce accident rates. The reconstruction involves a major realignment of the highway such that a 100-km/h design speed is attained. At one point on the highway, a

245-m equal tangent crest vertical curve exists. Measurements show that at 0 + 107.290 station from the *PVC*, the vertical curve offset is 1 m. Assess the adequacy of this existing curve in light of the reconstruction design speed of 100 km/h and, if the existing curve is inadequate, compute a satisfactory curve length. (Consider both minimum and desirable SSDs.)

3.8. Two level sections of an east-west highway ($G = 0$) are to be connected. Currently, the two sections of highway are separated by a 1300-m (horizontal distance), 2.0% grade. The westernmost section of highway is the higher of the two and is at elevation 44 m. If the highway has a 100-km/h design speed, determine, for the crest and sag vertical curves required, the stationing and elevation of the *PVCs* and *PVTs* given that the *PVC* of the crest curve (on the westernmost level highway section) is at station 0 + 000.000 and elevation 44 m. In solving this problem, assume desirable SSDs and that the curve *PVIs* are at the intersection of $G = 0$ and the 2.0% grade (i.e., $A = 2$).

3.9. Consider Problem 3.8. Suppose it is necessary to keep the entire alignment within the 1300 m that currently separate the two level sections. It is determined that the crest and sag curves should be connected (i.e., the *PVT* of the crest and *PVC* of the sag) with a constant-grade section that has the lowest grade possible. Again using 100-km/h design speed, determine, for the crest and sag vertical curves, the stationing and elevation of the *PVCs* and *PVTs* given that the westernmost level section ends at station 0 + 000.000 and elevation 44 m. (Note that A must now be determined and will not be equal to 2.)

3.10. An equal tangent crest vertical curve is designed for 100 km/h desirable. The initial grade is +4.0% and the final grade is negative. What is the elevation difference between the *PVC* and the high point of the curve?

3.11. An equal tangent crest curve connects a +1.0% and a −0.5% grade. The *PVC* is at station 14 + 890.000 and the *PVI* is at station 15 + 150.200. Is this curve long enough to provide passing-sight distance at a 100-km/h design speed?

3.12. Due to accidents at a railroad crossing, an overpass (with a roadway surface 7.5 m above the existing road) is to be constructed on an existing level highway. The existing highway has a design speed of 80 km/h (desirable SSD). The overpass structure is to be level, centered above the railroad, and 70 m long. What length of the existing level highway must be reconstructed to provide an appropriate vertical alignment?

3.13. A section of a freeway ramp has a +4.0% grade and ends at station 0 + 122.000 and elevation 163 m. It must be connected to another section of the ramp (which has a 0.0% grade) that is at station 1 + 322.000 and elevation 150 m. It is determined that the crest and sag curves required to connect the ramp should be connected (i.e., the *PVT* of the crest and *PVC* of the sag) with a constant-grade section that has the lowest grade possible. Design a vertical alignment to connect between these two stations using a

60-km/h design speed (desirable). Provide the lengths of the curves and constant-grade section.

3.14. A 400-m equal tangent crest vertical curve is currently designed for 80 km/h (desirable). A civil engineering student contends that 100 km/h is safe in a van because of a higher driver's eye height. If all other design inputs are standard, how high must the van driver's eye height be for the student's claim to be valid?

3.15. A roadway has a design speed of 90 km/h. At station 2 + 000.000 a +3.0% grade roadway section ends, and at station 3 + 090.000 a +2.0% grade roadway section begins. The +3.0% grade section (at station 2 + 000.000) is at a higher elevation than the +2.0% grade section (at station 3 + 090.000). If a −5.0% constant-grade section is used to connect the crest and sag vertical curves that are needed to link the +3.0% and +2.0% grade sections, what is the elevation difference between stations 2 + 000.000 and 3 + 090.000 if the roadway is designed for desirable SSD? (The entire alignment, crest and sag curves, and constant-grade section must fit between stations 2 + 000.000 and 3 + 090.000).

3.16. You are asked to design a horizontal curve for a two-lane road. The road has 3.6-m lanes. Due to expensive excavation, it is determined that a maximum of 9 m can be cleared from the road's centerline toward the inside lane to provide for stopping-sight distance. Also, local guidelines dictate a maximum superelevation of 8.0%. What is the highest possible design speed for this curve?

3.17. A horizontal curve on a single-lane highway has its *PC* at station 1 + 346.200 and its *PI* at station 1 + 568.700. The curve has a superelevation of 6.0% and is designed for 120 km/h (desirable). What is the station of the *PT*?

3.18. A horizontal curve through mountainous terrain is being designed for a four-lane road with lanes that are 3.0 m wide. The central angle (Δ) is 40 degrees, the tangent distance is 155 m, and the stationing of the tangent intersection (*PI*) is 7 + 270.000. Under specified conditions and vehicle speed, the roadway surface is determined to have a coefficient of side friction of 0.08, and the curve's superelevation is 9.0%. What is the stationing of the *PC* and *PT*? What is the safe vehicle speed?

3.19. A new interstate highway is being built with a design speed of 120 km/h. For one of the horizontal curves, the radius (measured to the innermost vehicle path) is tentatively planned as 300 m. What rate of superelevation is required for this curve at 120-km/h design speed?

3.20. A developer is having a single-lane raceway constructed with a 180-km/h design speed. A curve on the raceway has a radius of 320 m, a central angle of 30 degrees, and *PI* stationing at 11 + 511.200. If the design coefficient of side friction is 0.20, determine the superelevation required at the design speed (do not ignore the normal component of the centripetal force). Also, compute the length of curve and stationing of the *PC* and *PT*.

3.21. A horizontal curve is being designed for a new two-lane highway (3.6-m lanes). The *PI* is at station 1 + 134.000, design speed is 110 km/h, and a maximum superelevation of 8.0% is to be used. If the central angle of the curve is 35 degrees, design a curve for the highway by computing the radius, and the stationing of the *PC* and *PT*.

3.22. You are asked to design a horizontal curve with a 40-degree central angle ($\Delta = 40$) for a two-lane road with 3.0-m lanes. The design speed is 120 km/h and superelevation is limited to 6.0%. Give the radius and length of curve that you would recommend.

3.23. A freeway exit ramp has a single lane and consists entirely of a horizontal curve with a central angle of 90 degrees and a length of 200 m. If the distance cleared from the centerline for sight distance is 6.95 m, what design speed was used? (Assume desirable SSD.)

3.24. For the horizontal curve in Problem 3.21, what distance must be cleared from the inside edge of the inside lane to provide adequate sight distance for minimum and desirable SSDs?

Chapter 4

Pavement Design and Highway Drainage

4.1 INTRODUCTION

The physical components of a highway system include the right of way, the pavement, shoulders, safety appurtenances, signs, signals, and markings. The pavement and shoulders represent the most costly items associated with highway construction and maintenance. In fact, the highway system in the United States is the most costly public works project undertaken by any society. Because the pavement and associated shoulder structures are the most expensive items to construct and maintain, it is important for highway engineers to have a basic understanding of pavement design principles.

In the United States, there are over 4.8 million kilometers of highways. Surprisingly, about 45% of these roads are not paved but are comprised of either gravel or a stabilized material (which consists of an aggregate material bound together with a cementing agent such as Portland cement, lime fly ash, or asphaltic cement). Highways that carry higher volumes of traffic with heavy axle loads require surfaces with asphalt concrete or Portland cement concrete to provide for all-weather operations and prevent permanent deformation of the highway surface. These types of pavements can cost upward of several million dollars per kilometer to construct. Some states, such as Pennsylvania, Texas, Illinois, and California, have pavement construction and rehabilitation budgets that approach a billion dollars per year and, when coupled with their maintenance budgets, it is easy to see why construction and maintenance of pavement infrastructure must be done in a cost-effective manner.

Pavement provides two basic functions. First, it helps guide the driver and delineate the roadway by giving a visual perspective of the horizontal and vertical alignment of the traveled path. Consequently, pavement gives the driver information about the driving task and the steering control of the vehicle. The second function of pavement is to support vehicle loads. This second function will be discussed in this chapter.

In addition to pavement design, this chapter also discusses highway drainage.

85

Drainage is an important concern in highway design because proper drainage can help keep pavement structures from experiencing premature deterioration and failure, and proper drainage prevents accumulation of water on the pavement surface that can lead to dangerous driving conditions resulting from tire hydroplaning and poor visibility from splash and spray.

4.2   PAVEMENT TYPES

In general, there are two types of pavement structures: flexible pavements and rigid pavements. There are, however, many variations of these pavement types including some with soil cement and stabilized bases that have cemented aggregate. Composite pavements (which are made of both rigid and flexible layers), continuously reinforced pavements, and post-tensioned pavements are other types of pavements, which usually require specialized designs and are not covered in this chapter.

As with any structure, the underlying soil must ultimately carry the load that is placed on it. A pavement's function is to distribute the traffic-load stresses to the soil (subgrade) at a magnitude that will not shear or distort the soil. Typical soil-bearing capacities can be less than 345 kPa and in some cases as low as 14 to 21 kPa. Also, when soil is saturated with water, the bearing capacity can be very low, and in these cases it is very important for pavement to distribute tire loads to the soil in such a way as to avoid failure of the pavement structure.

A typical automobile weighs approximately 12 kN, with tire pressures of 240 kPa. These loads are small when compared to a typical tractor semi-trailer truck that can weigh up to 355.8 kN, the legal limit in many states, on five axles with tire pressures of 690 kPa or higher. Truck loads such as these represent the standard type of loading used in pavement design. In this chapter, we give attention to an accepted procedure that can be used to design pavement structures for high-traffic-volume highway facilities subjected to heavy truck traffic. The design of lower-volume facilities, which may have stabilized-soil and gravel-surfaced pavements, can be found in other references (Yoder & Witczak, 1975).

4.2.1   Flexible Pavements

A flexible pavement is constructed with asphaltic cement and aggregates and usually consists of several layers, as shown in Fig. 4.1. The lower layer is called the subgrade (i.e., the soil itself). The upper 150 to 200 mm of the subgrade is usually scarified and blended to provide a uniform material before it is compacted to maximum density. The next layer is the subbase, which usually consists of crushed aggregate (rock). This material has better engineering properties (higher modulus values) than the subgrade material, in terms of its bearing capacity. The next layer (i.e., the base layer) is also often made of crushed aggregates (of a higher strength than those used in the subbase), which are either unstabilized or stabilized with a cementing material. The cementing material can be Portland cement, lime fly ash, or asphaltic cement.

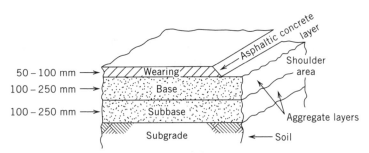

50 – 100 mm →
100 – 250 mm →
100 – 250 mm →

Figure 4.1 Typical flexible-pavement cross section.

The top layer of a flexible pavement is referred to as the wearing surface. It is usually made of asphaltic concrete, which is a mixture of asphalt cement and aggregates. The purpose of the wearing layer is to protect the base layer from wheel abrasion and to waterproof the entire pavement structure. It also provides a skid-resistant surface that is important for safe vehicle stops. Typical thicknesses of the individual layers are shown in Fig. 4.1. These thicknesses vary with the type of axle loading, available materials, and expected pavement design life, which is the number of years the pavement is expected to provide adequate service before it must undergo major rehabilitation.

4.2.2 Rigid Pavements

A rigid pavement is constructed with Portland cement concrete (PCC) and aggregates, as shown in Fig. 4.2. As is the case with flexible pavements, the subgrade (i.e., the lower layer) is often scarified, blended, and compacted to maximum density. In rigid pavements, the base layer (see Fig. 4.2) is optional, depending upon the engineering properties of the subgrade. If the subgrade soil is poor and erodable, then it is advisable to use a base layer. However, if the soil has good engineering properties and drains well, a base layer need not be used. The top layer (wearing surface) is the Portland cement concrete slab. Slab length varies from short slab spacing of 3 to 4 m to a spacing of 12 m or more.

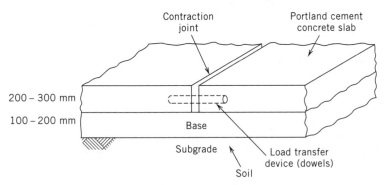

200 – 300 mm
100 – 200 mm

Figure 4.2 Typical rigid-pavement cross section.

Transverse contraction joints are built into the pavement to control cracking due to shrinkage of the concrete during the curing process. Load-transfer devices, such as dowel bars, are placed in the joints to minimize deflections and reduce stresses near the edges of the slabs. Slab thicknesses for PCC highway pavements usually vary from 200 to 300 mm, as shown in Fig. 4.2.

4.3 PAVEMENT SYSTEM DESIGN: PRINCIPLES FOR FLEXIBLE PAVEMENTS

The primary function of the pavement structure is to reduce and distribute the surface stresses (contact tire pressure) to an acceptable level at the subgrade (i.e., to a level that prevents permanent deformation). A flexible pavement reduces the stresses by distributing the traffic-wheel loads over greater and greater areas, through the individual layers, until the stress at the subgrade is at an acceptably low level. The traffic loads are transmitted to the subgrade by aggregate-to-aggregate particle contact. Confining pressures (lateral forces due to material weight) in the subbase and base layers increase the bearing strength of these materials. A cone of distributed loads reduces and spreads the stresses to the subgrade, as shown in Fig. 4.3.

4.3.1   Calculation of Flexible Pavement Stresses and Deflections

To design a pavement structure, an engineer must be able to calculate the stresses and deflections in the pavement system. In the simplest case, the wheel load can be assumed to consist of a point load on a single-layer system, as shown in Fig. 4.4. This type of load and configuration can be analyzed with the Boussinesq solutions that were derived for soils analysis. The Boussinesq theory assumes that the pavement is one layer thick and the material is elastic, homogeneous, and isotropic. The basic equation for the stress at a point in the system is

$$\sigma_z = 0.01 K \frac{P}{z^2} \qquad\qquad (4.1)$$

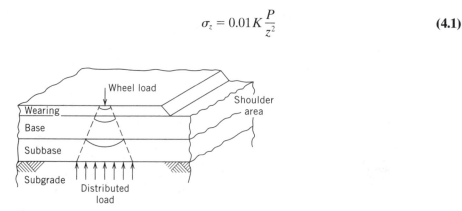

Figure 4.3 Distribution of load on a flexible pavement.

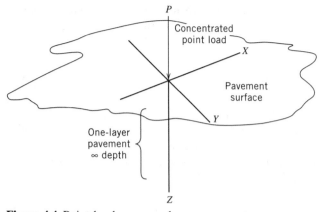

Figure 4.4 Point load on a one-layer pavement.

where σ_z is in kPa, P is the wheel load in newtons, z is the depth of the point in question in centimeters, and K is a variable defined as

$$K = \frac{3}{2\pi} \frac{1}{[1 + (r/z)^2]^{5/2}} \qquad (4.2)$$

where r is the radial distance in centimeters from the centerline of the point load to the point in question.

Although the Boussinesq theory is useful to begin the study of pavement stress calculations, it is not very representative of pavement-system loading and configuration because it applies to a point load on one layer. A more realistic approach is to expand the point load to an elliptical area that represents a tire footprint. The tire footprint can be defined by an equivalent circular area with a radius calculated by

$$a = \sqrt{\frac{P}{0.1p\pi}} \qquad (4.3)$$

where P is the tire load in newtons, p is the tire pressure in kPa, and a is the equivalent load radius of the tire footprint in centimeters. The integration of the load from a point to a circular area can be used to determine the stresses and deflections in a one-layer pavement system.

Several researchers have developed influence charts, graphical solutions, and equations for the calculation of stresses and deflections, which take into account a circular load. Ahlvin and Ulery provided solutions for the evaluation of stresses, strains, and deflections at any point in a homogeneous half-space (Ahlvin and Ulery 1962). Their work makes it easier to analyze a more complex pavement system than that considered in the Boussinesq example. The one-layer equations by Ahlvin and Ulery can be used for a material with any Poisson ratio, μ. (The

Poisson ratio describes the change in width to length when a load is applied along the vertical axis.) The Ahlvin and Ulery equation for the calculation of vertical stress, σ_z, is

$$\sigma_z = p(A + B) \tag{4.4}$$

where p is the pressure due to the load in kPa, and A and B are function values that depend on z/a and r/a, the depth in radii and offset distance in radii, respectively. The variables z, a, and r are as defined for the Boussinesq and circular load equations. The equation for radial-horizontal stress, σ_r (which is a cause of pavement cracking), is

$$\sigma_r = p[2\mu A + C + (1 - 2\mu)F] \tag{4.5}$$

and the equation for deflection, Δ_z, is

$$\Delta_z = \frac{p(1 + \mu)a}{E} \left[\frac{z}{a} A + (1 - \mu)H \right] \tag{4.6}$$

where E is the modulus of elasticity (known as Young's modulus, the ratio of stress to strain as a load is applied to a material), and C, F, and H are function values. The function values A, B, C, F, and H are presented in Table 4.1.

EXAMPLE 4.1

A tire with 689-kPa air pressure distributes a load over an area with a circular contact radius, a, of 12.5 cm. The pavement was constructed with a material that has a modulus of elasticity of 345.0 MPa and a Poisson ratio of 0.45. Calculate the radial horizontal stress and deflection at a point on the pavement surface under the center of the tire load. Also, calculate the radial-horizontal stress and deflection at a point at a depth of 50 cm and a radial distance of 25 cm from the center of the tire load. (Use Ahlvin and Ulery equations.)

SOLUTION

With $z = 0$ cm and $r = 0$ cm,

$$\frac{z}{a} = \frac{0}{12.5} = 0 \text{ and } \frac{r}{a} = \frac{0}{12.5} = 0$$

From Table 4.1, for the above values,

$$A = 1.0, B = 0, C = 0, F = 0.5, \text{ and } H = 2.0$$

Table 4.1 One-Layer Elastic Function Values

| | Function A | | | | | | | | | | | | | | | | |
| Depth (z/a) | Offset (r/a) | | | | | | | | | | | | | | | | |
	0	0.2	0.4	0.6	0.8	1	1.2	1.5	2	3	4	5	6	8	10	12	14
0	1.0	1.0	1.0	1.0	1.0	0.5	0	0	0	0	0	0	0	0	0	0	0
0.1	0.90050	0.89748	0.88679	0.86126	0.78797	0.43015	0.09645	0.02787	0.00856	0.00211	0.00084	0.00042					
0.2	0.80388	0.79824	0.77884	0.73483	0.63014	0.38269	0.15433	0.05251	0.01680	0.00419	0.00167	0.00083	0.00048	0.0020			
0.3	0.71265	0.70518	0.68316	0.62690	0.52081	0.34375	0.17964	0.07199	0.02440	0.00622	0.00250						
0.4	0.62861	0.62015	0.59241	0.53767	0.44329	0.31048	0.18709	0.08593	0.03118								
0.5	0.55279	0.54403	0.51622	0.46448	0.38390	0.28156	0.18556	0.09499	0.03701	0.01013	0.00407	0.00209	0.00118	0.0053			
0.6	0.48550	0.47691	0.45078	0.40427	0.33676	0.25588	0.17952	0.10010	0.04558								
0.7	0.42654	0.41874	0.39491	0.35428	0.29833	0.21727	0.17124	0.10228									
0.8	0.37531	0.36832	0.34729	0.31243	0.26581	0.19488	0.16206	0.10236									
0.9	0.33104	0.32492	0.30669	0.27707	0.23832	0.17868	0.15253	0.10094									
1	0.29289	0.28763	0.27005	0.24697	0.21468	0.15101	0.14329	0.09849	0.05185	0.01742	0.00761	0.00393	0.00226	0.00097	0.00025	0.00014	0.00009
1.2	0.23178	0.22795	0.21662	0.19890	0.17626	0.11892	0.12570	0.09192	0.05260	0.01935	0.00871	0.00459	0.00269	0.00115	0.00050	0.00029	0.00018
1.5	0.16795	0.16552	0.15877	0.14804	0.13436	0.08269	0.10296	0.08048	0.05116	0.02142	0.01013	0.00548	0.00325	0.00141	0.00073	0.00043	0.00027
2	0.10557	0.10453	0.10140	0.09647	0.09011	0.05974	0.07471	0.06275	0.04496	0.02221	0.01160	0.00659	0.00399	0.00180	0.00094	0.00056	0.00036
2.5	0.07152	0.07098	0.06947	0.06698	0.06373	0.04487	0.05555	0.04880	0.03787	0.02143	0.01221	0.00732	0.00463	0.00214	0.00115	0.00068	0.00043
3	0.05132	0.05101	0.05022	0.04886	0.04707	0.02749	0.04241	0.03839	0.03150	0.01980	0.01220	0.00770	0.00505	0.00242	0.00132	0.00079	0.00051
4	0.02986	0.02976	0.02907	0.02832	0.02802	0.01835	0.02651	0.02490	0.02193	0.01592	0.01109	0.00768	0.00536	0.00298	0.00160	0.00099	0.00065
5	0.01942	0.01938				0.01307			0.01573	0.01249	0.00949	0.00708	0.00527	0.00299	0.00179	0.00113	0.00075
6	0.01361					0.00976			0.01168	0.00983	0.00795	0.00628	0.00492	0.00291	0.00188	0.00124	0.00084
7	0.01005					0.00755			0.00894	0.00784	0.00661	0.00548	0.00445	0.00276	0.00193	0.00130	0.00091
8	0.00772					0.00600			0.00703	0.00635	0.00554	0.00472	0.00398	0.00256	0.00189	0.00134	0.00094
9	0.00612								0.00566	0.00520	0.00466	0.00409	0.00353	0.00241	0.00184	0.00133	0.00096
10								0.00477	0.00465	0.00438	0.00397	0.00352	0.00326				

Table 4.1 (Continued)

Function B

Offset (r/a)

Depth (z/a)	0	0.2	0.4	0.6	0.8	1	1.2	1.5	2	3	4	5	6	8	10	12	14
0	0	0	0	0	0	0	0	0	0	0	0	0	0	0	0	0	0
0.1	0.09852	0.10140	0.11138	0.13424	0.18796	0.05388	-0.07899	-0.02672	-0.00845	-0.00210	-0.00084	-0.00042	0	0	0	0	0
0.2	0.18857	0.19306	0.20772	0.23524	0.25983	0.08513	-0.07759	-0.04448	-0.01593	-0.00412	-0.00166	-0.00083	-0.00024	-0.00010			
0.3	0.28362	0.26787	0.28018	0.29483	0.27257	0.10757	-0.04316	-0.04999	-0.02166	-0.00599	-0.00245						
0.4	0.32016	0.32259	0.32748	0.32273	0.26925	0.12404	-0.00766	-0.04535	-0.02522								
0.5	0.35777	0.35752	0.35323	0.33106	0.26236	0.13591	0.02165	-0.03455	-0.02651	-0.00991	-0.00388	-0.00199	-0.00116	-0.00049	-0.0025	-0.0014	-0.00009
0.6	0.37831	0.37531	0.36308	0.32822	0.25411	0.14440	0.04457	-0.02101									
0.7	0.38487	0.37962	0.36072	0.31929	0.24638	0.14986	0.06209	-0.00702	-0.02329								
0.8	0.38091	0.37408	0.35133	0.30699	0.23779	0.15292	0.07530	0.00614									
0.9	0.36962	0.36275	0.33734	0.29299	0.22891	0.15404	0.08507	0.01795									
1	0.35355	0.34553	0.32075	0.27819	0.21978	0.15355	0.09210	0.02814	-0.01005	-0.01115	-0.00608	-0.00344	-0.00210	-0.00092	-0.00048	-0.00028	-0.00018
1.2	0.31485	0.30730	0.28481	0.24836	0.20113	0.14915	0.10002	0.04378	0.00023	-0.00995	-0.00632	-0.00378	-0.00236	-0.00107	-0.00068	-0.00040	-0.00026
1.5	0.25602	0.25025	0.23338	0.20694	0.17368	0.13732	0.10193	0.05745	0.01385	-0.00669	-0.00600	-0.00401	-0.00265	-0.00126	-0.00084	-0.00050	-0.00033
2	0.17889	0.18144	0.16644	0.15198	0.13375	0.11331	0.09254	0.06371	0.02836	0.00028	-0.00410	-0.00371	-0.00278	-0.00148	-0.00094	-0.00059	-0.00039
2.5	0.12807	0.12633	0.12126	0.11327	0.10298	0.09130	0.07869	0.06022	0.03429	0.00661	-0.00130	-0.00271	-0.00250	-0.00156	-0.00099	-0.00065	-0.00046
3	0.09487	0.09394	0.09099	0.08635	0.08033	0.07325	0.06551	0.05354	0.03511	0.01112	0.00157	-0.00134	-0.00192	-0.00151	-0.00094	-0.00068	-0.00050
4	0.05707	0.05666	0.05562	0.05383	0.05145	0.04773	0.04532	0.03995	0.03066	0.01515	0.00595	0.00155	-0.00029	-0.00109	-0.00070	-0.00068	-0.00049
5	0.03772	0.03760				0.03384			0.02474	0.01522	0.00810	0.00371	0.00132	-0.00043	-0.00037	-0.00047	-0.00045
6	0.02666					0.02468			0.01968	0.01380	0.00867	0.00496	0.00254	0.00028	-0.00002	-0.00029	-0.00037
7	0.01980					0.01868			0.01577	0.01204	0.00842	0.00547	0.00332	0.00093	0.00035	-0.00008	-0.00025
8	0.01526					0.01459			0.01279	0.01034	0.00779	0.00554	0.00372	0.00141	0.00066	0.00012	-0.00012
9	0.01212					0.01170			0.01054	0.00888	0.00705	0.00533	0.00386	0.00178			
10								0.00924	0.00879	0.00764	0.00631	0.00501	0.00382	0.00199			

Table 4.1 (Continued)

Function C

Depth (z/a)	Offset (r/a)																
	0	0.2	0.4	0.6	0.8	1	1.2	1.5	2	3	4	5	6	8	10	12	14
0	0	0	0	0	0	0	0	0	0	0	0	0	0	0	0	0	0
0.1	-0.04926	-0.05142	-0.05903	-0.07708	-0.12108	0.02247	0.12007	0.04475	0.01536	0.00403	0.00164	0.00082					
0.2	-0.09429	-0.09755	-0.10872	-0.12977	-0.14552	0.02419	0.14896	0.07892	0.02951	0.00796	0.00325	0.00164	0.00094	0.00039			
0.3	-0.13181	-0.13484	-0.14415	-0.15023	-0.12990	0.01988	0.13394	0.09816	0.04148	0.01169	0.00483						
0.4	-0.16008	-0.16188	-0.16519	-0.15985	-0.11168	0.01292	0.11014	0.10422	0.05067								
0.5	-0.17889	-0.17835	-0.17497	-0.15625	-0.09833	0.00483	0.08730	0.10125	0.05690	0.01824	0.00778	0.00399	0.00231	0.00098	0.00050	0.00029	0.00018
0.6	-0.18915	-0.18633	-0.17336	-0.14934	-0.08967	-0.00304	0.06731	0.09313	0.06129								
0.7	-0.19244	-0.18831	-0.17393	-0.14147	-0.08409	-0.01061	0.05028	0.08253									
0.8	-0.19046	-0.18481	-0.16784	-0.13393	-0.08066	-0.01744	0.03582	0.07114									
0.9	-0.18481	-0.17841	-0.16024	-0.12664	-0.07828	-0.02337	0.02359	0.05993									
1	-0.17678	-0.17050	-0.15188	-0.11995	-0.07634	-0.02843	0.01331	0.04939	0.05429	0.02726	0.01333	0.00726	0.00433	0.00188	0.00098	0.00057	0.00036
1.2	-0.15742	-0.15117	-0.13467	-0.10763	-0.07289	-0.03575	-0.00245	0.03107	0.04522	0.02791	0.01467	0.00824	0.00501	0.00221	0.00141	0.00083	0.00039
1.5	-0.12801	-0.12277	-0.11101	-0.09145	-0.06711	-0.04124	-0.01702	0.01088	0.03154	0.02652	0.01570	0.00933	0.00585	0.00266	0.00179	0.00107	0.00069
2	-0.08944	-0.08491	-0.07976	-0.06925	-0.05560	-0.04144	-0.02687	-0.00782	0.01267	0.02070	0.01527	0.01013	0.00321	0.00327	0.00128	0.00083	
2.5	-0.06403	-0.06068	-0.05839	-0.05259	-0.04522	-0.03605	-0.02800	-0.01536	0.00103	0.0134	0.00987	0.00707	0.00569	0.00209	0.00232	0.00145	0.00096
3	-0.04744	-0.04560	-0.04339	-0.04089	-0.03642	-0.03130	-0.02587	-0.01748	-0.00528	0.00792	0.01030	0.00888	0.00689	0.00392	0.00254	0.00168	0.00155
4	-0.02854	-0.02737	-0.02562	-0.02585	-0.02421	-0.02112	-0.01964	-0.01586	-0.00956	0.00038	0.00492	0.00602	0.00561	0.00389	0.00250	0.00177	0.00127
5	-0.01886	-0.01810				-0.01568			-0.00939	-0.00293	-0.00128	0.00329	0.00391	0.00341	0.00227	0.00173	0.00130
6	-0.01333					-0.01118			-0.00819	-0.00405	-0.00079	0.00129	0.00234	0.00272	0.00193	0.00161	0.00128
7	-0.00990					-0.00902			-0.00678	-0.00417	-0.00180	-0.00004	0.00113	0.00200	0.00157	0.00143	0.00120
8	-0.00763					-0.00699			-0.00552	-0.00393	-0.00225	-0.00077	0.00029	0.00134	0.00124	0.00122	0.00110
9	-0.00607					-0.00423			-0.00452	-0.00353	-0.00235	-0.00118	-0.00027	0.00082			
10								-0.00381	-0.00373	-0.00314	-0.00233	-0.00137	-0.00630	0.00040			

93

Table 4.1 (*Continued*)

Depth (z/a)	Function F Offset (r/a)																
	0	0.2	0.4	0.6	0.8	1	1.2	1.5	2	3	4	5	6	8	10	12	14
0	0.5	0.5	0.5	0.5	0.5	0	-0.34722	-0.22222	-0.12500	-0.05556	-0.03125	-0.02000	-0.01389	-0.00781	-0.00500	-0.00347	-0.00255
0.1	0.45025	0.44794	0.43981	0.41954	0.35789	0.03817	-0.20800	-0.17612	-0.10950	-0.05151	-0.02961	-0.01917	-0.01295	-0.00742			
0.2	0.40194	0.39781	0.38294	0.34823	0.26215	0.05466	-0.11165	-0.13381	-0.09441	-0.04750	-0.02798	-0.01835					
0.3	0.35633	0.35094	0.34508	0.29016	0.20503	0.06372	-0.05346	-0.09768	-0.08010	-0.04356	-0.02636						
0.4	0.31431	0.30801	0.28681	0.24469	0.17086	0.06848	-0.01818	-0.06835	-0.06684								
0.5	0.27639	0.26997	0.24890	0.20937	0.14752	0.07037	0.00388	-0.04529	-0.05479	-0.03595	-0.02320	-0.01590	-0.01154	-0.00681	-0.00450	-0.00318	-0.00237
0.6	0.24275	0.23444	0.21667	0.18138	0.13042	0.07068	0.01797	-0.02479									
0.7	0.21327	0.20762	0.18956	0.15903	0.11740	0.06963	0.02704	-0.01392	-0.03469								
0.8	0.18765	0.18287	0.16679	0.14053	0.10604	0.06774	0.03277	-0.00365									
0.9	0.16552	0.16158	0.14747	0.12528	0.09664	0.06533	0.03619	0.00408									
1	0.14645	0.14280	0.12395	0.11225	0.08850	0.06256	0.03819	0.00984	-0.01367	-0.01994	-0.01591	-0.01209	-0.00931	-0.00587	-0.00400	-0.00289	-0.00219
1.2	0.11589	0.11360	0.10460	0.09449	0.07486	0.05670	0.03913	0.01716	-0.00452	-0.01491	-0.01337	-0.01068	-0.00844	-0.00550	-0.00353	-0.00261	-0.00201
1.5	0.08398	0.08196	0.07719	0.06918	0.05919	0.04804	0.03686	0.02177	0.00413	-0.00879	-0.00995	-0.00870	-0.00723	-0.00495	-0.00307	-0.00233	-0.00183
2	0.05279	0.05348	0.04994	0.04614	0.04162	0.03593	0.03029	0.02197	0.01043	-0.00189	-0.00546	-0.00589	-0.00544	-0.00410	-0.00263	-0.00208	-0.00166
2.5	0.03576	0.03673	0.03459	0.03263	0.03014	0.02762	0.02406	0.01927	0.01188	0.00198	-0.00226	-0.00364	-0.00386	-0.00332	-0.00223	-0.00183	-0.00150
3	0.02566	0.02586	0.02255	0.02595	0.02263	0.02097	0.01911	0.01623	0.01144	0.00396	-0.00010	-0.00192	-0.00258	-0.00263	-0.00153	-0.00137	-0.00120
4	0.01493	0.01536	0.01412	0.01259	0.01386	0.01331	0.01256	0.01134	0.00912	0.00508	0.00209	0.00026	-0.00076	-0.00148	-0.00096	-0.00099	-0.00093
5	0.00971	0.01011				0.00905			0.00700	0.00475	0.00277	0.00129	0.00031	-0.00066	-0.00053	-0.00066	-0.00070
6	0.00680					0.00675			0.00538	0.00409	0.00278	0.00170	0.00088	-0.00010	-0.00020	-0.00041	-0.00049
7	0.00503					0.00483			0.00428	0.00346	0.00258	0.00178	0.00114	0.00027	0.00003	-0.00020	-0.00033
8	0.00386					0.00380			0.00350	0.00291	0.00229	0.00174	0.00125	0.00048	0.00020	-0.00005	-0.00019
9	0.00306					0.00374			0.00291	0.00247	0.00203	0.00163	0.00124	0.00062			
10								0.00267	0.00246	0.00213	0.00176	0.00149	0.00126	0.00070			

Table 4.1 (Continued)

| | Function H | | | | | | | | | | | | | | | | |
| | Offset (r/a) | | | | | | | | | | | | | | | | |
Depth (z/a)	0	0.2	0.4	0.6	0.8	1	1.2	1.5	2	3	4	5	6	8	10	12	14
0	2.0	1.97987	1.91751	1.80575	1.62553	1.27319	0.93676	0.71185	0.51671	0.33815	0.25200	0.20045	0.16626	0.12576	0.09918	0.08346	0.07023
0.1	1.80998	1.79018	1.72886	1.61961	1.44711	1.18107	0.92670	0.70888	0.51627	0.33794	0.25184	0.20081	0.16688	0.12512			
0.2	1.63961	1.62068	1.56242	1.46001	1.30614	1.09996	0.90098	0.70074	0.51382	0.33726	0.25162	0.20072					
0.3	1.48806	1.47044	1.40979	1.32442	1.19210	1.02740	0.86726	0.68823	0.50966	0.33638	0.25124						
0.4	1.35407	1.33802	1.28963	1.20822	1.09555	0.96202	0.83042	0.67238	0.50412								
0.5	1.23607	1.22176	1.17894	1.10830	1.01312	0.90298	0.79308	0.65429	0.49278	0.33293	0.24996	0.19982	0.16668	0.12493	0.09996	0.08295	0.07123
0.6	1.13238	1.11198	1.08350	1.02154	0.94120	0.84917	0.75653	0.63469	0.48061								
0.7	1.04131	1.03037	0.99794	0.91049	0.87742	0.80030	0.72143	0.61442									
0.8	0.96125	0.95175	0.92386	0.87928	0.82136	0.75571	0.68809	0.59398									
0.9	0.89072	0.88251	0.85856	0.82616	0.77950	0.71495	0.65677	0.57361									
1	0.82843	0.85005	0.80465	0.76809	0.72587	0.67769	0.62701	0.55364	0.45122	0.31877	0.24386	0.19673	0.16516	0.12394	0.09952	0.08292	0.07104
1.2	0.72410	0.71882	0.70370	0.67937	0.64814	0.61187	0.57329	0.51552	0.43013	0.31162	0.24070	0.19520	0.16369	0.12350	0.09876	0.08270	0.07064
1.5	0.60555	0.60233	0.57246	0.57633	0.55559	0.53138	0.50496	0.46379	0.39872	0.29945	0.23495	0.19053	0.16199	0.12281	0.09792	0.08196	0.07026
2	0.47214	0.47022	0.44512	0.45656	0.44502	0.43202	0.41702	0.39242	0.35054	0.27740	0.22418	0.18618	0.15846	0.12124	0.09700	0.08115	0.06980
2.5	0.38518	0.38403	0.38098	0.37608	0.36940	0.36155	0.35243	0.33698	0.30913	0.25550	0.21208	0.17898	0.15395	0.11928	0.09558	0.08061	0.06897
3	0.32457	0.32403	0.32184	0.31887	0.31464	0.30969	0.30381	0.29364	0.27453	0.23487	0.19977	0.17154	0.14919	0.11694	0.09300	0.07864	0.06848
4	0.24620	0.24588	0.24820	0.25128	0.24168	0.23932	0.23668	0.23164	0.22188	0.19908	0.17640	0.15596	0.13864	0.11172	0.08915	0.07675	0.06695
5	0.19805	0.19785				0.19455			0.18450	0.17080	0.15575	0.14130	0.12785	0.10585	0.08562	0.07452	0.06522
6	0.16554					0.16326			0.15750	0.14868	0.13842	0.12792	0.11778	0.09990	0.08197	0.07210	0.06377
7	0.14217					0.14077			0.13699	0.13097	0.12404	0.11620	0.10843	0.09387	0.07800	0.06928	0.06200
8	0.12448					0.12352			0.12112	0.11680	0.11176	0.10600	0.09976	0.08848			
9	0.11079					0.10989			0.10854	0.10548	0.10161	0.09702	0.09234	0.08298			
10								0.9900	0.09820	0.09510	0.09290	0.08980	0.08300	0.07710	0.07407	0.06678	0.05976

The radial–horizontal stress is calculated from Eq. 4.5:

$$\sigma_r = p[2\mu A + C + (1 - 2\mu)F]$$
$$= 689\{2(0.45)(1.0) + 0 + [1 - 2(0.45)](0.5)\}$$
$$= 654.55 \text{ kPa}$$

The deflection is calculated from Eq. 4.6:

$$\Delta_z = \frac{p(1 + \mu)a}{E}\left[\frac{z}{a}A + (1 - \mu)H\right]$$
$$= \frac{689(1 + 0.45)12.5}{345000}[0(1.0) + (1 - 0.45)2.0]$$
$$= 0.0398 \text{ cm or } 0.398 \text{ mm}$$

With $z = 50$ cm and $r = 25$ cm,

$$\frac{z}{a} = \frac{50}{12.5} = 4 \text{ and } \frac{r}{a} = \frac{25}{12.5} = 2$$

From Table 4.1, for the above values,

$$A = 0.02139, B = 0.03066, C = -0.00956, F = 0.00912, \text{ and } H = 0.22188$$

The radial–horizontal stress is

$$\sigma_r = 689\{2(0.45)(0.02193) + (-0.00956) + [1 - 2(0.45)](0.00912)\}$$
$$= 7.64 \text{ kPa}$$

The deflection is

$$\Delta_z = \frac{689(1 + 0.45)12.5}{345,000}\left[\frac{50}{12.5}(0.02193) + (1 - 0.45)(0.22188)\right]$$
$$= 0.00759 \text{ cm or } 0.0759 \text{ mm}$$

The Boussinesq theory and the Ahlvin and Ulery equations can be used to calculate the stresses and deflections in a simple pavement system. The utility of these is that they serve as a basis for more complex pavement analysis. With the availability of computers (both mainframe and personal), there have been significant advances in mechanistic pavement analysis methodologies. Pavement systems are represented as multilayer systems that are homogenous and isotropic with linear responses (i.e., the material returns to its original shape after the load is removed). Computer programs allow engineers to input various axle loads, tire pressures, and material properties. Program outputs consist of stresses, strains, and deflections. Some of the programs now in common use are BISAR, CHEVRON, KENLAYER, WESLEA, and ELSYM5. The ELSYM5 program is available on floppy disk from the Federal Highway Administration. The use of programs like ELSYM5 can greatly simplify the pavement analysis task.

4.4 THE AASHTO FLEXIBLE-PAVEMENT DESIGN PROCEDURE

There are several accepted flexible-pavement design procedures available including the Asphalt Institute method, the National Stone Association procedure, and the Shell procedure. Most of the procedures have been field verified and used by highway agencies for several years. The selection of one procedure over another is usually based on a highway agency's experience and satisfaction with design results.

A widely accepted flexible-pavement design procedure is presented in the "AASHTO Guide for Design of Pavement Structures," which is published by the American Association of State Highway and Transportation Officials (AASHTO). The procedure was first published in 1972, with the latest revisions in 1993. The 1993 AASHTO design procedure is the same as the 1986 AASHTO procedure except the new procedure has a revised section for overlay designs. Test data, used for the development of the design procedure, were collected at the AASHO Road Test in Illinois from 1958 to 1960 (AASHO, which stands for American Association of State Highway Officials, was the prior name of AASHTO). Unfortunately, the current AASHTO design procedure has not yet been converted to metric units and thus the material presented in this section is in U.S. customary units.

A pavement can be subjected to a number of detrimental effects including fatigue failures (cracking), which are the result of repeated loading caused by traffic passing over the pavement. The pavement is also placed in an uncontrolled environment that produces temperature extremes and moisture variations. The combination of the environment, traffic loads, material variations, and construction variations requires a comparatively complex set of design procedures to incorporate all of the variables. The AASHTO pavement design procedure meets most of the demands placed on a flexible-pavement design procedure. It considers environment, load, and materials in a methodology that is relatively easy to use. The AASHTO pavement design procedure has been widely accepted throughout the United States and around the world. Details of this procedure are presented in the following sections.

4.4.1 Serviceability Concept

Prior to the AASHO Road Test, there was no real consensus as to the definition of pavement failure. In the eyes of an engineer, pavement failure occurred whenever cracking, rutting, or other surface distresses became visible. In contrast, the motoring public usually associated pavement failure with poor ride quality. Pavement engineers conducting the ASSHO Road Test were faced with the task of combining the two failure definitions so that a single design procedure could be used to satisfy both critics. The Pavement Serviceability-Performance Concept was developed by Carey and Irick (1962) to handle the question concerning pavement failure. Carey and Irick considered pavement performance histories and noted that pavements usually begin their service life in excellent condition and deteriorate as traffic loading is applied in conjunction with prevailing environmental conditions. The performance curve is the historical record of the performance of the pavement. Pavement performance, at any point in time, is known as the present serviceability index or PSI. Examples of pavement performance (or PSI trends) are shown in Fig. 4.5.

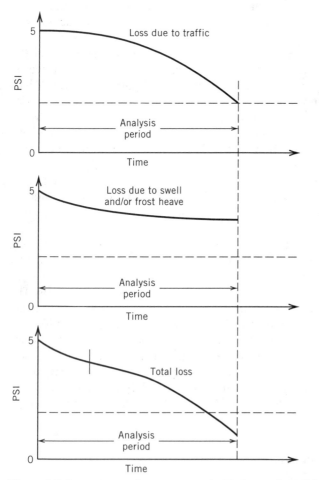

Figure 4.5 Pavement performance trends. Redrawn from "AASHTO Guide for Design of Pavement Structures," Washington, DC, The American Association of State Highway and Transportation Officials, copyright 1993. Used by permission.

At any time, the present serviceability index of a pavement can be measured. It is usually measured by a panel of raters who drive over the pavement section and rate the pavement performance on a scale of one to five with five being the smoothest ride. The accumulation of traffic loads causes the pavement to deteriorate and, as expected, the serviceability rating drops. At some point, a terminal serviceability index (TSI) is reached. At this point, most raters feel that the pavement can no longer perform in a serviceable manner.

Having raters evaluate all of the nation's highways on a continuous basis would be an overwhelming task. Instead, correlations have been made between panel opinions and measured variables such as pavement roughness, rutting, and cracking. Consequently, mechanical devices are now commonly used to determine PSI. It has been found that new pavements usually have an initial PSI rating of approximately 4.2 to 4.5. The point at which pavements are considered to have failed (i.e., the TSI) varies by type of highway. Highway facilities such as interstate highways or principal arterials usually have TSIs of 2.5 or 3.0, while local roads can have TSIs of 2.0.

4.4.2 Flexible-Pavement Design Equation

At the conclusion of the AASHO Road Test, a regression analysis (see Chapter 8 for a discussion on regression analysis) was performed to determine the interactions of traffic loadings, material properties, layer thickness, and climate. The relationship between axle loads and the thickness index of the pavement system is shown in Fig. 4.6. The thickness index represents a combination of layer

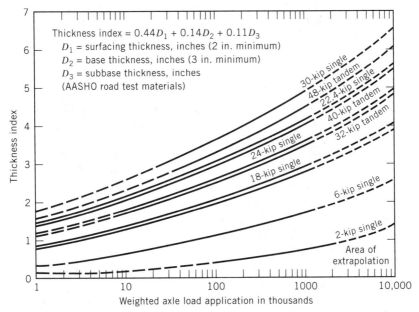

Figure 4.6 AASHO road test thickness index versus axle loads. Redrawn from "AASHO Road Test Report 5," Highway Research Board, Special Report 61E, Washington, DC, 1962. Used by permission.

thickness and strength coefficients. The term "thickness index" is the same as the term "structural number," which will be discussed later.

The relationship shown in Fig. 4.6 can be used to determine the required thickness index of a flexible pavement. For example, assume that a new pavement must withstand one million applications of a 12,000-lb (i.e., 12-kip, 53.4-kN) single-axle load. Based on the curves, it can be seen that a thickness index of approximately 2.5 is needed to sustain that type of loading. There are many combinations of pavement materials and thicknesses that can provide a thickness index value of 2.5, however, it is the responsibility of the pavement engineer to select a practical and economic combination of the materials to satisfy design inputs. Because of the large number of variables involved in real-world pavement design, the graph shown in Fig. 4.6 quickly loses its utility. Consequently, an equation was developed for flexible-pavement design, which replaces the graph in Fig. 4.6.

The basic equation for flexible-pavement design given in the 1993 AASHTO design guide permits engineers to determine a structural number (thickness index) required to carry a designated traffic loading. The AASHTO equation is

$$\log_{10} W_{18} = Z_R S_O + 9.36[\log_{10}(SN + 1)] - 0.20 + \frac{\log_{10}[\Delta PSI/(2.7)]}{0.40 + [1094/(SN + 1)^{5.19}]}$$

$$+ 2.32 \log_{10} M_R - 8.07 \tag{4.7}$$

where W_{18} is the 18-kp-equivalent single-axle load, Z_R is the reliability (z-statistic from the standard normal curve), S_O is the overall standard deviation of traffic, SN is the structural number, ΔPSI is the loss in serviceability from when the pavement is new until it reaches its TSI, and M_R is the soil resilient modulus of the subgrade in psi. A graphical solution to Eq. 4.7 is shown in Fig. 4.7. Details on the variables that serve as input to Eq. 4.7 and Fig. 4.7 are as follows:

W_{18} Automobiles and truck traffic provide a wide range of vehicle axle types and axle loads. If one were to attempt to account for the variety of traffic loadings encountered on a pavement, this input variable would require a significant amount of data collection and design evaluation. Instead, the problem of handling mixed traffic loading is solved with the adoption of a standard 18-kip (80.1-kN)-equivalent single-axle load (ESAL). The idea is to determine the impact of any axle load on the pavement in terms of the equivalent amount of pavement impact that an 18-kip single-axle load would have. For example, if a 44-kip tandem-axle (i.e., double-axle) load has 2.88 times the impact on pavement structure as an 18-kip single-axle load, 2.88 would be the W_{18} value assigned to this tandem-axle load. The AASHO Road Test also found that the 18-kip (80.1-kN)-equivalent axle load is a function of the terminal serviceability index of the pavement structure. The axle-load equivalency factors for flexible pavement design, with a TSI of 2.5, are presented in Tables 4.2 (for single axles), 4.3 (for tandem axles), and 4.4 (for triple axles).

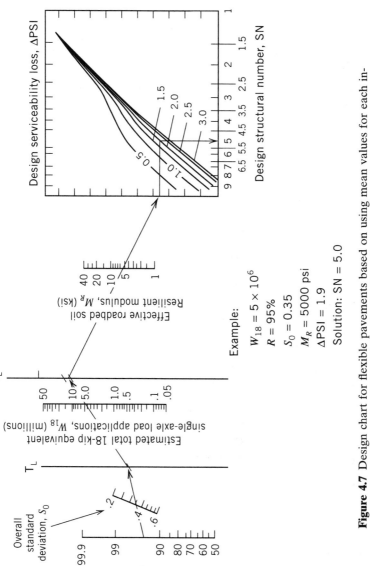

Figure 4.7 Design chart for flexible pavements based on using mean values for each input. Redrawn from "AASHTO Guide for Design of Pavement Structures," Washington, DC, The American Association of State Highway and Transportation Officials, copyright 1993. Used by permission.

Z_R is defined as the probability that serviceability will be maintained at adequate levels from a user's point of view throughout the design life of the facility. This factor estimates the probability that the pavement will perform at or above the TSI level during the design period, and it accounts for the inherent uncertainty in design. Equation 4.7 uses the z-statistic, which is obtained from the cumulative probabilities of the standard normal distribution (i.e., a normal distribution with mean equal to zero and variance equal to one). The z-statistics corresponding to various probability levels are given in Table 4.5. In the flexible-pavement-design nomograph (Fig. 4.7), the probabilities (in percent) are used directly (instead of the Z_R as in the case of Eq. 4.7) and these percent probabilities are denoted R, the reliability (see Table 4.5).

Table 4.2 Axle-Load Equivalency Factors for Flexible Pavements, Single Axles, and TSI = 2.5

Axle Load (kips)	Pavement Structural Number (SN)					
	1	2	3	4	5	6
2	0.0004	0.0004	0.0003	0.0002	0.0002	0.0002
4	0.003	0.004	0.004	0.003	0.002	0.002
6	0.011	0.017	0.017	0.013	0.010	0.009
8	0.032	0.047	0.051	0.041	0.034	0.031
10	0.078	0.102	0.118	0.102	0.088	0.080
12	0.168	0.198	0.229	0.213	0.189	0.176
14	0.328	0.358	0.399	0.388	0.360	0.342
16	0.591	0.613	0.646	0.645	0.623	0.606
18	1.00	1.00	1.00	1.00	1.00	1.00
20	1.61	1.57	1.49	1.47	1.51	1.55
22	2.48	2.38	2.17	2.09	2.18	2.30
24	3.69	3.49	3.09	2.89	3.03	3.27
26	5.33	4.99	4.31	3.91	4.09	4.48
28	7.49	6.98	5.90	5.21	5.39	5.98
30	10.3	9.5	7.9	6.8	7.0	7.8
32	13.9	12.8	10.5	8.8	8.9	10.0
34	18.4	16.9	13.7	11.3	11.2	12.5
36	24.0	22.0	17.7	14.4	13.9	15.5
38	30.9	28.3	22.6	18.1	17.2	19.0
40	39.3	35.9	28.5	22.5	21.1	23.0
42	49.3	45.0	35.6	27.8	25.6	27.7
44	61.3	55.9	44.0	34.0	31.0	33.1
46	75.5	68.8	54.0	41.4	37.2	39.3
48	92.2	83.9	65.7	50.1	44.5	46.5
50	112.0	102.0	79.0	60.0	53.0	55.0

Source: "AASHTO Guide for Design of Pavement Structures," The American Association of State Highway and Transportation Officials, Washington, DC, copyright 1993. Used by permission.

Table 4.3 Axle-Load Equivalency Factors for Flexible Pavements, Tandem Axles, and TSI = 2.5

Axle Load (kips)	Pavement Structural Number (SN)					
	1	2	3	4	5	6
2	0.0001	0.0001	0.0001	0.0000	0.0000	0.0000
4	0.0005	0.0005	0.0004	0.0003	0.0003	0.0002
6	0.002	0.002	0.002	0.001	0.001	0.001
8	0.004	0.006	0.005	0.004	0.003	0.003
10	0.008	0.013	0.011	0.009	0.007	0.006
12	0.015	0.024	0.023	0.018	0.014	0.013
14	0.026	0.041	0.042	0.033	0.027	0.024
16	0.044	0.065	0.070	0.057	0.047	0.043
18	0.070	0.097	0.109	0.092	0.077	0.070
20	0.107	0.141	0.162	0.141	0.121	0.110
22	0.160	0.198	0.229	0.207	0.180	0.166
24	0.231	0.273	0.315	0.292	0.260	0.242
26	0.327	0.370	0.420	0.401	0.364	0.342
28	0.451	0.493	0.548	0.534	0.495	0.470
30	0.611	0.648	0.703	0.695	0.658	0.633
32	0.813	0.843	0.889	0.887	0.857	0.834
34	1.06	1.08	1.11	1.11	1.09	1.08
36	1.38	1.38	1.38	1.38	1.38	1.38
38	1.75	1.73	1.69	1.68	1.70	1.73
40	2.21	2.16	2.06	2.03	2.08	2.14
42	2.76	2.67	2.49	2.43	2.51	2.61
44	3.41	3.27	2.99	2.88	3.00	3.16
46	4.18	3.98	3.58	3.40	3.55	3.79
48	5.08	4.80	4.25	3.98	4.17	4.49
50	6.12	5.76	5.03	4.64	4.86	5.28
52	7.33	6.87	5.93	5.38	5.63	6.17
54	8.72	8.14	6.95	6.22	6.47	7.15
56	10.3	9.6	8.1	7.2	7.4	8.2
58	12.1	11.3	9.4	8.2	8.4	9.4
60	14.2	13.1	10.9	9.4	9.6	10.7
62	16.5	15.3	12.6	10.7	10.8	12.1
64	19.1	17.6	14.5	12.2	12.2	13.7
66	22.1	20.3	16.6	13.8	13.7	15.4
68	25.3	23.3	18.9	15.6	15.4	17.2
70	29.0	26.6	21.5	17.6	17.2	19.2
72	33.0	30.3	24.4	19.8	19.2	21.3
74	37.5	34.4	27.6	22.2	21.3	23.6
76	42.5	38.9	31.1	24.8	23.7	26.1
78	48.0	43.9	35.0	27.8	26.2	28.8
80	54.0	49.4	39.2	30.9	29.0	31.7
82	60.6	55.4	43.9	34.4	32.0	34.8
84	67.8	61.9	49.0	38.2	35.3	38.1
86	75.7	69.1	54.5	42.3	38.8	41.7
88	84.3	76.9	60.6	46.8	42.6	45.6
90	93.7	85.4	67.1	51.7	46.8	49.7

Source: "AASHTO Guide for Design of Pavement Structures," The American Association of State Highway and Transportation Officials, Washington, DC, copyright 1993. Used by permission.

Table 4.4 Axle-Load Equivalency Factors for Flexible Pavements, Triple Axles, and TSI = 2.5

Axle Load (kips)	Pavement Structural Number (SN)					
	1	2	3	4	5	6
2	0.0000	0.0000	0.0000	0.0000	0.0000	0.0000
4	0.0002	0.0002	0.0002	0.0001	0.0001	0.0001
6	0.0006	0.0007	0.0005	0.0004	0.0003	0.0003
8	0.001	0.002	0.001	0.001	0.001	0.001
10	0.003	0.004	0.003	0.002	0.002	0.002
12	0.005	0.007	0.006	0.004	0.003	0.003
14	0.008	0.012	0.010	0.008	0.006	0.006
16	0.012	0.019	0.018	0.013	0.011	0.010
18	0.018	0.029	0.028	0.021	0.017	0.016
20	0.027	0.042	0.042	0.032	0.027	0.024
22	0.038	0.058	0.060	0.048	0.040	0.036
24	0.053	0.078	0.084	0.068	0.057	0.051
26	0.072	0.103	0.114	0.095	0.080	0.072
28	0.098	0.133	0.151	0.128	0.109	0.099
30	0.129	0.169	0.195	0.170	0.145	0.133
32	0.169	0.213	0.247	0.220	0.191	0.175
34	0.219	0.266	0.308	0.281	0.246	0.228
36	0.279	0.329	0.379	0.352	0.313	0.292
38	0.352	0.403	0.461	0.436	0.393	0.368
40	0.439	0.491	0.554	0.533	0.487	0.459
42	0.543	0.594	0.661	0.644	0.597	0.567
44	0.666	0.714	0.781	0.769	0.723	0.692
46	0.811	0.854	0.918	0.911	0.868	0.838
48	0.979	1.015	1.072	1.069	1.033	1.005
50	1.17	1.20	1.24	1.25	1.22	1.20
52	1.40	1.41	1.44	1.44	1.43	1.41
54	1.66	1.66	1.66	1.66	1.66	1.66
56	1.95	1.93	1.90	1.90	1.91	1.93
58	2.29	2.25	2.17	2.16	2.20	2.24
60	2.67	2.60	2.48	2.44	2.51	2.58
62	3.09	3.00	2.82	2.76	2.85	2.95
64	3.57	3.44	3.19	3.10	3.22	3.36
66	4.11	3.94	3.61	3.47	3.62	3.81
68	4.71	4.49	4.06	3.88	4.05	4.30
70	5.38	5.11	4.57	4.32	4.52	4.84
72	6.12	5.79	5.13	4.80	5.03	5.41
74	6.93	6.54	5.74	5.32	5.57	6.04
76	7.84	7.37	6.41	5.88	6.15	6.71
78	8.83	8.28	7.14	6.49	6.78	7.43
80	9.92	9.28	7.95	7.15	7.45	8.21
82	11.1	10.4	8.8	7.9	8.2	9.0
84	12.4	11.6	9.8	8.6	8.9	9.9
86	13.8	12.9	10.8	9.5	9.8	10.9
88	15.4	14.3	11.9	10.4	10.6	11.9
90	17.1	15.8	13.2	11.3	11.6	12.9

Source: "AASHTO Guide for Design of Pavement Structures," The American Association of State Highway and Transportation Officials, Washington, DC, copyright 1993. Used by permission.

Table 4.5 Cumulative Percent Probabilities of Reliability, R, of the Standard Normal Distribution, and Corresponding Z_R

R	0	1	2	3	4	5	6	7	8	9	9.5	9.9
90	−1.282	−1.341	−1.405	−1.476	−1.555	−1.645	−1.751	−1.881	−2.054	−2.326	−2.576	−3.080
80	−0.842	−0.878	−0.915	−0.954	−0.994	−1.036	−1.080	−1.126	−1.175	−1.227	−1.253	−1.272
70	−0.524	−0.553	−0.583	−0.613	−0.643	−0.675	−0.706	−0.739	−0.772	−0.806	−0.824	−0.838
60	−0.253	−0.279	−0.305	−0.332	−0.358	−0.385	−0.412	−0.440	−0.468	−0.496	−0.510	−0.522
50	0	−0.025	−0.050	−0.075	−0.100	−0.125	−0.151	−0.176	−0.202	−0.228	−0.241	−0.251

Example: To be 95% confident that the pavement will remain at or above its TSI (i.e., $R = 95$ for use in Fig. 4.7), a Z_R value of −1.645 would be used in Eq. 4.7 (and in Eq. 4.19).

Highways such as interstates and major arterials, which are costly to reconstruct (i.e., having their pavements rehabilitated) because of resulting traffic delay and disruption, require a high reliability level, whereas local roads, which will have lower impacts on users in the event of pavement rehabilitation, do not require high reliability levels. Typical reliability values for interstate highways are 90% or higher, whereas local roads can have a reliability as low as of 50%.

S_O The overall standard deviation, S_O, takes into account the designers' inability to accurately estimate the variation in future 18-kip (80.1-kN)-equivalent axle loads, and the statistical error in the equations resulting from variability in materials and construction practices. Typical values of S_O are in the order of 0.30 to 0.50.

SN The structural number, SN, represents the overall structural requirement needed to sustain the design's traffic loadings. The structural number is discussed further in Section 4.4.3.

ΔPSI The amount of serviceability loss, over the life of the pavement, ΔPSI, is determined during the pavement design process. The engineer must decide the final PSI level for a particular pavement. Recall that the loss of serviceability is caused by pavement roughness, cracking, patching, and rutting. As pavement distress increases, serviceability decreases. If the design is for a pavement with heavy traffic loads such as an interstate highway, then the serviceability loss may only be 1.2 (e.g., an initial PSI of 4.2 and a TSI of 3.0), whereas a low-volume road can be allowed to deteriorate further and a total serviceability loss of 2.7 or more is possible.

M_R The soil resilient modulus, M_R, is used to reflect the engineering properties of the subgrade (the soil). Each time a vehicle passes over pavement, stresses are developed in the subgrade. After the load passes, the subgrade soil relaxes and the stress is relieved. The resilient modulus test is used to determine the properties of the soil under this repeated load. The resilient modulus can be determined by AASHTO test method T274. Measurement of the resilient modulus is not performed by all transportation agencies; therefore, a relationship between M_R and the California bearing ratio, CBR, has been determined. The CBR has been widely used to determine the supporting characteristics of soils since the mid-1930s, and a significant amount of historical information is available. The CBR is the ratio of the load-bearing capacity of the soil to the load-bearing capacity of a high-quality aggregate, multiplied by 100. The relationship, used to provide a very basic approximation of M_R (in psi) from a known CBR, is

$$M_R = 1500 \times \text{CBR} \tag{4.8}$$

The coefficient of 1500 in Eq. 4.8 is used for CBR values less than 10. Caution must be exercised when applying this equation to higher CBRs because the coefficient (i.e., the 1500 shown in Eq. 4.8) has a range of 750 to 3000.

4.4.3 Structural Number

The objective of Eq. 4.7 and the nomograph in Fig. 4.7 is to determine a required structural number for given axle loadings, reliability, overall standard deviation, change in PSI, and soil resilient modulus. As previously mentioned, there are many pavement material combinations and thicknesses that will provide satisfactory pavement service life. The following equation can be used to relate individual material types and thickness to the structural number:

$$\text{SN} = a_1 D_1 + a_2 D_2 M_2 + a_3 D_3 M_3 \tag{4.9}$$

where a_1, a_2, and a_3 are structural-layer coefficients of the wearing surface, base, and subbase layers, respectively; D_1, D_2, and D_3 are the thicknesses of the wearing surface, base, and subbase layers (in inches), respectively; and M_2 and M_3 are drainage coefficients for the base and subbase, respectively. Values for the structural-layer coefficients for various types of material are presented in Table 4.6. Drainage coefficients are used to modify the thickness of the lower pavement layers (base and subbase) to take into account a material's drainage characteristics. A value of 1.0 for a drainage coefficient represents a material with good drainage characteristics (e.g., a sandy material). A soil such as clay does not drain very well and, consequently, will have a lower drainage coefficient (i.e., less than 1.0) than a sandy material. The reader is referred to AASHTO (1993) for further information on drainage coefficients.

Because there are many combinations of structural-layer coefficients and thicknesses that solve Eq. 4.9, there are some guidelines that can be used to narrow the number of solutions. Experience has shown that wearing layers are typically 2 to 4 in. (50.8 to 101.6 mm) thick, whereas subbases and bases range from 4 to 10 in. (101.6 to 254.0 mm) thick. Knowing which of the materials is the most costly per inch (mm) of depth will also assist with the solution of an initial layer thickness.

EXAMPLE 4.2

A pavement is to be designed to last 10 years. The initial PSI is 4.2 and the TSI (the final PSI) is determined to be 2.5. The subgrade has a soil resilient modulus of 15,000 psi (103.43 MPa). Reliability is 95% with an overall standard deviation

Table 4.6 Structural-Layer Coefficients

Pavement Component	Coefficient
Wearing Surface	
Sand-mix asphaltic concrete	0.35
Hot-mix asphaltic concrete	0.44
Base	
Crushed stone	0.14
Dense-graded crushed stone	0.18
Soil cement	0.20
Emulsion/aggregate-bituminous	0.30
Portland-cement/aggregate	0.40
Lime-pozzolan/aggregate	0.40
Hot-mix asphaltic concrete	0.40
Subbase	
Crushed stone	0.11

of 0.4. For design, the daily car, pickup truck, and light van traffic is 30,000, and the daily truck traffic consists of 1000 passes of single-unit trucks with two single axles and 350 passes of tractor semi-trailer trucks with single, tandem, and triple axles. The axle weights are

$$\text{cars, pickups, light vans} = \text{two 2000-lb (8.9-kN) single axles}$$

$$\text{single-unit truck} = \text{8000-lb (35.6-kN) steering, single axle}$$

$$= \text{22,000-lb (97.9-kN) drive, single axle}$$

$$\text{tractor semi-trailer truck} = \text{10,000-lb (44.5-kN) steering, single axle}$$

$$= \text{16,000-lb (71.2-kN) drive, tandem axle}$$

$$= \text{44,000-lb (195.7-kN) trailer, triple axle}$$

M_2 and M_3 are equal to 1.0 for the materials in the pavement structure. Four inches (10.16 cm) of hot mix asphalt is to be used as the wearing surface and 10 inches (25.40 cm) of crushed stone as the subbase. Determine the thickness required for the base if soil cement is the material to be used.

SOLUTION

Because the axle-load equivalency factors presented in Tables 4.2, 4.3, and 4.4 are a function of the structural number (SN), we have to assume an SN to start the problem (later we will arrive at a structural number and check to make sure that it is consistent with our assumed value). A typical assumption is to let SN = 4. Given this, 18-kip-equivalent single-axle loads (18-kip ESAL) for cars, pickups, and light vans are

$$\text{2-kip (8.9-kN) single-axle equivalent} = 0.0002 \text{ (Table 4.2)}$$

This gives an 18-kp ESAL total of 0.0004 for each vehicle. For single unit trucks,

8-kip (35.6-kN) single-axle equivalent = 0.041 (Table 4.2)

22-kip (97.9-kN) single-axle equivalent = 2.090 (Table 4.2)

This gives an 18-kip ESAL total of 2.131 for single-unit trucks. For tractor semi-trailer trucks,

10-kip (44.5-kN) single-axle equivalent = 0.102 (Table 4.2)

16-kip (71.2-kN) tandem-axle equivalent = 0.057 (Table 4.3)

44-kip (195.7-kN) triple-axle equivalent = 0.769 (Table 4.4)

This gives an 18-kip ESAL total of 0.928 for tractor semi-trailer trucks. Note the comparatively small effect of cars and other light vehicles in terms of 18-kip ESAL. This small effect underscores the nonlinear relationship between axle loads and pavement damage. For example, looking at Table 4.2 with SN = 4, a 36-kip (160.2-kN) single-axle load has 14.4 times the impact on pavement when compared to a single 18-kip (80.1-kN) axle (i.e., twice the weight has 14.4 times the impact).

Given the computed 18-kip ESAL, the daily traffic on this highway produces an 18-kip ESAL total of 2467.8 (0.0004 × 30,000 + 2.131 × 1000 + 0.928 × 350). Traffic (total axle accumulations) over the 10-year design period will be

$$2467.8 \times 365 \times 10 = 9,007,470 \text{ 18-kip ESAL}$$

With an initial PSI of 4.2 and a TSI of 2.5, $\Delta PSI = 1.7$. Solving Eq. 4.7 for SN (using an equation solver on a calculator or computer) with $Z_R = -1.645$ (which corresponds to $R = 95\%$ as shown in Table 4.5) gives SN = 3.94 (Fig. 4.7 can also be used to arrive at an approximate solution for SN). Note that this is very close to the value that was assumed (i.e., SN = 4.0) to get the load equivalency factors from Tables 4.2, 4.3, and 4.4. If Eq. 4.7 gave SN = 5, we would go back and recompute total axle accumulations using the SN of 5 to read the axle-load equivalency factors in Tables 4.2, 4.3, and 4.4. Usually one iteration of this type is all that is needed. (Later, Example 4.5 will provide a demonstration of this type of iteration.)

Given that SN = 3.94, Eq. 4.9 can be applied with $a_1 = 0.44$ (surface course, hot mix asphalt, Table 4.6), $a_2 = 0.20$ (base course, soil cement, Table 4.6), and $a_3 = 0.11$ (subbase, crushed stone, Table 4.6), $M_2 = 1.0$ (given), $M_3 = 1.0$ (given), $D_1 = 4.0$ in. (given), and $D_3 = 10.0$ in. (given).

$$SN = a_1D_1 + a_2D_2M_2 + a_3D_3M_3$$

$$3.94 = 0.44(4) + 0.20D_2(1.0) + 0.11(10.0)(1.0)$$

Solving for D_2 gives D_2 = 5.4 in. (13.72 cm). Using D_2 = <u>5.5 in. or 14 cm</u> would be a conservative estimate and allow for variations in construction. Rounding up to the nearest 0.5 in. (or 1 cm) is a safe practice.

EXAMPLE 4.3

A flexible pavement is constructed with 4 in. (10.16 cm) of hot mix asphalt wearing surface, 8 in. (20.32 cm) of emulsion/aggregate-bituminous base, and 8 in. (20.32 cm) of crushed stone subbase. The subgrade has a soil resilient modulus of 10,000 psi (68.95 MPa), and M_2 and M_3 are equal to 1.0 for the materials in the pavement structure. The overall standard deviation is 0.5, the initial PSI is 4.5, and the TSI is 2.5. The daily traffic has 1080 20-kip (89.0-kN) single axles, 400 24-kip (106.8-kN) single axles, and 680 40-kip (177.9-kN) tandem axles. How many years would you estimate this pavement would last (i.e., how long before its PSI drops below a TSI of 2.5) if you wanted to be 90% confident that your estimate was not too high, and if you wanted to be 99% confident that your estimate was not too high?

SOLUTION

The pavement's structural number is determined from Eq. 4.9 (using Table 4.6) as

$$SN = a_1D_1 + a_2D_2M_2 + a_3D_3M_3$$

$$5.04 = 0.44(4) + 0.30(8)(1.0) + 0.11(8.0)(1.0)$$

For the daily axle loads, the equivalency factors (reading axle equivalents from Tables 4.2 and 4.3 while using SN = 5, which is very close to the 5.04 computed above) are

20-kip (89.0-kN) single-axle equivalent = 1.51 (Table 4.2)

24-kip (106.8-kN) single-axle equivalent = 3.03 (Table 4.2)

40-kip (177.9-kN) tandem-axle equivalent = 2.08 (Table 4.3)

Thus total daily 18-kip ESAL is

daily W_{18} = 1.51(1080) + 3.03(400) + 2.08(680)

= 4257.2   18-kip ESAL

Applying Eq. 4.7, with S_0 = 0.5, SN = 5.04, ΔPSI = 2.0 (4.5 − 2.5), and M_R = 10,000 psi, we find that at R = 90% (Z_R = −1.282 for using Eq. 4.7, as shown in Table 4.5), W_{18} is 26,128,077. Therefore, the number of years is

$$\text{years} = \frac{26,128,077}{365 \times 4257.2}$$

$$= \underline{\underline{16.82 \text{ years}}}$$

Similarly, with $R = 99\%$ ($Z_R = -2.326$ for using Eq. 4.7, as shown in Table 4.5), W_{18} is 7,854,299. Therefore, the number of years is

$$\text{years} = \frac{7,854,299}{365 \times 4257.2}$$

$$= \underline{\underline{5.05 \text{ years}}}$$

These results show that you would be 99% confident that the pavement will last (have a PSI above 2.5) at least 5.05 years, and you would be 90% confident that it would have a PSI above 2.5 for 16.82 years. This example demonstrates the large impact that the chosen reliability values can have on pavement design.

4.5 PAVEMENT SYSTEM DESIGN: PRINCIPLES FOR RIGID PAVEMENTS

Rigid pavements distribute wheel loads by the beam action of the Portland cement concrete (PCC) slab, which is made of a material that has a high modulus of elasticity, on the order of 27.6 to 34.5 GPa (4 to 5 million psi). This beam action (see Fig. 4.8) distributes the wheel loads over a large area of the pavement, thus reducing the high stresses experienced at the surface of the pavement to a level that is acceptable to the subgrade soil.

4.5.1 Calculation of Rigid-Pavement Stresses and Deflections

H. M. Westergaard (1926) presented an important theoretical analysis for rigid pavements. He assumed that the PCC slabs act as homogeneous, isotropic, and elastic solids, and that subgrades react much like a liquid. As a result, the deflections of the slab at any point are a function of the load and the modulus of the subgrade reaction (which is related to the subgrade CBR and soil resilient modulus). Using U.S. customary units (the system in which the equations were originally derived),

$$\Delta = \frac{p}{k} \qquad (4.10)$$

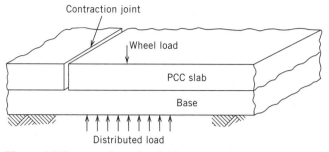

Figure 4.8 Beam action of a rigid pavement.

where k is the modulus of subgrade reaction in pounds per cubic inch (pci), p is the reactive pressure in psi, and Δ is the slab deflection in inches. The modulus k is assumed to be constant at each point under the slab and independent of deflection. The Westergaard equations were developed for three loading cases: an interior load, an edge load, and a corner load, as shown in Fig. 4.9. The equations have been referenced throughout the pavement design literature to the point where some equations have been miscopied and misused. Even Dr. Westergaard had to correct his equations in later publications. In more recent work, Ioannides, Thompson, and Barenberg (1985) reconsidered the Westergaard solutions and compared them with finite element analysis. Their work has refined and corrected the original solutions. These closed-form solutions provide the foundation for rigid-pavement analysis.

The Westergaard solutions for stresses and deflections as revised by Ioannides, Thompson, and Barenberg are presented here.

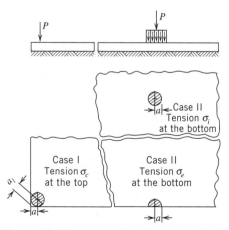

Figure 4.9 Westergaard loading cases. Redrawn from "Computation of Stresses in Concrete Roads," by H. M. Westergaard, *Highway Research Board Proceedings of the Fifth Annual Meeting,* Washington, DC, 1926. Used by permission.

For the interior loading:

$$\sigma_i = \frac{3P(1+\mu)}{2\pi h^2}\left[\ln\left(\frac{2l}{a}\right) + 0.5 - \gamma\right] + \frac{3P(1+\mu)}{64h^2}\left(\frac{a}{l}\right)^2 \tag{4.11}$$

$$\Delta_i = \frac{P}{8kl^2}\left\{1 + \left(\frac{1}{2\pi}\right)\left[\ln\left(\frac{a}{2l}\right) + \gamma - \frac{5}{4}\right]\left(\frac{a}{l}\right)^2\right\} \tag{4.12}$$

where σ_i is the bending stress in psi, P is the total load in lb, μ is the Poisson ratio, h is the slab thickness in inches, k is the modulus of subgrade reaction in pci, a is the radius of circular load in inches (e.g., the tire footprint radius), γ is Euler's constant and is equal to 0.577215, and l is the radius of relative stiffness (a measure of the slab thickness) and is defined as

$$l = \left(\frac{Eh^3}{12(1-\mu^2)k}\right)^{0.25} \tag{4.13}$$

where E is the modulus of elasticity in psi, and Δ_i is the slab deflection in inches. Note that a, the radius of the tire footprint, in U.S. customary units is

$$a = \sqrt{\frac{P}{p\pi}} \tag{4.14}$$

where P is the load in pounds, and p is the tire pressure in psi (compare this to Eq. 4.3).

For the edge loading:

$$\sigma_e = 0.529(1 + 0.54\mu)\left(\frac{P}{h^2}\right)\left[\log_{10}\left(\frac{Eh^3}{ka^4}\right) - 0.71\right] \tag{4.15}$$

$$\Delta_e = 0.408(1 + 0.4\mu)\left(\frac{P}{kl^2}\right) \tag{4.16}$$

For the corner loading:

$$\sigma_c = \frac{3P}{h^2}\left[1 - \left(\frac{a_l}{l}\right)^{0.72}\right] \tag{4.17}$$

$$\Delta_c = \frac{P}{kl^2}\left[1.205 - 0.69\left(\frac{a_l}{l}\right)\right] \tag{4.18}$$

where a_l is the distance to the point of action of the resulting load on a common angle bisection at the slab corner as shown in Fig. 4.9, and is equal to $\sqrt{2}\,a$.

EXAMPLE 4.4

A 15,000-lb (66.72-kN) wheel load is placed on a Portland cement concrete (PCC) slab that is 10.0 in. (254.0 mm) thick. The concrete has a modulus of elasticity of 4.5 million psi (31.0 GPa) with a Poisson ratio of 0.18. The modulus of subgrade reaction is 200 pci (54.3 N/cm^3). Tire pressure is 100 psi (689.5 kPa). Using the revised Westergaard equations, calculate the stress and deflection if the load is placed on the corner of the slab.

SOLUTION

We begin by computing the radius of relative stiffness, l, using Eq. 4.13:

$$l = \left(\frac{Eh^3}{12(1 - \mu^2)k}\right)^{0.25}$$

$$= \left(\frac{4,500,000(10)^3}{12(1 - (0.18)^2)200}\right)^{0.25}$$

$$= 37.31$$

To determine a_l, Eq. 4.14 is used:

$$a = \sqrt{\frac{P}{p\pi}} = \sqrt{\frac{15,000}{100\pi}} = 6.91 \text{ in. (17.55 cm)}$$

which gives $a_l = 9.77$ in. (i.e., $\sqrt{2} \times 6.91$) or 24.8 cm. Substituting values into Eq. 4.17 to get the corner stress, we obtain

$$\sigma_c = \frac{3P}{h^2}\left[1 - \left(\frac{a_l}{l}\right)^{0.72}\right]$$

$$= \frac{3(15,000)}{(10)^2}\left[1 - \left(\frac{12}{37.31}\right)^{0.72}\right]$$

$$= \underline{\underline{278.51 \text{ psi } (1.92 \text{ MPa})}}$$

Substituting values into Eq. 4.18 to get the corner deflection, we obtain

$$\Delta_c = \frac{P}{kl^2}\left[1.205 - 0.69\left(\frac{a_l}{l}\right)\right]$$

$$= \frac{15,000}{200(37.31)^2}\left[1.205 - 0.69\left(\frac{9.77}{37.31}\right)\right]$$

$$= \underline{\underline{0.055 \text{ in. (1.402 mm)}}}$$

4.6 THE AASHTO RIGID-PAVEMENT DESIGN PROCEDURE

The design procedure for rigid pavements presented in the AASHTO design guide is also based on the field results of the AASHO Road Test. The AASHTO design procedure is applicable to jointed plain (pavements that do not have steel reinforcement in the slab), reinforced (pavements with welded wire fabric reinforcement), and continuously reinforced pavements (pavements that have steel bars for longitudinal and transverse reinforcement). Because faulting, which is a distress due to different slab elevations, was not a failure consideration in the AASHO Road Test, the design of nondowelled joints must be checked with a design procedure other than the one presented herein.

The design procedure for rigid pavements is based on a selected reduction in serviceability and is similar to the design procedure followed for flexible pavements. However, instead of measuring pavement strength by using a structural number, the thickness of the PCC slab is the measure of strength. The regression equation that is used (also in U.S. customary units) to determine the thickness of a rigid-pavement PCC slab is

$$\log_{10} W_{18} = Z_R S_O + 7.35[\log_{10}(D+1)] - 0.06 + \frac{\log_{10}[\Delta \mathrm{PSI}/(3.0)]}{1 + [1.624 \times 10^7/(D+1)^{8.46}]}$$

$$+ (4.22 - 0.32\mathrm{TSI})\log_{10}\left(\frac{S_c' C_d[D^{0.75} - 1.132]}{215.63 J\{D^{0.75} - [18.42/(E_c/k)^{0.25}]\}}\right) \quad \textbf{(4.19)}$$

where W_{18} is the 18-kip-equivalent single-axle load, Z_R is the reliability (z-statistic from the standard normal curve), S_O is the overall standard deviation of traffic, D is the PCC slab thickness in inches, TSI is the pavement's terminal serviceability index, ΔPSI is the loss in serviceability from when the pavement is new until it reaches its TSI, S_c' is the concrete modulus of rupture in psi, C_d is a drainage coefficient, J is the load transfer coefficient, E_c is concrete elastic modulus in psi, and k is the modulus of subgrade reaction. A graphic solution to Eq. 4.19 is shown in Figs. 4.10 and 4.11. The terms used in Eq. 4.19 and Figs. 4.10 and 4.11 are presented here.

W_{18} The 18-kip (80.1-kN)-equivalent single-axle load is the same concept as that discussed in the flexible-pavement design procedure. However, instead of being a function of the structural number, these values are a function of slab thickness. The axle-load-equivalency factors used in rigid-pavement design are presented in Tables 4.7 (for single axles), 4.8 (for tandem axles), and 4.9 (for triple axles).

Z_R As was the case in flexible-pavement design, reliability, Z_R, is defined as the probability that serviceability will be maintained at adequate levels from a user's point of view throughout the design life of the facility (i.e., the PSI staying above the TSI). In the rigid-pavement design nomograph (Figs. 4.10 and 4.11), the probabilities (in percent) are used directly (instead of the Z_R as in the case of Eq. 4.19) and these percent probabilities are denoted R (see Table 4.5, which still applies).

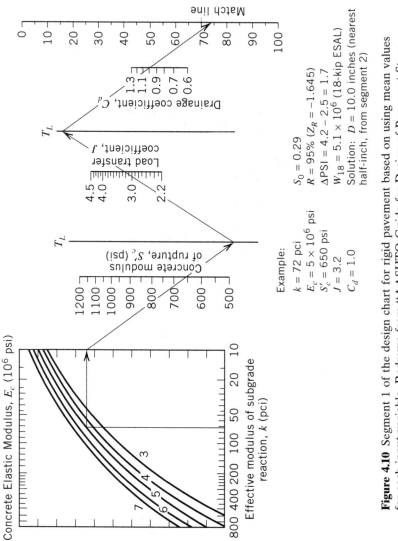

Figure 4.10 Segment 1 of the design chart for rigid pavement based on using mean values for each input variable. Redrawn from "AASHTO Guide for Design of Pavement Structures," The American Association of State Highway and Transportation Officials, Washington, DC, copyright 1993. Used by Permission.

Concrete Elastic Modulus, E_c (10^6 psi)

Effective modulus of subgrade reaction, k (pci)

800 400 200 100 50 20 10

Concrete modulus of rupture, S'_c (psi)

1200 1100 1000 900 800 700 600 500

Load transfer coefficient, J

4.5 4.0 3.0 2.2

Drainage coefficient, C_d

1.3 1.1 0.9 0.7 0.6

Match line

0 10 20 30 40 50 60 70 80 90 100

Example:

$k = 72$ pci
$E_c = 5 \times 10^6$ psi
$S'_c = 650$ psi
$J = 3.2$
$C_d = 1.0$

$S_0 = 0.29$
$R = 95\%$ ($Z_R = -1.645$)
$\Delta PSI = 4.2 - 2.5 = 1.7$
$W_{18} = 5.1 \times 10^6$ (18-kip ESAL)
Solution: $D = 10.0$ inches (nearest half-inch, from segment 2)

115

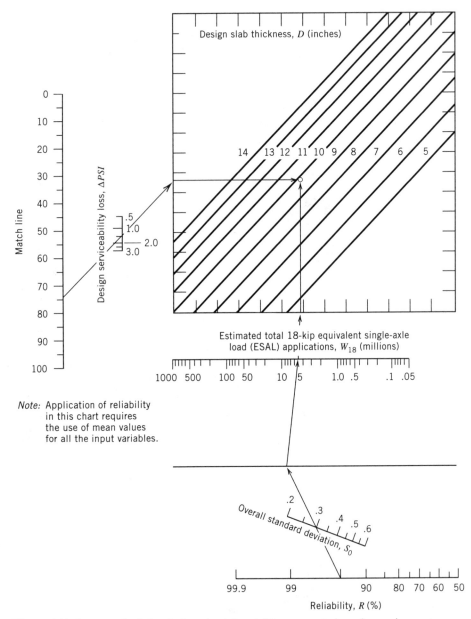

Figure 4.11 Segment 2 of the design chart for rigid pavements based on using mean values for each input variable. *Note*: Application of reliability in this chart requires the use of mean values for all the input variables. Redrawn from "AASHTO Guide for Design of Pavement Structures," The American Association of State Highway and Transportation Officials, Washington, DC, copyright 1993. Used by permission.

Table 4.7 Axle-Load Equivalency Factors for Rigid Pavements, Single Axles, and TSI = 2.5

Axle Load (kips)	Slab Thickness, D (inches)								
	6	7	8	9	10	11	12	13	14
2	0.0002	0.0002	0.0002	0.0002	0.0002	0.0002	0.0002	0.0002	0.0002
4	0.003	0.002	0.002	0.002	0.002	0.002	0.002	0.002	0.002
6	0.012	0.011	0.010	0.010	0.010	0.010	0.010	0.010	0.010
8	0.039	0.035	0.033	0.032	0.032	0.032	0.032	0.032	0.032
10	0.097	0.089	0.084	0.082	0.081	0.080	0.080	0.080	0.080
12	0.203	0.189	0.181	0.176	0.175	0.174	0.174	0.174	0.173
14	0.376	0.360	0.347	0.341	0.338	0.337	0.336	0.336	0.336
16	0.634	0.623	0.610	0.604	0.601	0.599	0.599	0.599	0.598
18	1.00	1.00	1.00	1.00	1.00	1.00	1.00	1.00	1.00
20	1.51	1.52	1.55	1.57	1.58	1.58	1.59	1.59	1.59
22	2.21	2.20	2.28	2.34	2.38	2.40	2.41	2.41	2.41
24	3.16	3.10	3.22	3.36	3.45	3.50	3.53	3.54	3.55
26	4.41	4.26	4.42	4.67	4.85	4.95	5.01	5.04	5.05
28	6.05	5.76	5.92	6.29	6.61	6.81	6.92	6.98	7.01
30	8.16	7.67	7.79	8.28	8.79	9.14	9.35	9.46	9.52
32	10.8	10.1	10.1	10.7	11.4	12.0	12.3	12.6	12.7
34	14.1	13.0	12.9	13.6	14.6	15.4	16.0	16.4	16.5
36	18.2	16.7	16.4	17.1	18.3	19.5	20.4	21.0	21.3
38	23.1	21.1	20.6	21.3	22.7	24.3	25.6	26.4	27.0
40	29.1	26.5	25.7	26.3	27.9	29.9	31.6	32.9	33.7
42	36.2	32.9	31.7	32.2	34.0	36.3	38.7	40.4	41.6
44	44.6	40.4	38.8	39.2	41.0	43.8	46.7	49.1	50.8
46	54.5	49.3	47.1	47.3	49.2	52.3	55.9	59.0	61.4
48	66.1	59.7	56.9	56.8	58.7	62.1	66.3	70.3	73.4
50	79.4	71.7	68.2	67.8	69.6	73.3	78.1	83.0	87.1

Source: "AASHTO Guide for Design of Pavement Structures," The American Association of State Highway and Transportation Officials, Washington, DC, copyright 1993. Used by permission.

S_O As was the case in flexible-pavement design, the overall standard deviation, S_O, takes into account designers' inability to accurately estimate future 18-kip (80.1-kN)-equivalent axle loads and the statistical error in the equations resulting from variability in materials and construction practices.

TSI The pavement's terminal serviceability index, TSI, is the point at which the pavement can no longer perform in a serviceable manner, as discussed previously in the flexible-pavement design procedure.

ΔPSI The amount of serviceability loss, ΔPSI, over the life of the pavement is the difference between the initial PSI and the TSI as discussed in the flexible-pavement design procedure.

S_c' The concrete modulus of rupture, S_c', is a measure of the tensile strength of the concrete and is determined by loading a beam specimen, at the third points, to failure. The test method is ASTM C 78, Flexural Strength of Concrete. Because concrete gains strength with age, the average 28-day strength is used for design purposes. Typical values are 500 to 1200 psi (3.45 to 8.27 MPa).

Table 4.8 Axle-Load Equivalency Factors for Rigid Pavements, Tandem Axles, and TSI = 2.5

Axle Load (kips)	Slab Thickness, D (inches)								
	6	7	8	9	10	11	12	13	14
2	0.0001	0.0001	0.0001	0.0001	0.0001	0.0001	0.0001	0.0001	0.0001
4	0.0006	0.0006	0.0005	0.0005	0.0005	0.0005	0.0005	0.0005	0.0005
6	0.002	0.002	0.002	0.002	0.002	0.002	0.002	0.002	0.002
8	0.007	0.006	0.006	0.005	0.005	0.005	0.005	0.005	0.005
10	0.015	0.014	0.013	0.013	0.012	0.012	0.012	0.012	0.012
12	0.031	0.028	0.026	0.026	0.025	0.025	0.025	0.025	0.025
14	0.057	0.052	0.049	0.048	0.047	0.047	0.047	0.047	0.047
16	0.097	0.089	0.084	0.082	0.081	0.081	0.080	0.080	0.080
18	0.155	0.143	0.136	0.133	0.132	0.131	0.131	0.131	0.131
20	0.234	0.220	0.211	0.206	0.204	0.203	0.203	0.203	0.203
22	0.340	0.325	0.313	0.308	0.305	0.304	0.303	0.303	0.303
24	0.475	0.462	0.450	0.444	0.441	0.440	0.439	0.439	0.439
26	0.644	0.637	0.627	0.622	0.620	0.619	0.618	0.618	0.618
28	0.855	0.854	0.852	0.850	0.850	0.850	0.849	0.849	0.849
30	1.11	1.12	1.13	1.14	1.14	1.14	1.14	1.14	1.14
32	1.43	1.44	1.47	1.49	1.50	1.51	1.51	1.51	1.51
34	1.82	1.82	1.87	1.92	1.95	1.96	1.97	1.97	1.97
36	2.29	2.27	2.35	2.43	2.48	2.51	2.52	2.52	2.53
38	2.85	2.80	2.91	3.03	3.12	3.16	3.18	3.20	3.20
40	3.52	3.42	3.55	3.74	3.87	3.94	3.98	4.00	4.01
42	4.32	4.16	4.30	4.55	4.74	4.86	4.91	4.95	4.96
44	5.26	5.01	5.16	5.48	5.75	5.92	6.01	6.06	6.09
46	6.36	6.01	6.14	6.53	6.90	7.14	7.28	7.36	7.40
48	7.64	7.16	7.27	7.73	8.21	8.55	8.75	8.86	8.92
50	9.11	8.50	8.55	9.07	9.68	10.14	10.42	10.58	10.66
52	10.8	10.0	10.0	10.6	11.3	11.9	12.3	12.5	12.7
54	12.8	11.8	11.7	12.3	13.2	13.9	14.5	14.8	14.9
56	15.0	13.8	13.6	14.2	15.2	16.2	16.8	17.3	17.5
58	17.5	16.0	15.7	16.3	17.5	18.6	19.5	20.1	20.4
60	20.3	18.5	18.1	18.7	20.0	21.4	22.5	23.2	23.6
63	23.5	21.4	20.8	21.4	22.8	24.4	25.7	26.7	27.3
64	27.0	24.6	23.8	24.4	25.8	27.7	29.3	30.5	31.3
66	31.0	28.1	27.1	27.6	29.2	31.3	33.2	34.7	35.7
68	35.4	32.1	30.9	31.3	32.9	35.2	37.5	39.3	40.5
70	40.3	36.5	35.0	35.3	37.0	39.5	42.1	44.3	45.9
72	45.7	41.4	39.6	39.8	41.5	44.2	47.2	49.8	51.7
74	51.7	46.7	44.6	44.7	46.4	49.3	52.7	55.7	58.0
76	58.3	52.6	50.2	50.1	51.8	54.9	58.6	62.1	64.8
78	65.5	59.1	56.3	56.1	57.7	60.9	65.0	69.0	72.3
80	73.4	66.2	62.9	62.5	64.2	67.5	71.9	76.4	80.2
82	82.0	73.9	70.2	69.6	71.2	74.7	79.4	84.4	88.8
84	91.4	82.4	78.1	77.3	78.9	82.4	87.4	93.0	98.1
86	102.0	92.0	87.0	86.0	87.0	91.0	96.0	102.0	108.0
88	113.0	102.0	96.0	95.0	96.0	100.0	105.0	112.0	119.0
90	125.0	112.0	106.0	105.0	106.0	110.0	115.0	123.0	130.0

Source: "AASHTO Guide for Design of Pavement Structures," The American Association of State Highway and Transportation Officials, Washington, DC, copyright 1993. Used by permission.

Table 4.9 Axle-Load Equivalency Factors for Rigid Pavements, Triple Axles, and TSI = 2.5

Axle Load (kips)	Slab Thickness, D (inches)								
	6	7	8	9	10	11	12	13	14
2	0.0001	0.0001	0.0001	0.0001	0.0001	0.0001	0.0001	0.0001	0.0001
4	0.0003	0.0003	0.0003	0.0003	0.0003	0.0003	0.0003	0.0003	0.0003
6	0.001	0.001	0.001	0.001	0.001	0.001	0.001	0.001	0.001
8	0.003	0.002	0.002	0.002	0.002	0.002	0.002	0.002	0.002
10	0.006	0.005	0.005	0.005	0.005	0.005	0.005	0.005	0.005
12	0.011	0.010	0.010	0.009	0.009	0.009	0.009	0.009	0.009
14	0.020	0.018	0.017	0.017	0.016	0.016	0.016	0.016	0.016
16	0.033	0.030	0.029	0.028	0.027	0.027	0.027	0.027	0.027
18	0.053	0.048	0.045	0.044	0.044	0.043	0.043	0.043	0.043
20	0.080	0.073	0.069	0.067	0.066	0.066	0.066	0.066	0.066
22	0.116	0.107	0.101	0.099	0.098	0.097	0.097	0.097	0.097
24	0.163	0.151	0.144	0.141	0.139	0.139	0.138	0.138	0.138
26	0.222	0.209	0.200	0.195	0.194	0.193	0.192	0.192	0.192
28	0.295	0.281	0.271	0.265	0.263	0.262	0.262	0.262	0.262
30	0.384	0.371	0.359	0.354	0.351	0.350	0.349	0.349	0.349
32	0.490	0.480	0.468	0.463	0.460	0.459	0.458	0.458	0.458
34	0.616	0.609	0.601	0.596	0.594	0.593	0.592	0.592	0.592
36	0.765	0.762	0.759	0.757	0.756	0.755	0.755	0.755	0.755
38	0.939	0.941	0.946	0.948	0.950	0.951	0.951	0.951	0.951
40	1.14	1.15	1.16	1.17	1.18	1.18	1.18	1.18	1.18
42	1.38	1.38	1.41	1.44	1.45	1.46	1.46	1.46	1.46
44	1.65	1.65	1.70	1.74	1.77	1.78	1.78	1.78	1.78
46	1.97	1.96	2.03	2.09	2.13	2.15	2.16	2.16	2.16
48	2.34	2.31	2.40	2.49	2.55	2.58	2.59	2.60	2.60
50	2.76	2.71	2.81	2.94	3.02	3.07	3.09	3.10	3.11
52	3.24	3.15	3.27	3.44	3.56	3.62	3.66	3.68	3.68
54	3.79	3.66	3.79	4.00	4.16	4.26	4.30	4.33	4.34
56	4.41	4.23	4.37	4.63	4.84	4.97	5.03	5.07	5.09
58	5.12	4.87	5.00	5.32	5.59	5.76	5.85	5.90	5.93
60	5.91	5.59	5.71	6.08	6.42	6.64	6.77	6.84	6.87
62	6.80	6.39	6.50	6.91	7.33	7.62	7.79	7.88	7.93
64	7.79	7.29	7.37	7.82	8.33	8.70	8.92	9.04	9.11
66	8.90	8.28	8.33	8.83	9.42	9.88	10.17	10.33	10.42
68	10.1	9.4	9.4	9.9	10.6	11.2	11.5	11.7	11.9
70	11.5	10.6	10.6	11.1	11.9	12.6	13.0	13.3	13.5
72	13.0	12.0	11.8	12.4	13.3	14.1	14.7	15.0	15.2
74	14.6	13.5	13.2	13.8	14.8	15.8	16.5	16.9	17.1
76	16.5	15.1	14.8	15.4	16.5	17.6	18.4	18.9	19.2
78	18.5	16.9	16.5	17.1	18.2	19.5	20.5	21.1	21.5
80	20.6	18.8	18.3	18.9	20.2	21.6	22.7	23.5	24.0
82	23.0	21.0	20.3	20.9	22.2	23.8	25.2	26.1	26.7
84	25.6	23.3	22.5	23.1	24.5	26.2	27.8	28.9	29.6
86	28.4	25.8	24.9	25.4	26.9	28.8	30.5	31.9	32.8
88	31.5	28.6	27.5	27.9	29.4	31.5	33.5	35.1	36.1
90	34.8	31.5	30.3	30.7	32.2	34.4	36.7	38.5	39.8

Source: "AASHTO Guide for Design of Pavement Structures," The American Association of State Highway and Transportation Officials, Washington, DC, copyright 1993. Used by permission.

Table 4.10 Relationship Between California Bearing Ratio (CBR) and Modulus of Subgrade Reaction, k

CBR Value	k Value, pci
2	100
10	200
20	250
25	290
40	420
50	500
75	680
100	800

C_d The drainage coefficient, C_d, is slightly different from those used in flexible-pavement design. In rigid-pavement design, it accounts for the drainage characteristics of the subgrade. A value of 1.0 for a drainage coefficient represents a material with good drainage characteristics (e.g., a sandy material). Other soils, with less than ideal drainage characteristics, will have drainage coefficents that are less than 1.0.

J The load transfer coefficient, J, is a factor that is used to account for the ability of pavement to transfer a load from one PCC slab to another across the slab joints. Many rigid pavements have dowel bars across the joints to transfer loads between slabs. Pavements with dowel bars at the joints are typically designed with a J value of 3.2.

E_c The concrete modulus of elasticity, E_c, is derived from the stress-strain curve as taken in the elastic region. As discussed previously, the modulus of elasticity is also known as Young's modulus. Typical values of E_c for Portland cement concrete are between 3 and 7 million psi (20.7 and 48.3 GPa).

k The modulus of subgrade reaction, k, depends upon several different factors including the moisture content and density of the soil. It should be noted that most highway agencies do not perform testing to measure the k value of the soil. At best, the agency will have a CBR value for the subgrade. Typical values for k range from 100 to 800 pci (27.1 to 216.5 N/cm³). Table 4.10 provides a relationship between CBR and k values.

EXAMPLE 4.5

A rigid pavement is to be designed to provide a service life of 20 years and has an initial PSI of 4.4 and a TSI of 2.5. The modulus of subgrade reaction is determined to be 300 pci (81.3 N/cm³). For design, the daily car, pickup truck, and light van traffic is 20,000, and the daily truck traffic consists of 200 passes of a single-unit truck with a single and tandem axle, and 410 passes of a tractor semi-trailer truck with a single, tandem, and triple axle. The axle weights are

$$\text{cars, pickups, light vans} = \text{two 2,000-lb (8.9-kN) single axles}$$
$$\text{single-unit truck} = \text{10,000-lb (44.5-kN) steering, single axle}$$
$$= \text{22,000-lb (97.9-kN) drive, tandem axle}$$
$$\text{tractor semi-trailer truck} = \text{12,000-lb (53.4-kN) steering, single axle}$$
$$= \text{18,000-lb (80.1-kN) drive, tandem axle}$$
$$= \text{50,000-lb (222.4-kN) trailer, triple axle}$$

Reliability is 95%, overall standard deviation is 0.45, the concrete's modulus of elasticity is 4.5 million psi (31.03 GPa), the concrete's modulus of rupture is 900 psi (6.21 MPa), the load transfer coefficient is 3.2, and the drainage coefficient in 1.0. Determine the required slab thickness.

SOLUTION

Because the axle-load equivalency factors presented in Tables 4.7, 4.8, and 4.9 are a function of the slab thickness (D), we have to assume a D to start the problem (later we will arrive at a slab thickness and check to make sure that it is consistent with our assumed value). A typical assumption is to let $D = 10$ in. Given this, 18-kip-equivalent single-axle loads (18-kip ESAL) for cars, pickups, and light vans are

$$\text{2-kip (8.9-kN) single-axle equivalent} = 0.0002 \text{ (Table 4.7)}$$

This gives an 18-kip ESAL total of 0.0004 for each vehicle. For single-unit trucks,

$$\text{10-kip (44.5-kN) single-axle equivalent} = 0.081 \text{ (Table 4.7)}$$
$$\text{22-kip (97.9-kN) tandem-axle equivalent} = 0.305 \text{ (Table 4.8)}$$

This gives an 18-kip ESAL total of 0.386 for single-unit trucks. For tractor semi-trailer trucks,

$$\text{12-kip (53.4-kN) single-axle equivalent} = 0.175 \text{ (Table 4.7)}$$
$$\text{18-kip (80.1-kN) tandem-axle equivalent} = 0.132 \text{ (Table 4.8)}$$
$$\text{50-kip (222.4-kN) triple-axle equivalent} = 3.020 \text{ (Table 4.9)}$$

This gives an 18-kip ESAL total of 3.327 for tractor semi-trailer trucks.
    Given the computed 18-kip ESAL, the daily traffic on this highway produces an 18-kip ESAL total of 1449.27 (0.0004 × 20,000 + 0.386 × 200 + 3.327 × 410). Traffic (total axle accumulations) over the 20-year design period will be

$$1449.27 \times 365 \times 20 = 10{,}579{,}671 \text{ 18-kip ESAL}$$

With an initial PSI of 4.4 and a TSI of 2.5, ΔPSI = 1.9. Solving Eq. 4.18 for D (using an equation solver on a calculator or computer) with $Z_R = -1.645$ (which corresponds to $R = 95\%$ as shown in Table 4.5) gives $D = 9.21$ in. (23.39 cm). (Figs. 4.10 and 4.11 can also be used to arrive at an approximate solution for D.) Note that this value differs from the slab thickness assumed to get the axle-load equivalency factors. Recomputing the axle-load equivalency factors with $D = 9$ in. (for Tables 4.7, 4.8, and 4.9) for cars, pickups, and light vans gives

2-kip (8.9-kN) single-axle equivalent = 0.0002 (Table 4.7)

This gives an 18-kip ESAL total of 0.0004 (same as before) for each vehicle. For single-unit trucks,

10-kip (44.5-kN) single-axle equivalent = 0.082 (Table 4.7)

22-kip (97.9-kN) tandem-axle equivalent = 0.308 (Table 4.8)

This gives an 18-kip ESAL total of 0.390 (up from 0.386) for single-unit trucks. For tractor semi-trailer trucks,

12-kip (53.4-kN) single-axle equivalent = 0.176 (Table 4.7)

18-kip (80.1-kN) tandem-axle equivalent = 0.133 (Table 4.8)

50-kip (222.4-kN) triple-axle equivalent = 2.940 (Table 4.9)

This gives an 18-kip ESAL total of 3.249 (down from 3.327) for tractor semi-trailer trucks.

Given the computed 18-kip ESAL, the daily traffic on this highway produces an 18-kip ESAL of 1418.09 (0.0004 × 20,000 + 0.390 × 200 + 3.249 × 410). Traffic (total axle accumulations) over the 20-year design period will be

$$1418.09 \times 365 \times 20 = 10{,}352{,}057 \text{ 18-kip ESAL}$$

Again solving Eq. 4.18 for D gives $D = 9.17$ in. (23.29 cm). (Figs. 4.10 and 4.11 can also be used to arrive at an approximate solution for D.) This is very close to the assumed $D = 9.0$ in. and is only a minor change from the 9.21 in. (23.39 cm) previously obtained. To be conservative, we would round up to the nearest 0.5 in. and make the slab <u>9.5 in.</u> (approximately 24 cm).

The final point to be covered with regard to pavement design relates to the case where there are multiple lanes of a highway (such as an interstate) in one direction. Because traffic tends to be distributed among the lanes, in some instances the pavement can be designed using a fraction of the total directional W_{18}. However, because traffic tends to concentrate in the right lane (particularly heavy vehicles), this fraction is not as simple as dividing the W_{18} by the number of lanes. In equation form,

$$\text{design-lane } W_{18} = PDL(\text{directional } W_{18}) \tag{4.20}$$

Table 4.11 Proportion of Directional W_{18} Assumed to Be in the Design Lane

Number of Directional Lanes	Proportion of Directional W_{18} in the Design Lane (*PDL*)
1	1.00
2	0.80–1.00
3	0.60–0.80
4	0.50–0.75

where *PDL* is the proportion of directional W_{18} assumed to be in the design lane. AASHTO-recommended values for *PDL* are given in Table 4.11.

As an example, suppose the computed directional W_{18} is an 18-kip ESAL of 10,000,000 and that there are three lanes in the direction of travel. If the highway is conservatively designed, Table 4.11 shows that 80% of the axle loads can be assumed to be in the design lane (i.e., PDL = 0.8). So the design W_{18} would be 8,000,000 (0.8 × 10,000,000) and this value would be used in the equations and nomographs. This design procedure applies to both flexible and rigid pavements.

EXAMPLE 4.6

In 1986, a rigid pavement on a northbound section of interstate highway was designed with a 12-in. (30.48-cm) PCC slab, an E_c of 6 × 10⁶ psi (48.27 GPa), a concrete modulus of rupture of 800 psi (5.52 MPa), a load transfer coefficient of 3.0, an initial PSI of 4.5, and a terminal serviceability index of 2.5. The overall standard deviation was 0.45, the modulus of subgrade reaction was 190 pci (51.49 N/cm³), and a reliability of 95% was used along with a drainage coefficient of 1.0. The pavement was designed for a 20-year life, and traffic was assumed to be composed entirely of tractor semi-trailer trucks with one 16-kip (71.2-kN) single axle, one 20-kip (88.9-kN) single axle, and one 35-kip (155.7-kN) tandem axle (the effect of all other vehicles was ignored). The interstate has four northbound lanes and was conservatively designed. How many tractor semi-trailer trucks, per day, were assumed to be traveling in the northbound direction?

SOLUTION

Given that the slab thickness D is 12 in. (30.48 cm), for the tractor semi-trailer trucks we have

16-kip (71.2-kN) single-axle equivalent = 0.599 (Table 4.7)

20-kip (89.0-kN) single-axle equivalent = 1.590 (Table 4.7)

35-kip (155.7-kN) tandem-axle equivalent = 2.245 (Table 4.8)

Note that the 2.245 value for the 35,000-lb (155.7-kN) tandem-axle linear interpolation uses 34-kip and 36-kip values [(1.97 + 2.52)/2]. The summation of these axle equivalents gives 4.434 18-kip ESAL per truck.

With an initial PSI of 4.5 and a TSI of 2.5, ΔPSI = 2.0. Solving Eq. 4.18 for W_{18} with $Z_R = -1.645$ (which corresponds to $R = 95\%$ as shown in Table 4.5) gives $W_{18} = 39{,}740{,}309$ 18-kip ESAL (Figs. 4.10 and 4.11 can also be used to arrive at an approximate solution for W_{18}.) Thus the total daily truck traffic in the design lane is

$$= \frac{39{,}740{,}309 \text{ 18-kip ESAL}}{365 \text{ days/year} \times 20 \text{ years} \times 4.434 \text{ 18-kip ESAL/truck}}$$

$$= 1227.76 \text{ trucks/day}$$

To determine the total directional volume (i.e., total number of northbound trucks), we note (from Table 4.11) that the *PDL* for a conservative design on a four-lane highway is 0.75, and the application of Eq. 4.20 gives

$$\text{directional } W_{18} = \frac{\text{design-lane } W_{18}}{PDL}$$

$$= \frac{1227.76}{0.75}$$

$$= \underline{\underline{1637.01 \text{ trucks/day}}}$$

4.7   HIGHWAY DRAINAGE

One of the most important elements of highway design, operation, and maintenance is adequate and functional drainage. Both rural and urban highways need good drainage if they are to operate safely and require minimum maintenance throughout their service life. Poor drainage can cause operational effects such as splash and spray creating limited visibility, water films that can lead to tire hydroplaning, and flooded areas that become impassable to vehicle traffic. Furthermore, water entering the pavement structure can weaken the layer materials, and cause pumping of fine particles from beneath the pavement and phenomena such as asphalt stripping.

In urban areas, water drains from the pavement surface and flows to curbs and drain inlets that allow the water to enter a storm water system. In most cases the storm water system carries the water to a natural water drainage source such as a stream, river, or lake. In rural areas, highway water on the pavement surface is directed to the shoulders and into side ditches or drainage swales. The side ditches usually run parallel with the highway. In cases where water must cross the highway, a culvert or a short-span bridge is used. The purpose of rural-highway drainage, as in urban areas, is to direct the water to a natural water source such as a river or stream.

This book provides a basic understanding of surface water flow in rural areas.

The reader is referred to other sources for further details on highway drainage (AASHTO, 1992).

4.7.1 Pavement Cross Slopes

Highway pavements are normally crowned or sloped to facilitate the movement of water from the traveled portion of the highway. A compromise is made to move water quickly and at the same time allow for proper operation of vehicles. Typically, a paved-surfaced highway, such as one with an asphalt cement concrete wearing layer, should have a cross slope of 1.5% to 2.0% with an additional 0.5% to 1.0% for added multiple-lane facilities (three or more lanes). Pavements with surface treatments or gravel/soil surfaces have cross slopes in the range of 2.0% to 6.0%.

After water leaves the travel lanes, it flows over shoulder areas, which usually have steeper cross slopes than the travel lanes. Shoulder cross slopes of 4.0% to 6.0% are common. Finally, the water flows to a side drainage ditch, away from the highway and pavement structure.

4.7.2 Surface Runoff Quantities

The first step in a drainage analysis is to determine how much water can be expected to collect in the drainage area. Water flow evaluations in streams, rivers, and large drainage areas can be very complicated. Often statistical analyses combined with measured stream flows and other records are used to determine the quantity of water runoff for drainage-structure design and analysis. However, for small drainage areas, a less rigorous analysis procedure can be used. This procedure, known as the rational method, is well suited to the analysis of small drainage areas (i.e., drainage areas up to 80 hectares or 197.7 acres). While the rational method is popular and simple to use, the reader is cautioned that it has many shortcomings, and engineering judgment must be applied during the analysis.

The rational method estimates the peak discharge of water expected at a particular location (e.g., a culvert inlet or drainage ditch) for a particular rainstorm. The rational equation is

$$Q = 0.00278(C_r I A_d) \tag{4.21}$$

where Q is the peak discharge in m^3/s, C_r is the runoff factor, I is the rainfall intensity in mm/h, and A_d is the drainage area in hectares. These variables are explained as follows.

Q The peak discharge of water, Q, is the maximum expected quantity of water that will accumulate at a particular location for a given storm. The peak discharge is the water that runs over the surface and is not absorbed by the soil and vegetation. This is the water that must be carried by the culvert or inlet without overflow.

C_r The runoff factor, C_r, accounts for the dissipated water during a storm. During a storm, some of the water will dissipate due to evaporation, transpiration, absorption by vegetation, ponding, and percolation into the soil. The runoff factor is a ratio that represents the proportion of surface runoff to the total amount of water falling on the area. Areas with paved surfaces will have runoff factors close to 1.0, whereas areas with grass and vegetation will have runoff factors in the order of 0.4 to 0.7. Surfaces with steep cross slopes usually have higher runoff factors than surfaces composed of the same material but with flatter slopes. Sandy soils and gravels typically absorb more water than clay soils, and dense urban areas tend to have higher runoff factors than rural residential areas. The exact ratio of runoff is difficult to predict; however, engineering judgment can provide a reasonable estimate of C_r. Typical runoff factors are presented in Table 4.12.

I In order to determine the peak discharge of water, information about the rainfall intensity, I, is needed. Some storms provide light rainfall for only a few minutes whereas others provide heavy rain for extended periods of time. For design purposes, the quantity of water considered is a function of the duration of the storm, its intensity (i.e., rainfall in mm/h), and the probability of storm occurrence. Storm occurrence probabilities are accounted for by using the return period, which incorporates the probability of a peak discharge (water flow) being equaled or exceeded in any given year. The return period and storm probability are reciprocals. For example, a peak water discharge having a 50-year return period (often called the 50-year design storm) has an occurrence probability of 0.02. A 50-year return period has a 2% chance of being equaled or exceeded in any given year. Likewise, a 100-year return period has a 1% chance of being equaled or exceeded in any given year. Return periods (storm frequencies) used for highway culvert designs are presented in Table 4.13.

A_d The drainage area, A_d, is the total area from which runoff can be expected. This area can be determined from topographical maps, satellite reconnaissance, or other sources.

Table 4.12 Runoff Coefficients for the Rational Formula

Type of Drainage Surface	Coefficient of Runoff, C_r
Pavement, Portland cement or asphaltic cement	0.75–0.95
Pavement, surface treatment	0.65–0.80
Pavement, gravel surface	0.25–0.60
Sandy soil, cultivated, brush or woods	0.15–0.30
Clay soil, cultivated, brush or woods	0.30–0.75
Urban business district	0.60–0.80
Urban residential area	0.50–0.70
Rural residential area	0.35–0.60
Parks, golf courses, grassy meadows	0.15–0.30

Table 4.13 Storm Frequency for Highway Culvert Design

Type of Highway Facility	Return Period (Storm Frequency) in Years
Interstate highways	50
Limited access freeways and arterials	25
Collectors	10
Locals	10
City streets	10

The storm frequency is selected for the type of highway facility in conjunction with the duration of the storm, which is selected to provide maximum surface runoff. The storm or rainfall duration used is equal to time of concentration. The time of concentration, used with the rational formula, is defined as the time it takes for water to flow from the most remote location in the drainage areas to a specified location. The time of concentration is a function of terrain type and the slope of the drainage area. If there are several different surface types in a drainage area, the time of flow can be added for each surface. In order to determine the time of concentration, the distance the water must flow and speed of the water are needed. Table 4.14 provides average water speeds over different surfaces.

Accurate estimates of the duration, frequency, and intensity of storms are needed for analysis of the peak flow discharge. Storm characteristics are a function of geographic location and, over time, there have been sufficient data collected within the United States to develop storm intensity-rainfall curves. State departments of transportation have prepared standard curves for their design manuals. An example of a typical curve is presented in Figure 4.12 (note that these curves vary depending on geographic location). It is suggested that the designer obtain standard curves for the location of the design.

Table 4.14 Average Runoff Water Speed Over Surface Conditions for Time of Concentration (in Meters per Second)

Surface Condition	Slope (%)						
	0–3	4–7	8–10	11–15	16–20	21–25	26–30
Woodland	0.15	0.30	0.45	0.50	0.60	0.80	1.10
Pastureland	0.25	0.45	0.65	0.80	0.90	1.25	1.35
Cultivated (row crop)	0.30	0.60	0.90	1.10	1.20	1.35	1.50
Pavement	1.50	3.65	4.70	5.50	—	—	—
Natural draw (not well defined)	0.25	0.75	1.20	1.85	—	—	—

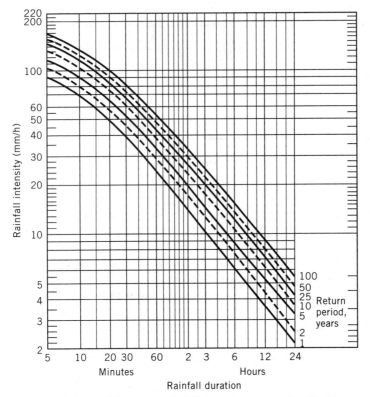

Figure 4.12 Rainfall intensity-rainfall duration curves (applicable to central Pennsylvania).

EXAMPLE 4.7

An engineer plans to install a culvert under a collector-type highway to reduce flooding in the area. The drainage area is 9 hectares of pastureland in a rural residential area in central Pennsylvania. Water in the drainage area flows on an approximate 5% slope for 1215 m before reaching the culvert. Estimate the peak discharge that can be expected at the culvert.

SOLUTION

Because the drainage area is small (less than 80 hectares), the rational formula can be used to estimate maximum water runoff. From Table 4.14, the speed of the water for a 5% slope on pastureland is 0.45 m/s. Therefore, time of concentration is

$$\frac{1215\text{ m}}{0.45\text{ m/s}} = 2700\text{ s} = 45\text{ min}$$

Using the time of concentration as the required rainfall duration and the fact that a collector roadway requires a 10-year return period, Figure 4.12 can be used to find a rainfall intensity of 47 mm/h. In rural residential areas, a runoff coefficient of 0.35 to 0.60 is suggested (see Table 4.12). For maximum runoff (i.e., a conservative design), a factor of 0.60 is selected.

Using Eq. 4.21, the peak discharge at the culvert site is

$$Q = 0.00278(C_r I A_d)$$
$$= 0.00278(0.60)(47)(9)$$
$$= \underline{\underline{0.706 \text{ m}^3/\text{s}}}$$

4.7.3 Basic Highway Culvert Design

Highway culverts are used to transport water under a highway to allow water to continue to a natural water source (e.g., a river or lake). Usually, a structure that is 6 m or less in span length is considered to be a culvert. Structures greater than 6 m in span length are considered to be a bridge. A culvert will usually cross the roadway at right angles or at a slight skew. However, the culvert should be aligned with the natural flow of the water.

Culverts are usually constructed with corrugated metal (galvanized steel) or reinforced Portland cement concrete. Plastic culverts are also available for smaller-sized applications. In the past, stone or Portland cement concrete headwalls were often used at the end of the culvert. Today, culvert ends are designed to be flush with the roadside cross slope to avoid obstructing the path of errant vehicles.

The complete design of a culvert is beyond the scope of this book. The reader is directed elsewhere for detailed culvert design information (AASHTO 1992); however, this text will provide a basic understanding of culvert design.

To begin, note that water flow in a culvert is a function of the energy between the inlet and the outlet of the culvert. Thus the capacity of the culvert is controlled by the water at the inlet or the outlet with respect to whether the culvert is flowing full or only partially full of water. An inlet-controlled culvert has water above the culvert inlet (i.e., water is ponded before it enters the culvert) and an outlet-controlled culvert has water ponded at the end of the culvert.

Although analysis of the water flow in a culvert is complex, a few basic principles should be kept in mind. The slope of a culvert is significant in determining whether the culvert is inlet or outlet controlled. A culvert on a steep slope is usually inlet controlled whereas a culvert on a flat slope is usually outlet controlled. Regardless of the control characteristics, the slope should be steep enough to provide scour velocity to ensure that sediment and debris do not build up in the culvert and cause a flow restriction. On the other hand, if the water flows too fast, it can damage the culvert and headwalls. Typically, a 2% slope is sufficient to keep the culvert clean while at the same time keeping the water speed at a reasonable level. Water speed should be in the range of 1.5 to 3.0 m/s. By keeping the speed of the water to a reasonable level, the following basic equation can be used to estimate the size of culvert needed for a peak discharge:

$$Q = VA_c \qquad (4.22)$$

where Q is the discharge in m³/s, V is the water speed in m/s, and A_c is the cross-sectional area in m².

EXAMPLE 4.8

In Example 4.7, the peak discharge was found to be 0.706 m³/s. It has been determined that a water flow speed of 1.5 m/s is needed to maintain scour velocity. Calculate the approximate diameter of a circular culvert needed to carry 0.706 m³/s if the culvert is flowing full. Assume that the flow in the culvert is not complex and water does not pond at either end of the culvert.

SOLUTION

Using Eq. 4.22, the cross-sectional area needed is

$$A_c = \frac{Q}{V} = \frac{0.706}{1.5} = 0.471 \text{ m}^2$$

Converting the area of the culvert to a diameter, d,

$$d = 2\sqrt{\frac{A_c}{\pi}} = 2\sqrt{\frac{0.471}{3.1416}} = \underline{\underline{0.774 \text{ m}}}$$

Because culvert pipe is available only in even increments, a 0.8-m diameter pipe would be used.

4.7.4 Highway Drainage Ditch Design

The majority of water coming from highway facilities flows in open channels (highway ditches or swales) that run parallel to the highway. These ditches range from very small (less than a meter wide) to large channels exceeding 5 meters in width. As the size of the ditch increases, the complexity of design also increases. Besides the need to understand water flow in the ditches, there is a need to ensure that the roadside is safe for errant vehicles. Cross slopes, channel shape, guiderail, and other safety features must also be evaluated for a complete design. The reader is directed to the AASHTO *Roadside Design Guide* (1996) for additional information concerning the design of a safe roadside environment.

Table 4.15 Roughness Coefficient, *n*, for Manning's Formula

Type of Channel Lining	*n*
Smooth concrete	0.012
Smooth asphalt	0.015
Earth	0.020
Rock	0.035
Grass and brush	0.050
Ductile iron pipe	0.013
Corrugated steel pipe	0.024
Corrugated plastic pipe	0.024

As was shown in Equation 4.22, the capacity of a drainage facility is a function of the cross-sectional area and the speed of the water. The area is easy to obtain from the geometric shape and dimensions of the facility. The speed of the water is determined from the classic equation known as Manning's formula. The equation applies to steady flow in a uniform open channel. Manning's formula is written as

$$V = \frac{R_h^{2/3} S^{1/2}}{n} \qquad (4.23)$$

where V is the water speed in m/s, S is the slope of the channel in m/m, n is Manning's roughness coefficient (values of n are given in Table 4.15), and R_h is the hydraulic radius,

$$R_h = \frac{A_c}{WP} \qquad (4.24)$$

where A_c is the area of cross section in m^2, and WP is the wetted perimeter (the wetted length in the cross section in meters). The hydraulic elements of various cross sections (i.e., area, wetted perimeter, and hydraulic radius) are presented in Table 4.16.

EXAMPLE 4.9

A highway agency plans to construct a rock-lined drainage channel next to a highway. The channel will be rectangular in shape with a 1.0-m bottom and a water depth of 0.5 m. The slope of the channel is determined to be 3.0%. Calculate the speed of the water and the flow rate.

Table 4.16 Hydraulic Elements of Various Cross Sections

Cross Section	Area ($\mathcal{A}_c$)	Wetted Perimeter ($\mathcal{W}P$)	Hydraulic Radius ($\mathcal{R}_h$)
Trapezoid	$bd + zd^2$	$b + 2d\sqrt{z^2 + 1}$	$\dfrac{bd + zd^2}{b + 2d\sqrt{z^2 + 1}}$
Rectangle	bd	$b + 2d$	$\dfrac{bd}{b + 2d}$
Triangle	zd^2	$2d\sqrt{z^2 + 1}$	$\dfrac{zd}{2\sqrt{z^2 + 1}}$
Parabola	$\dfrac{2}{3}dT$	$T + \dfrac{8d^2}{3T}$	$\dfrac{2dT^2}{3T^2 + 8d^2}$
Circle < ½ full	$\dfrac{D^2}{8}\left(\dfrac{\pi\theta}{180} - \sin\theta\right)$	$\dfrac{\pi D\theta}{360}$	$\dfrac{45D}{\pi\theta}\left(\dfrac{\pi\theta}{180} - \sin\theta\right)$
Circle > ½ full	$\dfrac{D^2}{8}\left(2\pi - \dfrac{\pi\theta}{180} + \sin\theta\right)$	$\dfrac{\pi D(360 - \theta)}{360}$	$\dfrac{45D}{\pi(360-\theta)}\left(2\pi - \dfrac{\pi\theta}{180} + \sin\theta\right)$

1. Satisfactory approximation for the interval $0 < \dfrac{d}{T} \leqq 0.25$

    When $\dfrac{d}{T} > 0.25$, use $WP = \dfrac{1}{2}\sqrt{16d^2 + T^2} + \dfrac{T^2}{8d}\sinh^{-1}\dfrac{4d}{T}$

2. $\theta = 4\sin^{-1}\sqrt{d/D}$ Insert θ in degrees in equations

3. $\theta = 4\cos^{-1}\sqrt{d/D}$

132

SOLUTION

The area of flow in the channel is

$$A_c = (1.0)(0.5) = 0.5 \text{ m}^2$$

The wetted perimeter, *WP*, is

$$WP = 1.0 + (2)(0.5) = 2.0 \text{ m}$$

The hydraulic radius, R_h, is calculated with Eq. 4.24 and is

$$R_h = \frac{0.5}{2} = 0.25 \text{ m}$$

Table 4.15 shows that Manning's roughness coefficient for a rock-lined channel is 0.035. Applying Manning's formula (Eq. 4.23) gives

$$V = \frac{R_h^{2/3} S^{1/2}}{n}$$

$$= \frac{(0.25)^{2/3}(0.03)^{1/2}}{0.035}$$

$$= \underline{\underline{1.98 \text{ m/s}}}$$

Using Eq. 4.22, the flow rate is

$$Q = VA_c = 1.98 \, (0.5) = \underline{\underline{0.99 \text{ m}^3/\text{s}}}$$

NOMENCLATURE FOR CHAPTER 4

a	equivalent tire radius
a_l	distance to point of action of the resulting load on a common angle bisection
a_1, a_2, a_3	structural-layer coefficients for wearing surface, base, and subbase
A_c	cross-sectional area
A_d	drainage area
C_d	drainage coefficient for rigid-pavement design
C_r	runoff coefficient
CBR	California bearing ratio

D	slab thickness, AASHTO design equation
D_1, D_2, D_3	structural-layer thicknesses for wearing surface, base, and subbase
E	modulus of elasticity (Young's modulus)
E_c	concrete modulus of elasticity
h	slab thickness, Westergaard solutions
I	rainfall intensity
l	radius of relative stiffness
k	modulus of subgrade reaction
M_2, M_3	drainage coefficients for base and subbase
n	Manning's roughness coefficient
pci	pounds per cubic inch
p	tire pressure
P	wheel load
PDL	proportion of directional W_{18} assumed in the design lane
psi	pounds per square inch
PSI	present serviceability index
Q	peak discharge
r	radial distance from the center of the load
R	reliability
R_h	hydraulic radius
S	slope of channel
S'_c	concrete modulus of rupture
S_O	overall standard deviation for AASHTO design equations
SN	structural number
TSI	terminal serviceability index
V	water speed
WP	wetted perimeter
W_{18}	18-kip (80.1-kN)-equivalent single-axle load
z	depth in pavement
Z_R	reliability for AASHTO design equations
Δ	deflection
Δ_c	corner deflection
Δ_e	edge deflection

Δ_i	interior deflection
Δ_z	deflection at depth z
ΔPSI	change in the present serviceability index
σ_c	bending stress, corner
σ_e	bending stress, edge
σ_i	bending stress, interior
σ_r	radial horizontal stress
σ_z	vertical stress
μ	Poisson ratio
γ	Euler's constant

REFERENCES

AASHTO (American Association of State Highway and Transportation Officials). "AASHTO Guide for Design of Pavement Structures." Washington, DC, 1993.

―――. *Highway Drainage Guidelines.* Washington, DC, 1992.

―――. *Roadside Design Guide.* Washington, DC, 1996.

Ahlvin, R. G., and H. H. Ulery. "Tabulated Values for Determining the Complete Pattern of Stresses, Strains, and Deflections Beneath a Uniform Circular Load on a Homogeneous Half Space." *Highway Research Board Bulletin* 342, 1962.

Carey, W., and P. Irick. "The Pavement Serviceability-Performance Concept." Highway Research Board Special Report 61E, AASHO Road Test, 1962.

Highway Research Board. "AASHTO Road Test Report 5." Special Report G. Washington, DC, 1962.

Ioannides, A. M., M. R. Thompson, and E. J. Barenberg. "Westergaard Solutions Reconsidered." *Transportation Research Record* 1043. Washington, DC, 1985.

Westergaard, H. M. "Computation of Stresses in Concrete Roads." *Proceedings of the Fifth Annual Meeting of the Highway Research Board.* Washington, DC, 1926.

Yoder, E. J., and M. W. Witczak. *Principles of Pavement Design,* 2d ed. New York: Wiley, 1975.

PROBLEMS

4.1. A tire carries a 22.2-kN load and has a pressure of 706.8 kPa. The pavement that the tire is on is constructed with a modulus of elasticity of 300 MPa.

A deflection of 0.41 mm is observed at a point at the pavement surface, 2 cm from the center of the tire load. Using Ahlvin and Ulery equations, what is the radial horizontal stress at this point?

4.2. A wheel carrying a 30-kN load generates a deflection of 0.855 mm, 5 cm below the center of the tire load. The contact area is measured at 530.7 cm^2 and the Poisson ratio is 0.5. Using Ahlvin and Ulery equations, what is the pavement's modulus of elasticity?

4.3. A wheel load produces a circular contract radius of 9 cm and the pavement material has a modulus of elasticity of 300 MPa. At a point on the pavement surface under the center of the tire load, the radial-horizontal stress is 600 kPa and the deflection is 0.42 mm. Using Ahlvin and Ulery equations, what is the load applied to the wheel?

4.4. A pavement is 10 cm thick and has a modulus of elasticity of 250 MPa with a Poisson ratio of 0.40. A wheel load is applied 20 cm from the edge of the pavement. The wheel's tire has a pressure of 700 kPa and a circular contact radius of 5 cm. Using Ahlvin and Ulery equations, determine the vertical stress, radial-horizontal stress, and deflection at a point at the bottom of the pavement, at the pavement's edge.

4.5. Truck A has two single axles. One axle weighs 12,000 lb (53.38 kN) and the other weighs 23,000 lb (102.3 kN). Truck B has an 8000-lb (35.6-kN) single axle and a 43,000-lb (191.3-kN) tandem axle. On a flexible pavement with a 3-in. (7.62-cm) hot-mix asphalt wearing surface, a 6-in. (15.24-cm) soil-cement base, and an 8-in. (20.32-cm) crushed stone subbase, which truck will cause more pavement damage? (Assume drainage coefficients are 1.0.)

4.6. A flexible pavement has a 4-in. (10.16-cm) hot-mix asphalt wearing surface, a 7-in. (17.78-cm) dense-graded crushed stone base, and a 10-in. (25.4-cm) crushed stone subbase. The pavement is on a soil with a resilient modulus of 5000 psi (34.48 MPa). The pavement was designed with 90% reliability, an overall standard deviation of 0.4, and a ΔPSI of 2.0 (a TSI of 2.5). The drainage coefficients are 0.9 and 0.8 for the base and subbase, respectively. How many 25-kip (111.3-kN) single-axle loads can be carried before the pavement reaches its TSI (with given reliability)?

4.7. A highway has the following pavement design traffic:

Daily Count	Axle Load
300 (single axle)	10,000 lb (44.5 kN)
120 (single axle)	18,000 lb (80.1 kN)
100 (single axle)	23,000 lb (102.3 kN)
100 (tandem axle)	32,000 lb (142.3 kN)
30 (single axle)	32,000 lb (142.3 kN)
100 (triple axle)	40,000 lb (177.9 kN)

A flexible pavement is designed to have 4 in. (10.16 cm) of sand-mix asphalt wearing surface, 6 in. (15.24 cm) of soil–cement base, and 7 in. (17.78 cm) of crushed stone subbase. The pavement has a 10-year design life, a reliability of 85%, an overall standard deviation of 0.30, drainage coefficients that are 1.0, an initial PSI of 4.7, and a TSI of 2.5. What is the minimum acceptable soil resilient modulus?

4.8. Consider the conditions in Problem 4.7. Suppose the state has relaxed truck weight limits and the impact has been to reduce the number of 18,000-lb (80.1-kN) single-axle loads from 120 to 20 and increase the number of 32,000-lb (142.3-kN) single-axle loads from 30 to 90 (all other traffic is unaffected). Under these revised daily counts, what is the minimum acceptable soil resilient modulus?

4.9. A flexible pavement was designed for the following traffic with a 12-year design life:

Daily Count	Axle Load
1300 (single axle)	8000 lb (35.6 kN)
900 (tandem axle)	15,000 lb (66.7 kN)
20 (single axle)	40,000 lb (177.9 kN)
200 (tandem axle)	40,000 lb (177.9 kN)

The highway was designed with 4 in. (10.16 cm) of hot-mix asphalt wearing surface, 4 in. (10.16 cm) of hot-mix asphaltic base, and 8 in. (20.32 cm) of crushed stone subbase. The reliability was 70%, overall standard deviation was 0.5, ΔPSI was 2.0 (with a TSI of 2.5), and all drainage coefficients were 1.0. What was the soil resilient modulus of the subgrade used in design?

4.10. An engineer plans to replace the rigid pavement in Example 4.5 with a flexible pavement. The chosen design has 6 in. (17.78 cm) of sand-mix asphalt wearing surface, 9 in. (22.86 cm) of soil-cement base, and 10 in. (25.4 cm) of crushed stone subbase. All drainage coefficients are 1.0 and the soil resilient modulus is 5000 psi. If the highways's traffic is the same (same axle loadings per vehicle as in Example 4.5), how many years would you be 95% sure that this pavement will last? (Assume any parameters not given in this problem are the same as those given in Example 4.5.)

4.11. You have been asked to design a flexible pavement and the following traffic is expected for design:

Daily Count	Axle Load
5000 (single axle)	10,000 lb (44.5 kN)
400 (single axle)	24,000 lb (106.8 kN)
1000 (tandem axle)	30,000 lb (133.4 kN)
100 (tandem axle)	50,000 lb (222.4 kN)

There are 3 lanes in the design direction (conservative design is to be used). Reliability is 90%, overall standard deviation is 0.40, ΔPSI is 1.8, and the design life is 15 years. The soil has a resilient modulus of 13,750 psi (94.81 MPa). If the TSI is 2.5, what is the required structural number?

4.12. A flexible pavement is designed with 5 in. (12.7 cm) of hot-mix asphalt wearing surface, 6 in (15.24 cm) of hot-mix asphaltic base, and 10 in. (25.4 cm) of crushed stone subbase. All drainage coefficients are 1.0. Daily traffic is 200 passes of a 20-kip (89.0-kN) single axle, 200 passes of a 40-kip (177.9-kN) tandem axle, and 80 passes of a 22-kip (97.9-kN) single axle. If the initial minus the terminal PSI is 2.0 (the TSI is 2.5), the soil resilient modulus is 3000 psi (20.69 MPa), and overall standard deviation is 0.6, what is the probability (i.e., reliability) that this pavement will last 20 years before reaching its terminal serviceability?

4.13. A flexible pavement is designed with 4 in. (10.16 cm) of sand-mix asphalt wearing surface, 6 in. (15.24 cm) of dense-graded crushed stone base, and 8 in. (20.32 cm) of crushed stone subbase. All drainage coefficients are 1.0. The pavement is designed for 18-kip (80.1-kN) single-axle loads (1290 per day). The initial PSI is 4.5 and the TSI is 2.5. The soil has a resilient modulus of 12,000 psi (82.74 MPa). If the overall standard deviation is 0.40, what is the probability that this pavement will have a PSI greater than 2.5 after 20 years?

4.14. A tire carries a 10,000-lb (44.5-kN) load and the tire's pressure is 90 psi (620.55 kPa). The tire is on a pavement with a modulus of elasticty of 4.2 million psi (28.96 GPa) and a Poisson ratio of 0.25. The modulus of subgrade reactions is 150 pci (40.72 N/cm³). If, for edge loading, the edge stress is 218.5 psi (1.51 MPa), what is the pavement's slab thickness? (Use the revised Westergaard equations.)

4.15. A pavement has an 8-in. (20.32-cm) slab with a modulus of elasticity of 3.5 million psi (24.13 GPa) and a Poisson ratio of 0.30. The radius of relative stiffness is 30.106. A 12,000-lb (53.4-kN) wheel load is applied (interior loading) and produces an interior deflection of 0.008195 in. (0.21 mm). What is the interior stress? (Use the revised Westergaard equations.)

4.16. A pavement has a 10-in. (25.4-cm) slab with a Poisson ratio of 0.36. The pavement is on a subgrade with a soil resilient modulus of 250 pci (67.88 N/cm³). A 17,000-lb (75.6-kN) load is applied (corner loading), and a_l is 7 in. (17.78 cm). If the corner deflection is 0.05 in. (0.127 cm), what is the modulus of elasticity of the pavement? (Use the revised Westergaard equations.)

4.17. A 12-in. (30.48-cm) pavement slab has a modulus of elasticity of 4 million psi (27.58 GPa) and a Poisson ratio of 0.40. The pavement is on a soil with a modulus of subgrade reaction equal to 300 pci (81.45 N/cm³). A wheel load of 9000 lb (22.86 kN) is applied and the radius of circular load is 5 in. (12.7 cm). What would the interior and edge stresses be, and what would the interior and edge slab deflections be? (Use the revised Westergaard equations.)

4.18. Consider the two trucks in Problem 4.5. Which truck will cause more pavement damage on a rigid pavement with a 10-in. (25.4-cm) slab?

4.19. You have been asked to design the pavement for an access highway to a major truck terminal. The design daily truck traffic consists of the following:

Daily Count	Axle Load
80 (single axle)	22,500 lb (100.1 kN)
570 (tandem axle)	25,000 lb (111.2 kN)
50 (tandem axle)	39,000 lb (173.5 kN)
80 (triple axle)	48,000 lb (213.5 kN)

The highway is to be designed with rigid pavement having a modulus of rupture of 600 psi (4.14 MPa) and a modulus of elasticity of 5 million psi (34.48 GPa). The reliability is to be 95%, the overall standard deviation is 0.4, the drainage coefficient is 0.9, ΔPSI is 1.7 (with a TSI of 2.5), and the load transfer coefficient is 3.2. The modulus of subgrade reaction is 200 pci (54.2 N/cm³). If a 20-year design life is to be used, determine the required slab thickness.

4.20. A rigid pavement is being designed with the same parameters as those used in Problem 4.9. The modulus of subgrade reaction is 300 pci (81.3 N/cm³) and the slab thickness is determined to be 8.5 in. (21.59 cm). The load transfer coefficient is 3.0, the drainage coefficient is 1.0, and the modulus of elasticity is 4 million psi (27.58 GPa). What is the design modulus of rupture? (Assume any parameters not given in this problem are the same as those given in Problem 4.9.)

4.21. A rigid pavement is designed with a 10-in. (25.40-cm) slab, an E_c of 6 × 10^6 psi (48.27 GPa), a concrete modulus of rupture of 432 psi (2.98 MPa), a load transfer coefficient of 3.0, an initial PSI of 4.7, and a terminal serviceability index of 2.5. The overall standard deviation is 0.35, the modulus of subgrade reaction is 190 pci (51.49 N/cm³), and a reliability of 90% is used along with a drainage coefficient of 0.8. The pavement is designed assuming traffic is composed entirely of trucks (100 per day). Each truck has one 20-kip (89.0-kN) single axle, and one 42-kip (186.82-kN) tandem axle (the effect of all other vehicles is ignored). A section of this road is to be replaced (due to different subgrade characteristics) with a flexible pavement having a structural number of 4 and is expected to last the same number of years as the rigid pavement. What is the assumed soil resilient modulus? (Assume all other factors are the same as for the rigid pavement.)

4.22. Consider the loading conditions in Problem 4.7. A rigid pavement is used with a modulus of subgrade reaction of 200 pci (54.2 N/cm³), a slab thickness of 8 in. (20.32 cm), a load transfer coefficient of 3.2, a modulus of elasticity of 5 million psi (34.48 GPa), a modulus of rupture of 600 psi (4.14 MPa), and a drainage coefficient of 1.0. How many years would the pavement be expected to last using the same reliability as in Problem 4.7? (Assume all other factors as in Problem 4.7.)

4.23. Consider Problem 4.22. How long would the rigid pavement be expected to last if you wanted to be 95% sure that the pavement would stay above the 2.5 TSI?

4.24. Consider the traffic conditions in Example 4.5. Suppose a 10-in. (25.4-cm) slab was used and all other parameters are as described in Example 4.5. What would the design life be if the drainage coefficient was 0.8, and what would it be if it was 0.6?

4.25. Consider the conditions in Example 4.6. Suppose all of the parameters are the same but further soil tests found that the modulus of subgrade reaction was only 150 pci (40.65 N/cm³). In light of this new soil finding, how would the design life of the pavement change?

4.26. Consider the conditions in Example 4.6. Suppose all of the parameters are the same but a quality control problem resulted in a modulus of rupture of 600 psi (4.14 MPa) instead of 800 psi (5.52 MPa). How would the design life of the pavement change?

4.27. A parabolic channel is on a 3.5% slope. Water in the channel has a surface width of 1.6 m and the channel is lined with smooth concrete. If the water flow is 3.68 m³/s, what is the water depth in the channel?

4.28. Consider the conditions in Problem 4.27. What would the water depth be if the channel had a rectangular cross section instead of a parabolic one?

4.29. A culvert is being designed for an arterial highway. The drainage area is determined to be 24 hectares in an urban business district and the time of concentration is 30 minutes. If a water speed of 2 m/s is needed to maintain scour velocity and a conservative design is used, what are the dimensions of a square culvert needed to carry the water flow? (Assume that the flow in the culvert is not complex and does not pond at either end, and that Figure 4.12 applies.)

4.30. A trapezoidal channel has a base of 2 m and 45-degree walls. A smooth asphalt lining is used and the channel slopes at 2.5%. If the water speed is measured at 10 m/s, what is the water depth in the channel?

4.31. A circular culvert on an interstate highway has a 1-m diameter and during the design storm water flows at 2.0 m/s. The drainage area is 17 hectares in an urban residential area and a conservative design was used. What time of concentration was used in design? (Assume that the flow in the culvert is not complex and does not pond at either end, and that Figure 4.12 applies.)

4.32. Consider the conditions in Problem 4.31. Suppose an additional 12 hectares are added to the drainage area and the time of concentration becomes 70 minutes. What diameter of culvert is needed to handle the additional drainage area and maintain the 2.0 m/s water speed? (Assume that the flow in the culvert is not complex and does not pond at either end, and that Figure 4.12 applies.)

Chapter 5

Elements of Traffic Analysis

5.1 INTRODUCTION

It is important to realize that the primary function of a highway is to provide a transportation service. In an engineering context this service is measured in terms of the highway's ability to accommodate vehicle traffic safely and with some acceptable level of performance (i.e., providing acceptable vehicle speeds). We have already discussed many safety-related aspects of highway design (see Chapter 3) and now begin to focus on measures of performance.

The analysis of vehicle traffic provides the basis for measuring the operating performance of highways. In undertaking such an analysis, the various dimensions of traffic, such as number of vehicles per unit time (traffic volume), vehicle types, vehicle speeds, and the variation in traffic volumes over time, must be addressed because they will influence highway design (the selection of the number of lanes, pavement types, and geometric design) and highway operations (selection of traffic control devices including signs, markings, and traffic signals), both of which impact the performance of the highway in terms of its ability to handle vehicle traffic. In light of this, it is important for the analysis of traffic to begin with theoretically consistent quantitative techniques that can be used to model traffic volumes, speeds, and temporal fluctuations. The intent of this chapter is to focus on such models of traffic and thus provide the groundwork for quantifying measures of performance (i.e., levels of service as will be discussed in Chapter 7).

5.2 TRAFFIC FLOW, SPEED, AND DENSITY

Traffic flow, speed, and density are variables that form the underpinnings of traffic analysis. To begin to study these variables, the basic definitions of traffic flow, speed, and density must be presented. Traffic flow, q, is simply defined as the number of vehicles, n, passing some designated highway point in a time interval of duration, t, or

$$q = \frac{n}{t} \tag{5.1}$$

An example of units for q would be vehicles per hour. In addition to the total number of vehicles passing a point in some time interval, the amount of time between the passing of successive vehicles (or, time between the arrival of successive vehicles) is of interest. The time between the passage of the front bumpers of successive vehicles, at some designated highway point, is known as the time headway. These time headways are related to t, as defined in Eq. 5.1, by

$$t = \sum_{i=1}^{n} h_i \tag{5.2}$$

where h_i is the time headway of the ith vehicle (the time that has transpired between the arrival of vehicles i and i-1). Substituting Eq. 5.2 into Eq. 5.1 gives

$$q = \frac{n}{\sum_{i=1}^{n} h_i} \tag{5.3}$$

or

$$q = \frac{1}{\bar{h}} \tag{5.4}$$

where $\bar{h}$ is the average headway ($\Sigma h_i/n$). The importance of time headways in traffic analysis will be given additional attention in forthcoming sections of this chapter.

Average traffic speed is defined in two ways. The first is the arithmetic mean of the speeds observed at some designated point along the highway. This is referred to as the time-mean speed, $\bar{u}_t$, and is expressed as

$$\bar{u}_t = \frac{1}{n} \sum_{i=1}^{n} u_i \tag{5.5}$$

where u_i is the spot speed (i.e., the speed of the vehicle at the designated point on the highway, as might be obtained using a radar gun) of the ith vehicle. The second definition of average traffic speed is more useful in the context of traffic analysis and is determined on the basis of the time necessary for a vehicle to traverse some known length of highway, l. This measure of average traffic speed is known as the space-mean speed (denoted simply as u) and is

$$u = \frac{(1/n) \sum_{i=1}^{n} l_i}{\bar{t}} \tag{5.6}$$

where l_i is the length of highway used for the speed measurement of vehicle i, and

$$\bar{t} = \frac{1}{n} [t_1(l_1) + t_2(l_2) + \cdots + t_n(l_n)] \tag{5.7}$$

where $t_n(l_n)$ is the time necessary for vehicle n to traverse a section of highway of length l. Note that if all vehicle speeds are measured over the same length of highway ($L = l_1 = l_2 = \cdots l_n$),

$$u = \frac{1}{(1/n) \sum_{i=1}^{n} [1/(L/t_i)]} \tag{5.8}$$

which is the harmonic mean of speed (space-mean speed). This space-mean speed is the average speed used in traffic models.

Finally, traffic density, k, refers to the number of vehicles occupying some length of highway at some specified time, and is simply

$$k = \frac{n}{l} \tag{5.9}$$

5.3 BASIC TRAFFIC STREAM MODELS

Based on the definitions presented in the preceding section, a simple identity provides the basic relationship among traffic flow, speed (space-mean speed), and density:

$$q = uk \tag{5.10}$$

with typical units of flow (q), speed (u), and density (k) being vehicles per hour (veh/h), kilometers per hour (km/h), and vehicles per kilometer (veh/km), respectively. As will be shown in the forthcoming discussion, Eq. 5.10 will serve the important function of linking specific models of traffic into a consistent generalized model.

5.3.1   Speed-Density Model

The most intuitive starting point for developing a consistent, generalized traffic model is to focus on the relationship between speed and density. To begin, consider a section of highway with only a single vehicle on it. Under these conditions, the density will be very low (veh/km) and the driver will be able to travel freely at a speed close to the design speed of the highway. This speed is referred to as the free-flow speed (denoted here as u_f) because vehicle speed is not inhibited by the presence of other vehicles. As more and more vehicles begin to use a section of highway, the traffic density will increase and the average operating speed of vehicles will decline from the free-flow value as drivers slow to allow for the maneuvers of other vehicles. Eventually, the highway section will become so congested (i.e., will have such a high density) that the traffic will come to a stop ($u = 0$) and the density will be determined by the length of the vehicles and the spaces that drivers leave between them. This high-density condition is referred to as the jam density, k_j.

One possible representation of the process just described is the linear relationship shown in Fig. 5.1. Mathematically, such a relationship can be expressed as

$$u = u_f\left(1 - \frac{k}{k_j}\right) \tag{5.11}$$

The advantage of using a linear representation of the speed-density relationship is that it provides a basic insight into the relationships among traffic flow, speed, and density interactions without having these insights clouded by the additional complexity that a nonlinear speed-density relationship introduces. However, it is important to note that field studies have shown that the speed-density relationship tends to be nonlinear at low densities and high densities (i.e., those that approach the jam density). In fact, the overall speed-density relationship is better repre-

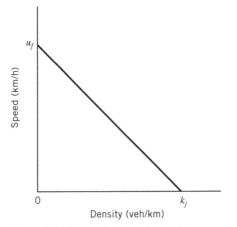

Figure 5.1 Illustration of a typical linear speed-density relationship.

sented by three relationships: (1) a nonlinear relationship at low densities that has speed slowly declining from the free-flow value, (2) a linear relationship over the large medium-density region (i.e., speed declining linearly with density as shown in Eq. 5.11), and (3) a nonlinear relationship near the jam density as the speed asymptotically approaches zero with increasing density. For the purposes of exposition, we only present traffic stream models that are based on the assumption of a linear speed-density relationship. Examples of nonlinear speed-density relationships are provided elsewhere (Drew 1965; Pipes 1967).

5.3.2 Flow-Density Model

Using the assumption of a linear speed-density relationship as shown in Eq. 5.11, a parabolic flow-density model can be obtained by substituting Eq. 5.11 into Eq. 5.10:

$$q = u_f \left(k - \frac{k^2}{k_j} \right) \tag{5.12}$$

The general form of Eq. 5.12 is shown in Fig. 5.2. Note in this figure that the maximum flow rate, q_m, represents the highest rate of traffic flow that the highway is capable of handling. This is referred to as the traffic flow at capacity or simply the capacity of the highway. The traffic density that corresponds to this maximum flow rate is k_m, and the corresponding speed is u_m. Equations for q_m, k_m, and u_m can be derived by differentiating Eq. 5.12 because at maximum flow,

$$\frac{dq}{dk} = u_f \left(1 - \frac{2k}{k_j} \right) = 0 \tag{5.13}$$

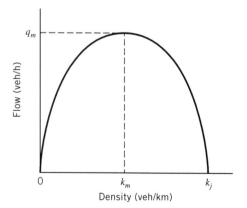

Figure 5.2 Illustration of the parabolic flow-density relationship.

and because the free-flow speed (u_f) is not equal to zero,

$$k_m = \frac{k_j}{2}$$ (5.14)

Substituting Eq. 5.14 into Eq. 5.11 gives

$$u_m = u_f\left(1 - \frac{k_j}{2k_j}\right)$$

$$= \frac{u_f}{2}$$ (5.15)

and using Eq. 5.14 and Eq. 5.15 in Eq. 5.10 gives

$$q_m = u_m k_m$$

$$= \frac{u_f k_j}{4}$$ (5.16)

5.3.3 Speed-Flow Model

Returning to the linear speed-density model (Eq. 5.11), a corresponding speed-flow model can be developed by rearranging Eq. 5.11 to

$$k = k_j\left(1 - \frac{u}{u_f}\right)$$ (5.17)

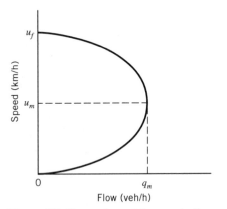

Figure 5.3 Illustration of the parabolic speed-flow relationship.

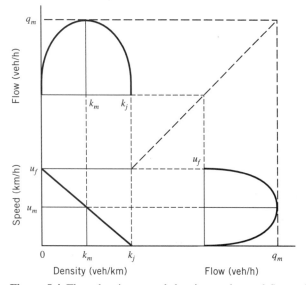

Figure 5.4 Flow-density, speed-density, and speed-flow relationships (assuming a linear speed-density model).

By substituting Eq. 5.17 into Eq. 5.10, we obtain

$$q = k_j \left(u - \frac{u^2}{u_f} \right) \tag{5.18}$$

The speed-flow model defined by Eq. 5.18 again gives a parabolic function as shown in Fig. 5.3. Note that Fig. 5.3 shows that two speeds are possible for flows, q, up to the highway's capacity q_m (this follows from the two densities possible for given flows as shown in Fig. 5.2). It is desirable, for any given flow, to keep the average space-mean speed on the upper portion of the speed-flow curve (i.e., above u_m). When speeds drop below u_m, traffic is in a highly congested and unstable condition. This concept will be discussed in more detail in Chapter 7.

All of the flow, speed, and density relationships and their interactions, as we have discussed, are graphically represented in Fig. 5.4.

EXAMPLE 5.1

A section of highway is known to have a free-flow speed of 90 km/h and a capacity of 3300 veh/h. In a given hour, 2100 vehicles were counted at a specified point along this highway section. If the linear speed-density relationship shown in Eq. 5.11 applies, what would you estimate the space-mean speed of these 2100 vehicles to be?

SOLUTION

The jam density is first determined from Eq. 5.16 as

$$k_j = \frac{4q_m}{u_f}$$

$$= \frac{4 \times 3300}{90}$$

$$= 146.67 \text{ veh/km}$$

Rearranging Eq. 5.18 to solve for u,

$$\frac{k_j}{u_f} u^2 - k_j u + q = 0$$

Substituting,

$$\frac{146.67}{90} u^2 - 146.67u + 2100 = 0$$

which gives u = 72.12 km/h or 17.87 km/h. Both of these speeds are feasible as shown in Fig. 5.3.

5.4 MODELS OF TRAFFIC FLOW

With the basic relationships among traffic flow, speed, and density formalized, we now turn attention to a more microscopic view of traffic flow. That is, instead of simply modeling the number of vehicles passing a specified point on a highway in some time interval, there is considerable analytic value in modeling the time between the arrivals of successive vehicles (i.e., the notion of vehicle time headways presented earlier). The most simplistic approach to vehicle arrival modeling is to assume that all vehicles are equally or uniformly spaced. This results in what is termed a deterministic, uniform arrival pattern. Under this assumption, if the traffic flow is 360 veh/h, the number of vehicles arriving in any 5-minute time interval is 30, and the headway between all vehicles is 10 seconds (because h will equal $3600/q$). However, actual observations show that such uniformity of traffic flow is not always realistic because some 5-minute intervals are likely to have more or less traffic flow than other 5-minute intervals. Thus a representation of vehicle arrivals that goes beyond the deterministic, uniform assumption is often warranted.

5.4.1 Poisson Models

Models that account for the nonuniformity of flow are derived by assuming that the pattern of vehicle arrivals corresponds to some random process. The problem then becomes one of selecting a probability distribution that is a reasonable representation of observed traffic arrival patterns. An example of such a distribution is the Poisson distribution (the limitations of which will be discussed later), which is expressed as

$$P(n) = \frac{(\lambda t)^n e^{-\lambda t}}{n!} \tag{5.19}$$

where t is the duration of the time interval over which vehicles are counted, $P(n)$ is the probability of having n vehicles arrive in time t, and λ is the average vehicle flow or arrival rate (in vehicles per unit time).

EXAMPLE 5.2

An engineer counts 360 veh/h at a specific highway location. Assuming that the arrival of vehicles at this highway location is Poisson distributed, estimate the probabilities of having 0, 1, 2, 3, 4, and 5 or more vehicles arriving over a 20-second interval.

SOLUTION

The average arrival rate, λ, is 360 veh/h or 0.1 vehicles per second (veh/s). Using this in Eq. 5.19 with $t = 20$ seconds, we find the probabilities of having 0, 1, 2, 3, and 4 vehicles are

$$P(0) = \frac{(0.1 \times 20)^0 e^{-0.1(20)}}{0!} = 0.135$$

$$P(1) = \frac{(0.1 \times 20)^1 e^{-0.1(20)}}{1!} = 0.271$$

$$P(2) = \frac{(0.1 \times 20)^2 e^{-0.1(20)}}{2!} = 0.271$$

$$P(3) = \frac{(0.1 \times 20)^3 e^{-0.1(20)}}{3!} = 0.180$$

$$P(4) = \frac{(0.1 \times 20)^4 e^{-0.1(20)}}{4!} = 0.090$$

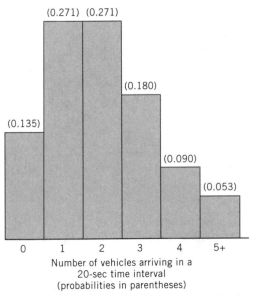

Figure 5.5 Histogram of the Poisson distribution for $\lambda = 0.1$ vehicles per second.

For five or more vehicles,

$$P(5) = 1 - P(n < 5)$$
$$= 1 - 0.135 - 0.271 - 0.271 - 0.180 - 0.090$$
$$= \underline{\underline{0.053}}$$

A histogram of these probabilities is shown in Fig. 5.5

EXAMPLE 5.3

Traffic data are collected in fifteen 60-second intervals at a specific highway location as shown in Table 5.1. Assuming the traffic is Poisson distributed and continues to arrive at the same rate as that observed in the first fifteen time periods, what is the probability that six or more vehicles will arrive in each of the next three 60-second time intervals (i.e., 12:15 P.M. to 12:16 P.M, 12:16 P.M. to 12:17 P.M., and 12:17 P.M. to 12:18 P.M.)?

SOLUTION

Table 5.1 shows that a total of 101 vehicles arrive in the 15-minute period from 12:00 P.M. to 12:15 P.M. Thus the average arrival rate, λ, is 0.112 veh/s (101/900). As in Example 5.2, Eq. 5.19 is applied to find the probabilities of exactly 0, 1, 2, 3, 4, and 5 vehicles arriving. With $\lambda = 0.112$ veh/s and $t = 60$ seconds ($\lambda t =$

Table 5.1 Observed Traffic Data for Example 5.3

Time Period	Observed Number of Vehicles
12:00 P.M. to 12:01 P.M.	3
12:01 P.M. to 12:02 P.M.	5
12:02 P.M. to 12:03 P.M.	4
12:03 P.M. to 12:04 P.M.	10
12:04 P.M. to 12:05 P.M.	7
12:05 P.M. to 12:06 P.M.	4
12:06 P.M. to 12:07 P.M.	8
12:07 P.M. to 12:08 P.M.	11
12:08 P.M. to 12:09 P.M.	9
12:09 P.M. to 12:10 P.M.	5
12:10 P.M. to 12:11 P.M.	3
12:11 P.M. to 12:12 P.M.	10
12:12 P.M. to 12:13 P.M.	9
12:13 P.M. to 12:14 P.M.	7
12:14 P.M. to 12:15 P.M.	6

6.733), the application of Eq. 5.19 gives probabilities 0.0012, 0.008, 0.027, 0.0606, 0.102, and 0.137 for having 0, 1, 2, 3, 4, and 5 vehicles, respectively, arriving in a 60-second time interval. The summation of these probabilities (0.3358) is the probability that 0 to 5 vehicles will arrive in any given 60-second time interval (i.e., $P(n \leq 5)$), so 1 minus 0.3358 (0.6642) is the probability that 6 or more vehicles will arrive in any 60-second time interval (i.e., $P(n \geq 6)$). The probability that 6 or more vehicles will arrive in three successive time intervals (i's) is simply the product of probabilities:

$$P(n \geq 6) \text{ for three successive time periods} = \prod_{i=1}^{3} P(n \geq 6)$$

$$= (0.6642)^3$$

$$= \underline{\underline{0.293}}$$

The assumption of Poisson vehicle arrivals also implies a distribution of the time intervals between the arrivals of successive vehicles (i.e., time headway). To show this, let the average arrival rate, λ, be in units of vehicles per second so that

$$\lambda = \frac{q}{3600} \tag{5.20}$$

where q is the flow in vehicles per hour. Substituting Eq. 5.20 into Eq. 5.19 gives

$$P(n) = \frac{(qt/3600)^n e^{-qt/3600}}{n!} \tag{5.21}$$

Note that the probability of having no vehicles arrive in a time interval of length t (i.e., $P(0)$) is equivalent to the probability of a vehicle headway, h, being greater than or equal to the time interval t. So from Eq. 5.21,

$$P(0) = P(h \geq t)$$
$$= e^{-qt/3600}$$

(5.22)

This distribution of vehicle headways is known as the negative exponential distribution and is often simply referred to as the exponential distribution.

EXAMPLE 5.4

Consider the traffic situation in Example 5.2 (i.e., 360 veh/h). Again assume that the vehicle arrivals are Poisson distributed. What is the probability that the gap between successive vehicles will be less than 8 seconds, and what is the probability that the gap between successive vehicles will be between 8 and 10 seconds?

SOLUTION

By definition, $P(h < t) = 1 - P(h \geq t)$. Therefore, Eq. 5.22 gives the probability that the gap will be less than 8 seconds as

$$P(h < t) = 1 - e^{-qt/3600}$$
$$= 1 - e^{-360(8)/3600}$$
$$= \underline{\underline{0.551}}$$

To determine the probability that the gap will be between 8 and 10 seconds, we compute the probability that the gap will be greater than or equal to 10 seconds as

$$P(h \geq t) = e^{-qt/3600}$$
$$= e^{-360(10)/3600}$$
$$= \underline{\underline{0.368}}$$

So the probability that the gap will be between 8 and 10 seconds is $\underline{\underline{0.081}}$ ($1 - 0.551 - 0.368$).

To help visualize the shape of the exponential distribution, Fig. 5.6 shows the probability distribution implied by Eq. 5.22 with the flow, q, equal to 360 veh/h as in Example 5.4.

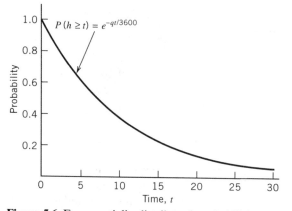

Figure 5.6 Exponentially distributed probabilities of headways greater than or equal to t, with $q = 360$ veh/h.

5.4.2   Limitations of Poisson Models

Empirical observations have shown that the assumption of Poisson-distributed traffic arrivals is most realistic in lightly congested traffic conditions. As traffic flows become heavily congested or when traffic signals cause cyclical traffic stream disturbances, other distributions of traffic flow become more appropriate. The primary limitation of Poisson models of vehicle arrivals is the constraint imposed by the Poisson distribution that the mean of period observations equals the variance. For example, the mean of period-observed traffic in Example 5.3 is 6.733 while the corresponding variance, σ^2, is 6.729. Because these two are virtually equal, the Poisson model was appropriate for this example. If the variance is significantly greater than the mean, data are said to be overdispersed, and if the variance is significantly less than the mean, data are said to be underdispersed. In both cases the Poisson distribution is no longer appropriate and other distributions should be used (e.g., the negative binomial). Such distributions are discussed in detail in more specialized sources (Poch and Mannering 1996; Transportation Research Board 1975).

5.5   QUEUING THEORY AND TRAFFIC FLOW ANALYSIS

The formation of traffic queues in congested periods is a source of considerable time delay and results in the loss of highway performance. Under some extreme conditions, queuing delay can account for 90% or more of a motorist's total trip travel time. Given this, it is essential in traffic analysis to develop a clear understanding of the characteristics of queue formation and dissipation along with mathematical formulations that can predict queuing-related elements.

As is well known, the problem of queuing is not unique to traffic analysis. Nontransportation fields, such as the design and operation of industrial plants, retail stores, and service-oriented industries, must give serious consideration to the problem of queuing. The impact that queues have on performance and productivity in manufacturing, retailing, and other fields has led to numerous theories of queuing behavior (i.e., the process by which queues form and dissipate). As will be shown, the models of traffic flow previously presented (i.e., uniform, deterministic arrivals, and Poisson arrivals) will form the basis for studying traffic queues within the more general context of queuing theory.

5.5.1 Dimensions of Queuing Models

The purpose of traffic queuing models is to provide a means to estimate important measures of highway performance including vehicle delay and traffic queue lengths. Such estimates are critical to highway design (e.g., the required length of left-turning bays, and the number of lanes at intersections) and traffic operations control including the timing of traffic signals at intersections.

Queuing models are derived from underlying assumptions regarding arrival patterns, departure characteristics, and queue disciplines. Section 5.4 explored traffic arrival patterns where, given an average vehicle arrival rate, two possible distributions of the time between the arrival of successive vehicles were considered: (1) equal time intervals (derived from the assumption of uniform, deterministic arrivals) and (2) exponentially distributed time intervals (derived from the assumption of Poisson-distributed arrivals). In addition to vehicle arrival assumptions, the derivation of traffic queuing models requires assumptions relating to vehicle departure characteristics. Of particular interest is the distribution of the amount of time it takes a vehicle to depart; for example, the time to pass through an intersection at the beginning of a green signal, the time required to pay a toll at a toll booth, or the time a driver takes before deciding to proceed after stopping at a stop sign. As was the case for arrival patterns discussed previously, given an average vehicle departure rate, the assumption of a deterministic or exponential distribution of departure times is appropriate. Another important aspect of queuing models is the number of available departure channels. For most traffic applications only one departure channel will exist, such as a highway lane or group of lanes passing through an intersection. However, multiple departure channels are encountered in some traffic applications, such as at toll booths at entrances to turnpikes or bridges.

The final necessary assumption relates to the queue discipline. In this regard, two options have been popularized in the development of queuing models: first-in-first-out (FIFO), indicating that the first vehicle to arrive is the first vehicle to depart, and last-in-first-out (LIFO), indicating that the last vehicle to arrive is the first to depart. For virtually all traffic-oriented queues, the FIFO queuing discipline is the more realistic of the two.

Queuing models are typically identified by three alphanumeric values: the first value indicates the arrival rate assumption; the second value gives the departure rate assumption; and the third value indicates the number of departure channels. For traffic arrival and departure assumptions, the uniform, deterministic distribution is denoted D, and the exponential distribution is denoted M. Thus a $D/D/1$

queuing model assumes deterministic arrivals and departures with one departure channel. Similarly, an $M/D/1$ queuing model assumes exponentially distributed arrival times, deterministic departure times, and one departure channel.

5.5.2 *D/D/1* Queuing

The case of deterministic arrivals and departures with one departure channel ($D/D/1$ queue) is an excellent starting point in understanding queuing models because of its simplicity. The $D/D/1$ queue lends itself to an intuitive graphical or mathematical solution that is best illustrated by example.

EXAMPLE 5.5

Vehicles arrive at an entrance to a recreational park. There is a single gate (at which all vehicles must stop), where a park attendant distributes free brochures. The park opens at 8:00 A.M., at which time vehicles begin to arrive at a rate of 480 veh/h. After 20 minutes, the arrival flow rate declines to 120 veh/h and continues at that level for the remainder of the day. If the time required to distribute the brochure is 15 seconds, and assuming $D/D/1$ queuing, describe the operational characteristics of the queue.

SOLUTION

We begin by putting arrival and departure rates into common units of vehicles per minute.

$$\lambda = \frac{480 \text{ veh/h}}{60 \text{ min/h}} = 8 \text{ veh/min} \qquad \text{for } t \leq 20 \text{ min}$$

$$\lambda = \frac{120 \text{ veh/h}}{60 \text{ min/h}} = 2 \text{ veh/min} \qquad \text{for } t > 20 \text{ min}$$

$$\mu = \frac{60 \text{ s/min}}{15 \text{ s/veh}} = 4 \text{ veh/min} \qquad \text{for all } t$$

Equations for the total number of vehicles that have arrived and departed up to a specified time, t, can now be written. Define t as the number of minutes after the start of the queuing process (in this case the number of minutes after 8:00 A.M.). The total number of vehicle arrivals at time t is equal to

$$8t \qquad \text{for } t \leq 20 \text{ min}$$

and

$$160 + 2(t - 20) \qquad \text{for } t > 20 \text{ min}$$

Similarly, the number of vehicle departures is

$$4t \qquad \text{for all } t$$

These equations can be illustrated graphically as shown in Fig. 5.7. When the arrival curve is above the departure curve, a queue condition will exist. The point at which the arrival curve falls below the departure curve is the moment when the queue dissipates (i.e., no more queue exists). In this example, the point of queue dissipation can be determined graphically by inspection of Fig. 5.7, or analytically by equating appropriate arrival and departure equations; that is,

$$160 + 2(t - 20) = 4t$$

Solving for t gives $t = 60$ minutes. Thus the queue that began to form at 8:00 A.M. will dissipate 60 minutes later (9:00 A.M.), at which time 240 vehicles will have arrived and departed (i.e., 4 veh/min × 60 min).

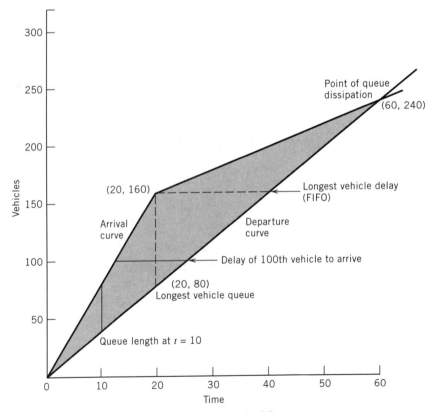

Figure 5.7 *D/D/*1 queuing diagram for Example 5.5.

Another aspect of interest is individual vehicle delay. Under the assumption of a FIFO queuing discipline, the delay of an individual vehicle is given by the horizontal distance between arrival and departure curves starting from the time of the vehicle's arrival in the queue. So, by inspection of Fig. 5.7, the 160th vehicle to arrive will have the longest delay of 20 minutes (the longest horizontal distance between arrival and departure curves), and vehicles arriving after the 239th vehicle will encounter no queue delay because the queue will have dissipated and the departure rate will continue to exceed the arrival rate. It follows that with the LIFO queuing discipline, the first vehicle to arrive will have to wait until the entire queue clears (i.e., 60 minutes of delay).

The total length of the queue at a specified time, expressed by the number of vehicles, is given by the vertical distance between arrival and departure curves at that time. For example, at 10 minutes after the start of the queuing process (8:10 A.M.) the queue is 40 vehicles long, and the longest queue (longest vertical distance between arrival and departure curves) will occur at $t = 20$ minutes and is 80 vehicles long (see Fig. 5.7).

Total vehicle delay, defined as the summation of the delays of each individual vehicle, is given by the total area between arrival and departure curves (see Fig. 5.7) and, in this case, is in units of vehicle-minutes. In this example, the areas between arrival and departure curves can be determined by summing triangular areas, giving total delay, D_t, as

$$D_t = \frac{1}{2}(80 \times 20) + \frac{1}{2}(80 \times 40)$$

$$= \underline{\underline{2400 \text{ veh-min}}}$$

Finally, because 240 vehicles encounter queuing delay (as previously determined), the average delay per vehicle is 10 minutes (2400 veh-min/240 veh).

EXAMPLE 5.6

After observing arrivals and departures at a highway toll booth over a 60-min period, an observer notes that the arrival and departure rates (or service rates) are deterministic but, instead of being uniform, change over time according to a known function. The arrival rate is given by the function $\lambda(t) = 2.2 + 0.17t - 0.0032t^2$, and the departure rate is given by $\mu(t) = 1.2 + 0.07t$, where t is in minutes after the beginning of the observation period and $\lambda(t)$ and $\mu(t)$ are in vehicles per minute. Determine the total vehicle delay at the toll booth and the longest queue assuming $D/D/1$ queuing.

SOLUTION

Note that this problem is an example of a time-dependent deterministic queue because the deterministic arrival and departure rates change over time. We begin by solving the time-to-queue dissipation by equating vehicle arrivals and departures:

$$\int_0^t 2.2 + 0.17t - 0.0032t^2 \, dt = \int_0^t 1.2 + 0.07t \, dt$$

$$2.2t + 0.085t^2 - 0.00107t^3 = 1.2t + 0.035t^2$$

$$-0.00107t^3 + 0.05t^2 + t = 0$$

which gives $t = 61.8$ minutes. Therefore, the total vehicle delay (which is the area between the arrival and departure functions) is

$$D_t = \int_0^{61.8} 2.2t + 0.085t^2 - 0.00107t^3 \, dt - \int_0^{61.8} 1.2t + 0.035t^2 \, dt$$

$$= 1.1t^2 + 0.0283t^3 - 0.0002675t^4 - 0.6t^2 - 0.0117t^3 \, |_0^{61.8}$$

$$= -0.0002675(61.8)^4 + 0.0166(61.8)^3 + 0.5(61.8)^2$$

$$= \underline{\underline{1925.8 \text{ veh-min}}}$$

The queue length (in vehicles) at any time t is given by the function

$$Q(t) = \int_0^t 2.2 + 0.17t - 0.0032t^2 \, dt - \int_0^t 1.2 + 0.07t \, dt$$

$$= -0.00107t^3 + 0.05t^2 + t$$

Solving for the maximum queue length gives

$$\frac{dQ(t)}{dt} = -0.00321t^2 + 0.1t + 1 = 0$$

$$t = \underline{\underline{39.12 \text{ min}}}$$

Substituting with $t = 39.12$ minutes gives the maximum queue length,

$$Q(39.12) = -0.00107t^3 + 0.05t^2 + t \, |_0^{39.12}$$

$$= -0.00107(39.12)^3 + 0.05(39.12)^2 + 39.12$$

$$= \underline{\underline{51.58 \text{ veh}}}$$

5.5.3 M/D/1 Queuing

The assumption of exponentially distributed times between the arrivals of successive vehicles (i.e., Poisson arrivals) will, in some cases, give a more realistic representation of traffic flow than the assumption of uniformly distributed arrival times. Therefore, the $M/D/1$ queue (exponentially distributed arrivals, deterministic departures, and one departure channel) has some important applications within the traffic analysis field. Although a graphical solution to an $M/D/1$ queue is difficult, a mathematical solution is straightforward. Defining a traffic intensity term, ρ, as the ratio of average arrival to departure rates (λ/μ), and assuming that ρ is less than 1, it can be shown that for an $M/D/1$ queue the average length of queue (in vehicles) is given as

$$\overline{Q} = \frac{\rho^2}{2(1 - \rho)} \qquad (5.23)$$

The average time waiting in the queue (for each vehicle) is

$$\overline{w} = \frac{\rho}{2\mu(1 - \rho)} \qquad (5.24)$$

The average time spent in the system (i.e., the summation of average queue waiting time and average departure time, or, as it is frequently referred to, service time) is

$$\overline{t} = \frac{2 - \rho}{2\mu(1 - \rho)} \qquad (5.25)$$

It is important to note that under the assumption that traffic intensity (ρ) is less than 1 (i.e., $\lambda < \mu$), the $D/D/1$ queue will predict no queue formation. However, a queuing model that is derived based on random arrivals or departures, such as the $M/D/1$ queuing model, will predict queue formations under such conditions. Also, note that the $M/D/1$ queuing model we have presented is based on steady-state conditions (i.e., constant average arrival and departure rates) with randomness arising from the assumed probability distribution of arrivals. This contrasts with the time-varying deterministic queuing case, as presented in Example 5.6, in which arrival and departure rates changed over time, but randomness was not present.

EXAMPLE 5.7

Consider the entrance to the recreational park described in Example 5.5 However, let the average arrival flow rate be 180 veh/h and Poisson distributed (exponential times between arrivals) over the entire period from park opening time (8:00 A.M.) until closing at dusk. Compute the average length of queue (in vehicles), average waiting time in queue, and average time spent in the system, assuming $M/D/1$ queuing.

SOLUTION

Putting arrival and departure rates into common units of vehicles per minute gives

$$\lambda = \frac{180 \text{ veh/h}}{60 \text{ min/h}} = 3 \text{ veh/min} \qquad \text{for all } t$$

$$\mu = \frac{60 \text{ s/min}}{15 \text{ s/veh}} = 4 \text{ veh/min} \qquad \text{for all } t$$

and

$$\rho = \frac{\lambda}{\mu} = \frac{3}{4} = 0.75$$

For the average length of queue (in vehicles), Eq. 5.23 is applied:

$$\overline{Q} = \frac{0.75^2}{2(1 - 0.75)}$$

$$= 1.125 \text{ veh}$$

For average waiting time in the queue, Eq. 5.24 gives

$$\overline{w} = \frac{0.75}{2(4)(1 - 0.75)}$$

$$= 0.375 \text{ min}$$

For average time spent in the system (queue time plus departure or service, time), Eq. 5.25 is used:

$$\overline{t} = \frac{2 - 0.75}{2(4)(1 - 0.75)}$$

$$= 0.625 \text{ min}$$

Alternatively, because the departure time (or service time) is $1/\mu$ (i.e., the 0.25 min it takes the park attendant to distribute the brochure),

$$\overline{t} = \overline{w} + \frac{1}{\mu}$$

$$= 0.375 + \frac{1}{4}$$

$$= 0.625 \text{ min}$$

5.5.4 *M/M/1* Queuing

A queuing model that assumes one departure channel and exponentially distributed departure time patterns in addition to exponentially distributed arrival times (i.e., an *M/M/1* queue) is also useful in some traffic applications. For example, exponentially distributed departure patterns might be a reasonable assumption at a toll booth where some arriving drivers have the correct toll and can be processed quickly, and others may not have the correct toll, thus producing a distribution of departures about some mean departure rate. Under standard *M/M/1* assumptions, it can be shown that the average length of queue (in vehicles), again assuming that ρ is less than 1, is

$$\overline{Q} = \frac{\rho^2}{1 - \rho} \qquad (5.26)$$

The average time waiting in the queue (for each vehicle) is

$$\overline{w} = \frac{\lambda}{\mu(\mu - \lambda)} \qquad (5.27)$$

The average time spent in the system (the summation of average queue waiting time and average departure time) is

$$\overline{t} = \frac{1}{\mu - \lambda} \qquad (5.28)$$

EXAMPLE 5.8

Assume that the park attendant in Examples 5.5 and 5.7 takes an average of 15 seconds to distribute brochures, but that the distribution time varies depending on whether or not park patrons have questions relating to park operating policies. Given an average arrival rate of 180 veh/h as in Example 5.7, compute the average length of queue (in vehicles), average waiting time in queue, and average time spent in the system, assuming *M/M/1* queuing.

SOLUTION

Using the average arrival rate, departure rate, and traffic intensity as determined in Example 5.7, Eq. 5.26 gives the average length of queue (in vehicles) as

$$\overline{Q} = \frac{0.75^2}{1 - 0.75}$$

$$= 2.25 \text{ veh}$$

The average waiting time in the queue (from Eq. 5.27) is

$$\overline{w} = \frac{3}{4(4-3)}$$

$$= \underline{\underline{0.75 \text{ min}}}$$

The average time spent in the system (from Eq. 5.28) is

$$\overline{t} = \frac{1}{4-3}$$

$$= \underline{\underline{1 \text{ min}}}$$

5.5.5   *M/M/N* Queuing

A more general formulation of the *M/M/1* queue is the *M/M/N* queue where *N* is the total number of departure channels. *M/M/N* queuing is a reasonable assumption at toll booths entering turnpikes or at toll bridges where there is often more than one departure channel available (i.e., more than one toll booth open). A parking lot is another example, with *N* being the number of parking stalls in the lot and the departure rate, μ, being the exponentially distributed times of parking duration. *M/M/N* queuing is also frequently encountered in nontransportation applications such as checkout lines at retail stores, security checks at airports, and so on.

The following equations describe the operational characteristics of *M/M/N* queuing. Note that, unlike the equations for *M/D/1* and *M/M/1*, which require traffic intensity, ρ, to be less than 1, the following equations allow ρ to be greater than 1 but apply only when ρ/N (which is called the utilization factor) is less than 1.

The probability of having no vehicles in the system is

$$P_0 = \frac{1}{\displaystyle\sum_{n_c=0}^{N-1} \frac{\rho^{n_c}}{n_c!} + \frac{\rho^N}{N!(1-\rho/N)}} \tag{5.29}$$

where n_c is the departure channel number. The probability of having *n* vehicles in the system is

$$P_n = \frac{\rho^n P_0}{n!} \qquad \text{for } n \le N \tag{5.30}$$

$$P_n = \frac{\rho^n P_0}{N^{n-N}N!} \qquad \text{for } n \ge N \tag{5.31}$$

The average length of queue (in vehicles) is

$$\overline{Q} = \frac{P_0\rho^{N+1}}{N!N}\left[\frac{1}{(1-\rho/N)^2}\right] \tag{5.32}$$

The average time spent in the system is

$$\overline{t} = \frac{\rho+\overline{Q}}{\lambda} \tag{5.33}$$

The average waiting time in the queue is

$$\overline{w} = \frac{\rho+\overline{Q}}{\lambda} - \frac{1}{\mu} \tag{5.34}$$

The probability of waiting in a queue (the probability of being in a queue, which is the probability that the number of vehicles in the system, n, is greater than the number of departure channels, N) is

$$P_{n>N} = \frac{P_0\rho^{N+1}}{N!N(1-\rho/N)} \tag{5.35}$$

EXAMPLE 5.9

At an entrance to a toll bridge, four toll booths are open. Vehicles arrive at the bridge at an average rate of 1200 veh/h, and, at the booths, drivers take an average of 10 seconds to pay their tolls. Both the arrival and departure rate can be assumed to be exponentially distributed. How would the average queue length, time in the system, and probability of waiting in a queue change if a fifth toll booth was opened?

SOLUTION

Using the equations for $M/M/N$ queuing, we first compute the four-booth case. Note that $\mu = 6$ veh/min, $\lambda = 20$ veh/min, and therefore $\rho = 3.333$. Also, because $\rho/N = 0.833$ (which is less than 1), Eqs. 5.29 to 5.35 can be used. The probability of having no vehicles in the system with four booths open (using Eq. 5.29) is

$$P_0 = \frac{1}{1 + \frac{3.333}{1!} + \frac{3.333^2}{2!} + \frac{3.333^3}{3!} + \frac{3.333^4}{4!(0.1667)}}$$

$$= 0.0213$$

The average queue length is (from Eq. 5.32)

$$\overline{Q} = \frac{0.213(3.333)^5}{4!4}\left[\frac{1}{(0.1667)^2}\right]$$

$$= 3.287 \text{ veh}$$

The average time spent in the system is (from Eq. 5.33)

$$\overline{t} = \frac{3.333 + 3.287}{20}$$

$$= 0.331 \text{ min}$$

The probability of having to wait in a queue is (from Eq. 5.35)

$$P_{n>N} = \frac{0.0213(3.333)^5}{4!4(0.1667)}$$

$$= 0.548$$

With a fifth booth open, the probability of having no vehicles in the system is (from Eq. 5.29)

$$P_0 = \frac{1}{1 + \dfrac{3.333}{1!} + \dfrac{3.333^2}{2!} + \dfrac{3.333^3}{3!} + \dfrac{3.333^4}{4!} + \dfrac{3.333^5}{5!(0.3333)}}$$

$$= 0.0318$$

The average queue length is (from Eq. 5.32)

$$\overline{Q} = \frac{0.0318(3.333)^6}{5!5}\left[\frac{1}{(0.3333)^2}\right]$$

$$= 0.654 \text{ veh}$$

The average time spent in the system is (from Eq. 5.33)

$$\overline{t} = \frac{3.333 + 0.654}{20}$$

$$= 0.199 \text{ min}$$

The probability of having to wait in a queue is (from Eq. 5.35)

$$P_{n>N} = \frac{0.0318(3.333)^6}{5!5(0.3333)}$$

$$= 0.218$$

So, opening a fifth booth reduces the average queue length by 2.633 veh (3.287 − 0.654), average time in the system by 0.132 min (0.331 − 0.199), and the probability of waiting by 0.33 (0.548 − 0.218).

EXAMPLE 5.10

A convenience store has four available parking spaces. The owner predicts that the duration of customer shopping (the time that a customer's vehicle will occupy a parking space) is exponentially distributed with a mean of 6 minutes. The owner knows that in the busiest hour customer arrivals are exponentially distributed with a mean arrival rate of 20 customers per hour. What is the probability that a customer will not have an open parking space available when arriving at the store?

SOLUTION

Putting mean arrival and departure rates in common units gives $\mu = 10$ veh/h, and $\lambda = 20$ veh/h. So $\rho = 2.0$ and, because $\rho/N = 0.5$ (which is less than 1), Eqs. 5.29 to 5.35 can be used. The probability of having no vehicles in the system with four parking spaces is (using Eq. 5.29)

$$P_0 = \frac{1}{1 + \dfrac{2}{1!} + \dfrac{2^2}{2!} + \dfrac{2^3}{3!} + \dfrac{2^4}{4!(0.5)}}$$

$$= 0.1304$$

Thus the probability of having not having an open parking space when arriving is (from Eq. 5.35)

$$P_{n>N} = \frac{0.1304(2)^5}{4!4(0.5)}$$

$$= 0.087$$

5.6 TRAFFIC ANALYSIS AT HIGHWAY BOTTLENECKS

Some of the most severe congestion problems occur at highway bottlenecks, which can be generally defined as a portion of highway with a lower capacity (q_m) than the incoming section of highway. This reduction in capacity can originate from a number of sources including a decrease in the number of highway lanes and reduced shoulder widths (which tend to cause drivers to slow and thus effectively reduce highway capacity, as will be discussed in Chapter 7). There are two general types of traffic bottlenecks—those that are recurring, and those that are incident-induced. Recurring bottlenecks exist where the highway itself limits capacity; for example, by a physical reduction in the number of lanes. Traffic congestion at such bottlenecks results from recurring traffic flows that exceed the vehicle capacity of the highway in the bottleneck area. In contrast, incident-induced bottlenecks occur as a result of vehicle breakdowns or accidents that effectively reduce highway capacity by restricting the through movement of traffic. Because incident-induced bottlenecks are unanticipated and temporary in nature, they have features that distinguish them from recurring bottlenecks, such as the possibility that the capacity resulting from an incident-induced bottleneck may change over time. For example, an accident may initially stop traffic flow completely, but as the wreckage is cleared, partial capacity (i.e., one lane open) may be provided for a period of time before full capacity is eventually restored. A feature shared by both recurring and incident-induced bottlenecks is the adjustment in traffic flow that may occur as travelers choose other routes and/or different trip departure times to avoid the bottleneck area, in response to visual information or traffic advisories.

The analysis of traffic flow at bottlenecks can be undertaken using the queuing models discussed in Section 5.5. The most intuitive approach from which traffic congestion at bottlenecks can be analyzed is to assume $D/D/1$ queuing.

EXAMPLE 5.11

An incident occurs on a freeway that has a capacity in the northbound direction, before the incident, of 4000 veh/h and a constant flow of 2900 veh/h during the morning commute (i.e., no adjustments to traffic flow result from the incident). At 8:00 A.M. a traffic accident closes the freeway to all traffic. At 8:12 A.M. the freeway is partially opened with a capacity of 2000 veh/h. Finally, the wreckage is removed and the freeway is restored to full capacity (4000 veh/h) at 8:31 A.M. Assume $D/D/1$ queuing to determine time of queue dissipation, longest queue length, total delay, average delay per vehicle, and longest wait of any vehicle (assuming FIFO).

SOLUTION

Let μ be the full-capacity departure rate and μ_r be the restrictive partial-capacity departure rate. Putting arrival and departure rates in common units of vehicles per minute,

$$\mu = \frac{4000 \text{ veh/h}}{60 \text{ min/h}} = 66.67 \text{ veh/min}$$

$$\mu_r = \frac{2000 \text{ veh/h}}{60 \text{ min/h}} = 33.33 \text{ veh/min}$$

$$\lambda = \frac{2900 \text{ veh/h}}{60 \text{ min/h}} = 48.33 \text{ veh/min}$$

The arrival rate is constant over the entire time period, and the total number of vehicles is equal to λt, where t is the number of minutes after 8:00 A.M. The total number of departing vehicles is

$$
\begin{array}{ll}
0 & \text{for } t \le 12 \text{ min} \\[4pt]
\mu_r(t - 12) & \text{for } 12 \text{ min} < t \le 31 \text{ min} \\[4pt]
633.33 + \mu(t - 31) & \text{for } t > 31 \text{ min}
\end{array}
$$

These arrival and departure rates can be represented graphically as shown in Fig. 5.8. As discussed in Section 5.5, for $D/D/1$ queuing, the queue will dissipate at the intersection point of arrival and departure curves, which can be determined as

$$
\lambda t = 633.33 + \mu(t - 31) \qquad \text{or} \qquad t = \underline{78.16 \text{ min}} \text{ (just after 9:18 A.M.)}
$$

At this time a total of 3777.5 vehicles (48.33×78.16) will have arrived and departed (for the sake of clarity, fractions of vehicles are used). The longest queue (longest vertical distance between arrival and departure curves) occurs at 8:31 A.M. and is

$$
\begin{aligned}
Q_m &= \lambda t - \mu_r(t - 12) \\
&= 48.33(31) - 33.33(19) \\
&= \underline{\underline{865 \text{ veh}}}
\end{aligned}
$$

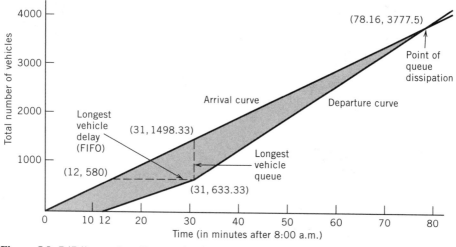

Figure 5.8 $D/D/1$ queuing diagram for Example 5.11.

Total vehicle delay, using equations for triangular and trapezoidal areas to calculate the total area between arrival and departure curves, is

$$D_t = \frac{1}{2}(12)(580) + \frac{1}{2}(580 + 1498.33)(19) - \frac{1}{2}(19)(633.33)$$

$$+ \frac{1}{2}(1498.33 - 633.33)(78.16 - 31)$$

$$= 37{,}604.2 \text{ veh-min}$$

The average delay per vehicle is 9.95 min (37,604.2/3777.5). The longest wait of any vehicle (the longest horizontal distance between arrival and departure curves), assuming a FIFO queuing discipline, will be the delay time of the 633.33rd vehicle to arrive. This vehicle will arrive 13.1 minutes (633.33/48.33) after 8:00 A.M. and will depart at 8:31 A.M., thus being delayed a total of 17.9 min.

NOMENCLATURE FOR CHAPTER 5

D deterministic arrivals or departures

D_t total vehicle delay

h time headway

k traffic density

k_j traffic jam density

k_m traffic density at maximum flow

l highway length

M exponentially distributed arrivals or departures

n number of vehicles

n_c departure channel number

N total number of departure channels

q traffic flow

q_m maximum traffic flow

Q length of queue

$\overline{Q}$ average length of queue

Q_m maximum number of vehicles in the queue

t time

$\bar{t}$ average time, average time spent in the system

u space-mean speed

u_f free-flow speed

u_m speed at maximum flow

$\overline{u}_t$ time-mean speed

$\overline{w}$ average time waiting in queue

λ arrival rate

μ departure rate

ρ traffic intensity

REFERENCES

Drew, D. R. "Deterministic Aspects of Freeway Operations and Control." *Highway Research Record* 99, 1965.

Pipes, L. A. "Car Following Models and the Fundamental Diagram of Road Traffic." *Transportation Research,* vol. 1, no. 1, 1967.

Poch, M., and F. Mannering. "Negative Binomial Analysis of Intersection-Accident Frequencies." *Journal of Transportation Engineering,* vol. 122, no. 2, March/April 1996.

Transportation Research Board. *Traffic Flow Theory: A Monograph.* Special Report 165, National Research Council, Washington, DC, 1975.

PROBLEMS

5.1. On a specific westbound section of highway, studies show that the speed-density relationship is

$$u = u_f \left[1 - \left(\frac{k}{k_j} \right)^{3.5} \right]$$

The highway's capacity is 3800 veh/h and the jam density is 140 veh/km. What is the space-mean speed of the traffic at capacity, and what is the free-flow speed?

5.2. A section of highway has a speed-flow relationship of the following form:

$$q = au^2 + bu$$

It is known that at capacity (which is 2900 veh/h) the space-mean speed of traffic is 50 km/h. Determine the space-mean speed when the flow is 1400 veh/h, and also the free-flow speed.

5.3. A section of highway has the following flow-density relationship:

$$q = 80k - 0.4k^2$$

What is the capacity of the highway section, the speed at capacity, and the density when the highway is at one-quarter of its capacity?

5.4. An observer has determined that the time headways between successive vehicles on a section of highway are exponentially distributed, and that 60% of the headways between vehicles are 13 seconds or greater. If the observer decides to count traffic in 30-second intervals, estimate the probability of the observer counting exactly four vehicles in an interval.

5.5 At a specified point on a highway, vehicles are known to arrive according to a Poisson process. Vehicles are counted in 20-second intervals, and vehicle counts are taken in 120 of these time intervals. It is noted that no cars arrive in 18 of these 120 intervals. Approximate the number of these 120 intervals in which exactly three cars arrive.

5.6. For the data collected in Problem 5.5, estimate the percentage of time headways that will be 10 seconds or greater, and those that will be less than 6 seconds.

5.7. A vehicle pulls out onto a single-lane highway that has a flow rate of 280 veh/h (Poisson distributed). The driver of the vehicle does not look for oncoming traffic. Road conditions and vehicle speeds on the highway are such that it takes 1.5 seconds for an oncoming vehicle to stop once the brakes are applied. Assuming a standard driver reaction time of 2.5 seconds, what is the probability that the vehicle pulling out will be in an accident with oncoming traffic?

5.8. Consider the conditions in Problem 5.7. How quick would the driver reaction times of oncoming vehicles have to be to have the probability of an accident equal to 0.15?

5.9. A toll booth on a turnpike is open from 8:00 A.M. to 12 midnight. Vehicles start arriving at 7:45 A.M. at a uniform deterministic rate of 6 per minute until 8:15 A.M. and from then on at 2 per minute. If vehicles are processed at a uniform deterministic rate of 6 per minute, determine when the queue will dissipate, total delay, longest queue length (in vehicles), longest vehicle delay under FIFO, and longest vehicle delay under LIFO.

5.10. Vehicles begin to arrive at a parking lot at 6:00 A.M. at a rate 8 per minute. Due to an accident on the access highway, no vehicles arrive from 6:20 to 6:30 A.M. From 6:30 A.M. on, vehicles arrive at a rate of 2 per minute. The parking lot attendant processes incoming vehicles (collects parking fees) at a rate of 4 per minute throughout the day. Assuming $D/D/1$ queuing, determine total vehicle delay.

5.11. Vehicles begin to arrive at a toll booth at 8:50 A.M., with an arrival rate of $\lambda(t) = 4.1 + 0.01t$ (with t in minutes and $\lambda(t)$ in vehicles per minute). The toll booth opens at 9:00 A.M. and processes vehicles at a rate of 12 per minute throughout the day. Assuming $D/D/1$ queuing, when will the queue dissipate and what will the total vehicle delay be?

5.12. At a parking lot, vehicles arrive according to a Poisson process and are processed (parking fee collected) at a uniform deterministic rate at a single station. The mean arrival rate is 4 veh/min and the processing rate is 5 veh/min. Determine the average length of queue (in vehicles), time spent in the system, and waiting time spent in the queue.

5.13. Consider the parking lot and conditions described in Problem 5.12. If the rate at which vehicles are processed became exponentially distributed (instead of deterministic) with a mean processing rate of 5 veh/min, what would be the average length of queue (in vehicles), time spent in the system, and waiting time spent in the queue?

5.14. Vehicles arrive at a toll booth with a mean arrival rate of 2 veh/min (the time between arrivals is exponentially distributed). The toll booth operator processes vehicles (collects tolls) at a uniform deterministic rate of one every 20 seconds. What is the average length of queue (in vehicles), time spent in the system, and waiting time spent in the queue?

5.15. To promote a business, the owner decides to pass out free transistor radios (along with a promotional brochure) at a booth in a parking lot. The owner begins giving the radios away at 9:15 A.M. and continues until 10:00 A.M. Vehicles start arriving for the radios at 8:45 A.M. at a uniform deterministic rate of 4 per minute and continue to arrive at this rate until 9:15 A.M. From 9:15 to 10:00 A.M. the arrival rate becomes 8 per minute. The radios and brochures are distributed at a uniform deterministic rate of 11 cars per minute over the 45-minute time period. Determine total delay, maximum queue length (in vehicles), and longest vehicle delay assuming FIFO and LIFO.

5.16. Consider the conditions described in Problem 5.15. Suppose the owner decides to accelerate the radio/brochure distribution rate (in veh/min) so that the queue that forms will dissipate at 9:45 A.M. What would this new distribution rate be?

5.17. A ferryboat queuing lane holds 30 vehicles. If vehicles are processed (tolls collected) at a uniform deterministic 4 vehicles per minute and processing begins when the lane reaches capacity, what is the uniform deterministic arrival rate if the vehicle queue dissipates 30 minutes after vehicles begin to arrive?

5.18. At a toll booth, vehicles arrive and are processed (tolls collected) at uniform deterministic rates λ and μ, respectively. The arrival rate is 2 veh/min. Processing begins 13 minutes after the arrival of the first vehicles and the queue dissipates t minutes after the arrival of the first vehicle. Letting the number of vehicles that must actually wait in a queue be x, develop an expression for determining processing rates in terms of x.

5.19. Vehicles arrive at a recreational park booth at a uniform deterministic rate of 4 veh/min. If uniform deterministic processing of vehicles (collecting fees) begins 30 minutes after the first arrival and the total delay is 3600 veh-min, how long after the arrival of the first vehicle will it take the queue to dissipate?

5.20. Trucks begin to arrive at a truck weighing station (with a single scale) at 6:00 A.M. at a deterministic but time-varying rate of $\lambda(t) = 4.3 - 0.22t$, with $\lambda(t)$ in veh/min and t in minutes. The departure rate is a constant 2 veh/min (i.e., time to weigh a truck is 30 seconds). When will the queue that forms dissipate, what will the total delay be, and what will the length (in vehicles) of the longest queue be?

5.21 Vehicles begin to arrive at a remote parking lot after the start of a major sporting event. They are arriving at a deterministic but time-varying rate of $\lambda(t) = 3.3 - 0.1t$, with $\lambda(t)$ in veh/min and t in minutes. The parking lot attendant processes vehicles (assigns spaces and collects fees) at a deterministic rate at a single station. A queue exceeding four vehicles will back up onto a congested street and is to be avoided. How many vehicles per minute must the attendant process to ensure that the queue does not exceed four vehicles?

5.22. A truck weighing station has a single scale. The time between truck arrivals at the station is exponentially distributed with a mean arrival rate of 1.5 veh/min. The time it takes vehicles to be weighed is exponentially distributed with a mean rate of 2 veh/min. When more than five trucks are in the system, the queue backs up onto the highway and interferes with through traffic. What is the probability that the number of trucks in the system will exceed five?

5.23. Consider the convenience store described in Example 5.10. The owner is concerned about customers not finding an available parking space when they arrive in the busiest hour. How many spaces must be provided for there to be less than a 1% chance of an arriving customer not finding an open parking space?

5.24. Vehicles arrive at a toll bridge at a rate of 430 veh/h (the time between arrivals is exponentially distributed). Two toll booths are open and each can process arrivals (collect tolls) at a mean of 10 seconds (the processing time is also exponentially distributed). What is the total time in the system spent by all vehicles in a one-hour period?

5.25. Vehicles leave an airport parking facility (i.e., arrive at parking fee collection booths) at a rate of 500 veh/h (the time between arrivals is exponentially distributed). The parking facility has a policy that the average time a patron spends in a queue while waiting to pay for parking is not to exceed 5 seconds. If the time required to pay for parking is exponentially distributed with a mean of 15 seconds, what is the fewest number of payment processing booths that must be open to keep the average time spent in a queue less than 5 seconds?

Chapter 6

Traffic Analysis at Signalized Intersections

6.1 INTRODUCTION

Due to conflicting traffic movements, highway intersections are a source of great concern to highway engineers. Intersections offer the potential for increased highway accident rates and considerable vehicle delay as vehicles yield to avoid conflicts with other vehicles. Most highway intersections are not signalized due to low traffic volumes and adequate sight distances. Furthermore, the economic expense of installing traffic signals and the possible increase in vehicle delay that signals may cause argue against their use. However, at some point, traffic volumes (and other factors) reach a level that make signalization economically feasible.

This chapter focuses only on signalized intersections, and the reader is referred to other sources for the study of unsignalized intersections (Transportation Research Board 1975, 1994). We begin by presenting definitions of terms and analytic techniques that can be used to calculate delay at signalized intersections. Attention is then turned to the practical aspects of traffic signal timing. Overall, this chapter provides the basic principles used in the analysis and design of signalized intersections.

6.2 DEFINITIONS

Following are a number of definitions (and associated notations where appropriate) of commonly used intersection-related terminology.

Approach A lane or group of lanes through which traffic enters an intersection.

Cycle One complete sequence (for all approaches) of signal indications (greens, yellows, reds).

Cycle length The total time for the signal to complete one cycle (identified by the symbol c and usually expressed in seconds).

Traffic Signal Phase The part of the cycle length allocated to a traffic movement that has the right of way, or any combination of traffic movements that receive the right of way simultaneously. The sum of the phase lengths (in seconds) is the cycle length.

Indication The illumination of one or more signal lenses (greens, yellows, reds) indicating a permitted or prohibited traffic movement.

Interval A period of time during which all signal indications (greens, yellows, reds) remain the same for all approaches.

Green Time The time within a cycle in which an approach has the green indication (expressed in seconds and given the symbol GT).

Red Time The time within a cycle in which an approach has the red indication (expressed in seconds and given the symbol RT).

Yellow Time The time within a cycle in which an approach has the yellow indication (expressed in seconds and given the symbol YT).

All-Red Time The time within a cycle in which all approaches have a red indication (expressed in seconds and given the symbol AR).

Change Interval The yellow time plus all-red time (the short period of time in which all approaches have a red signal) that provides for clearance of the intersection before conflicting traffic movements are given a signal indication that allows them to enter the intersection (expressed in seconds).

Lost Time Time during which the intersection is not effectively used by any approach. These times occur during the change interval (when the intersection is cleared) and at the beginning of each green indication as the first few vehicles in a standing queue experience start-up delays. The lost time is given the symbol LT.

Effective Green The time that is effectively used by the approach for traffic movement. This is generally taken to be the green time plus the change interval minus the lost time for the approach. Effective green is stated in seconds and given the symbol g.

Effective Red The time that is effectively not used by the approach for traffic movement. Stated in seconds, it is the cycle length minus the effective green time, and is given the symbol r.

Saturation Flow The maximum flow that could pass through an intersection from a given approach, if that approach were allocated all of the cycle time as effective green with no lost time (given the symbol s).

Approach Capacity The maximum flow that can pass through an intersection under prevailing highway and traffic conditions, given the effective green time allocated to the approach. It is equal to the saturation flow multiplied by the ratio of effective green to cycle length ($C = s \times g/c$).

Major Street The street at an intersection that has the higher traffic-volume approaches.

Minor Street The street at an intersection that has the lower traffic-volume approaches.

Protected Turn A turning movement made without the conflict of opposing traffic or pedestrians. This turn is made during an exclusive turning phase (e.g., a left-turn arrow).

Permitted Turn A turning movement that is made through opposing traffic flow or through conflicting pedestrian movement. This turn is made during gaps (time headways) in opposing traffic and conflicting pedestrian movements.

Signal Timing The operating characteristics of the signal with the parameters being the signal cycle length, green time, red time, yellow time, and all-red time (the settings produce, for all approaches, effective green and red times).

Pretimed Signal A signal whose timing (i.e., cycle length, green time, etc.) is fixed over specified time periods and does not change in response to changes in traffic flow at the intersection.

Semi-Actuated Signal A signal whose timing (i.e., cycle length, green time, etc.) is affected when vehicles are detected (e.g., by magnetic-loop detectors in the pavement) on some, but not all, approaches. These types of signals are usually found when a low-volume road intersects with a high-volume road. In such cases green time is allocated to the high-volume approaches until vehicles are detected on the low-volume approaches, then the green indication is briefly allocated to the low-volume approaches and then returned to the high-volume approaches.

Fully Actuated Signal A signal whose timing (i.e., cycle length, green time, etc.) is completely influenced by the traffic volumes, when detected, on all of the approaches. Fully actuated signals are most commonly used at intersections where substantial variations exist in approach traffic volumes over the course of a day.

6.3   ANALYSIS OF SIGNALIZED INTERSECTIONS WITH *D/D/*1 QUEUING

The assumption of $D/D/1$ queuing (as discussed in Chapter 5) provides a strong intuitive appeal that helps understanding of the analytic fundamentals underlying traffic analysis at signalized intersections. To begin applying $D/D/1$ queuing to signalized intersections, we consider the case where the approach capacity exceeds the approach arrivals. Under these conditions the assumption of $D/D/1$ queuing will result in a queuing system as shown in Fig. 6.1, where λ is the arrival rate (typically in vehicles per second), μ is the departure rate (in vehicles per second), g is the effective green (in seconds), r is the effective red (in seconds), t is the total transpired time (in seconds), the "arrivals λt" line gives the total number of vehicle arrivals at time t, the "departures μt" line gives the slope of vehicle departures (number of vehicles that depart) during effective greens, t_0 is the time from the start of the effective green until queue dissipation (in seconds), and c is the cycle length (in seconds). Note that the per-cycle approach arrivals

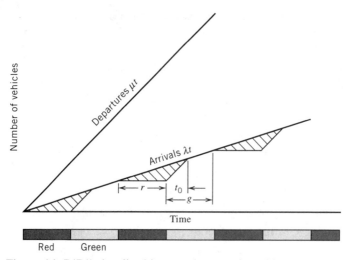

Figure 6.1 $D/D/1$ signalized intersection queuing with approach capacity (μg) exceeding arrivals (λc) for all cycles.

will be λc and the corresponding approach capacity (departures) per-cycle will be μg. Thus Fig. 6.1 is predicated on the assumption that μg exceeds λc for all cycles (i.e., no queues exist at the beginning or end of a cycle).

Given the properties of $D/D/1$ queues presented in Chapter 5, a number of general equations can be derived by simple inspection of Fig. 6.1.

1. For the time to queue dissipation after the start of the effective green, t_0, note that $\lambda(r + t_0) = \mu t_0$ and, with traffic intensity $\rho = \lambda/\mu$ (as in Chapter 5),

$$t_0 = \frac{\rho r}{(1 - \rho)} \tag{6.1}$$

2. The proportion of the cycle with a queue, P_q, is

$$P_q = \frac{r + t_0}{c} \tag{6.2}$$

3. The proportion of vehicles stopped, P_s, is

$$P_s = \frac{\lambda(r + t_0)}{\lambda(r + g)} = \frac{r + t_0}{c} = P_q$$

Also,

$$\tag{6.3}$$

$$P_s = \frac{\lambda(r + t_0)}{\lambda(r + g)} = \frac{\mu t_0}{\lambda c} = \frac{t_0}{\rho c}$$

4. The maximum number of vehicles in the queue, Q_m, is

$$Q_m = \lambda r \qquad \text{(6.4)}$$

5. The total vehicle delay per cycle, D_t, is

$$D_t = \frac{\lambda r^2}{2(1 - \rho)} \qquad \text{(6.5)}$$

6. The average vehicle delay per cycle, d, is

$$d = \frac{\lambda r^2}{2(1 - \rho)} \times \frac{1}{\lambda c}$$

$$= \frac{r^2}{2c(1 - \rho)} \qquad \text{(6.6)}$$

7. The maximum delay of any vehicle (assuming a FIFO queuing discipline), d_m, is

$$d_m = r \qquad \text{(6.7)}$$

EXAMPLE 6.1

An approach at a pretimed signalized intersection has a saturation flow of 2400 veh/h and is allocated 24 seconds of effective green in an 80-second signal cycle. If the flow at the approach is 500 veh/h, provide an analysis of the intersection assuming $D/D/1$ queuing.

SOLUTION

Putting arrival and departure rates into common units of vehicles per second,

$$\lambda = \frac{500 \text{ veh/h}}{3600 \text{ s/h}}$$

$$= 0.139 \text{ veh/s}$$

$$\mu = \frac{2400 \text{ veh/h}}{3600 \text{ s/h}}$$

$$= 0.667 \text{ veh/s}$$

This gives a traffic intensity of

$$\rho = \frac{0.139 \text{ veh/s}}{0.667 \text{ veh/s}}$$

$$= 0.208$$

Checking to make certain that capacity exceeds arrivals, we find that the capacity (μg) is 16 veh/cycle (0.667×24) which is greater than (permitting fractions of vehicles for the sake of clarity) the 11.12 arrivals ($\lambda c = 0.139 \times 80$). Therefore, Eqs. 6.1 to 6.7 are valid. By definition $r = c - g = 80 - 24 = 56$ s. Thus for

1. Time to queue dissipation after the start of the effective green (Eq. 6.1),

$$t_0 = \frac{0.208(56)}{(1 - 0.208)}$$

$$= 14.71 \text{ s}$$

2. Proportion of the cycle with a queue (Eq. 6.2),

$$P_q = \frac{56 + 14.71}{80}$$

$$= 0.884$$

3. Proportion of vehicles stopped (Eq. 6.3),

$$P_s = \frac{14.71}{0.208(80)}$$

$$= 0.884$$

4. Maximum number of vehicles in the queue (Eq. 6.4),

$$Q_m = 0.139(56)$$

$$= 7.78$$

5. Total vehicle delay per cycle (Eq. 6.5),

$$D_t = \frac{0.139(56)^2}{2(1 - 0.208)}$$

$$= 275.19 \text{ veh-s}$$

6. Average delay per vehicle (Eq. 6.6),

$$d = \frac{56^2}{2(80)(1 - 0.208)}$$

$$= 24.75 \text{ s/veh}$$

7. Maximum delay of any vehicle (Eq. 6.7),

$$d_m = r$$

$$= \underline{\underline{56 \text{ s}}}$$

Recall that Eqs. 6.1 through 6.7 are valid only when the approach capacity exceeds approach arrivals. For the case when approach arrivals exceed capacity for some signal cycles, $D/D/1$ queuing can again be used, as illustrated in the following example.

EXAMPLE 6.2

An approach to a pretimed signalized intersection has a saturation flow of 1700 veh/h. The signal's cycle length is 60 seconds and the approach's effective red is 40 seconds. During three consecutive cycles, 15, 8, and 4 vehicles arrive. Determine the total vehicle delay over the three cycles assuming $D/D/1$ queuing.

SOLUTION

For all cycles, the departure rate is

$$\mu = \frac{1700 \text{ veh/h}}{3600 \text{ s/h}}$$

$$= 0.472 \text{ veh/s}$$

During the first cycle, the number of vehicles that will depart from the signal is (permitting fractions for the sake of clarity)

$$\mu g = 0.472(20)$$

$$= 9.44 \text{ veh}$$

Therefore, 5.56 vehicles (15 − 9.44) will not be able to pass through the intersection on the first cycle even though they arrive during the first cycle. At the end of the second cycle, 23 vehicles (15 + 8) will have arrived but only 18.88 ($2\mu g$) will have departed, leaving 4.12 vehicles waiting at the beginning of the third cycle. At the end of the third cycle, a total of 27 vehicles will have arrived and as many as 28.32 ($3\mu g$) could have departed, so the queue that formed during the first cycle will dissipate at some time during the third cycle. This process is shown graphically in Fig. 6.2. From this figure, the total vehicle delay of the first

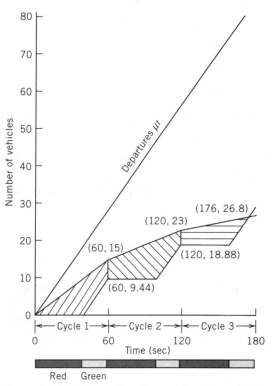

Figure 6.2 $D/D/1$ queuing diagram for Example 6.2.

cycle (the area between arrival and departure curves) is

$$D_1 = \frac{1}{2}(60)(15) - \frac{1}{2}(20)(9.44)$$

$$= 355.6 \text{ veh-s}$$

Similarly, the delay in the second cycle is

$$D_2 = \frac{1}{2}(60)(15 + 23) - (40)(9.44) - \frac{1}{2}(20)(9.44 + 18.88)$$

$$= 479.2 \text{ veh-s}$$

To determine the delay in the third cycle, it is necessary first to know exactly when, in this cycle, the queue dissipates. The time to queue dissipation after the start of the effective green, t_0, is (with λ_3 being the arrival rate during the third cycle, and n_3 being the number of vehicles in the queue at the start of the third cycle)

$$n_3 + \lambda_3(r + t_0) = \mu t_0$$

where

$$\lambda_3 = \frac{4 \text{ veh}}{60 \text{ s}}$$

$$= 0.067 \text{ veh/s}$$

Therefore,

$$(23 - 18.88) + 0.067(40 + t_0) = 0.472t_0$$

which gives $t_0 = 16.8$ seconds. Thus the queue will clear 56.8 seconds $(40 + 16.8)$ after the start of the third cycle, at which time a total of 26.8 vehicles $(0.067 \times 56.8 + 15 + 8)$ will have arrived and departed from the intersection. The vehicle delay for the third cycle is

$$D_3 = \frac{1}{2}(56.8)(23 + 26.8) - (40)(18.88) - \frac{1}{2}(16.8)(18.88 + 26.8)$$

$$= 275.41 \text{ veh-s}$$

giving the total delay over all the three cycles as

$$D_t = D_1 + D_2 + D_3$$

$$= 355.6 + 479.2 + 275.41$$

$$= \underline{\underline{1110.2 \text{ veh-s}}}$$

It is possible to handle the case where intersection arrivals and/or departures are deterministic but time varying in a fashion similar to that shown in Chapter 5. An example of time-varying arrivals is presented below.

EXAMPLE 6.3

The saturation flow of an approach to a pretimed signal is 12,000 veh/h. The signal has a 60-second cycle with 20 seconds of effective red allocated to the approach. At the beginning of an effective red (with no vehicles remaining in the queue from a previous cycle) vehicles start arriving at a rate $\lambda(t) = 0.88 + 0.02t + 0.00057t^2$ (where $\lambda(t)$ is in vehicles per second and t is the number of seconds from the beginning of the cycle). Thirty seconds into the cycle the arrival rate remains constant at its 30-second level and stays at that rate until the end of the cycle. What is the total vehicle delay over the cycle (in vehicle-seconds) assuming $D/D/1$ queuing?

SOLUTION

Vehicle arrivals for the first 30 seconds (again allowing fractions of vehicles for the sake of clarity) are

$$= \int_0^{30} 0.88 + 0.02t + 0.00057t^2$$

$$= 0.88t + 0.01t^2 + 0.00019t^3 \big|_0^{30}$$

$$= 40.53 \text{ veh}$$

The arrival rate after 30 seconds is 1.993 veh/s $[0.88 + 0.02(30) + 0.00057(30)^2]$ and is no longer time varying. During the effective green, the departure rate is 3.333 veh/s (12,000/3600). To determine when the queue will dissipate, let t' be the time after 30 seconds (i.e., the time after which the arrival rate is no longer time varying), so,

$$40.53 + 1.993t' = 3.333(t' + 10)$$

which gives $t' = 5.37$ seconds. Thus the queue clears at 35.37 seconds after the beginning of the cycle. For delay, the area under the arrival curve

$$= \int_0^{30} 0.88t + 0.01t^2 + 0.00019t^3 + \frac{1}{2}\{40.53 + [1.993(5.37) + 40.53]\}(5.37)$$

$$= 0.44t^2 + 0.0033t^2 + 0.0000475t^2 \big|_0^{30} + 246.38$$

$$= 523.58 + 246.38$$

$$= 769.96 \text{ veh-s}$$

The area under the departure curve

$$= \frac{1}{2}(51.23 \times 15.37)$$

$$= 393.7 \text{ veh-s}$$

So total vehicle delay over the cycle is <u>376.26 veh-s</u> (769.96 − 393.7).

6.4 ANALYSIS OF SIGNALIZED INTERSECTIONS WITH PROBABILISTIC ARRIVALS

The assumption of nonuniform traffic arrival, although often more realistic than the uniform (deterministic) arrival assumption, adds considerable complexity to the analysis of delay at signalized intersections. Over the past few decades, a number of delay formulas have been proposed to account for the randomness of vehicle arrivals at signalized intersections. Among the better known of these delay formulas are those developed by Webster (1958) and Allsop (1972). The Webster formula for approach delay at a pretimed signalized intersection is

$$d' = d + \frac{x^2}{2\lambda(1-x)} - 0.65 \left(\frac{c}{\lambda^2}\right)^{1/3} x^{2+5(g/c)} \tag{6.8}$$

where d' is the average vehicle delay, d is the average vehicle delay computed by assuming $D/D/1$ queuing (e.g., see Eq. 6.6), x is the ratio of approach arrivals to approach capacity (also referred to as the volume-to-capacity ratio and equal to $\lambda c/\mu g$), c is the cycle length, g is the effective green, and λ is the average vehicle arrival rate. The Webster formula was developed by using data from a computer simulation of intersection operations. Allsop noted that the second term in Webster's equation is obtained by assuming an additional queue interposed between the arriving traffic and the signal, and the third term is merely an empirical correction that ranges from 5% to 15% of the total mean delay. Based on these observations, Allsop proposed that average vehicle delay, based on the assumption of random vehicle arrivals, be computed as

$$d' = \frac{9}{10}\left[d + \frac{x^2}{2\lambda(1-x)}\right] \tag{6.9}$$

where all terms are as previously defined.

For information on other intersection delay formulas that are based on the assumption of random vehicle arrivals, the reader is referred to previously cited sources (Transportation Research Board 1975, 1994).

EXAMPLE 6.4

Compute the average approach per cycle using Webster's and Allsop's formulas given the conditions described in Example 6.1.

SOLUTION

As computed in Example 6.1, the delay, assuming uniform arrivals ($D/D/1$ queuing), d, is 24.75 s/veh. The ratio of approach arrivals to approach capacity is

$$x = \frac{\lambda c}{\mu g}$$

$$= \frac{0.139(80)}{0.667(24)}$$

$$= 0.695$$

Using Webster's formula (Eq. 6.8),

$$d' = 24.75 + \frac{0.695^2}{2(0.139)(1 - 0.695)} - 0.65 \left(\frac{80}{0.139^2}\right)^{1/3} 0.695^{2+5(24/80)}$$

$$= 24.72 + 5.697 - 2.921$$

$$= 27.426 \text{ s/veh}$$

Using Allsop's formula (Eq. 6.9),

$$d' = \frac{9}{10}\left[24.75 + \frac{0.695^2}{2(0.139)(1 - 0.695)}\right]$$

$$= \underline{27.402 \text{ s/veh}}$$

6.5   OPTIMAL TRAFFIC SIGNAL TIMING

Allocating effective green times to competing approaches in some optimal fashion has been a goal of highway engineers since traffic signals were first used. However, a drive on virtually any highway with signalized intersections shows that optimal signal timing has not yet been universally achieved. The problem of optimal timing is complicated by a number of factors. For example, although the distribution of traffic can be approximated by some arrival assumptions (deterministic or Poisson), to properly optimize signal timing the arrival pattern must be known with certainty. Traffic sensors embedded near intersections have resulted in traffic signal systems that respond to variations in traffic flow and thus have gone a long way to satisfy the traffic arrival information needed. However, a more fundamental question pervades signal optimization. Specifically, on what basis should traffic signal timings be optimized? Should vehicle delay be minimized? Should the number of vehicles stopped be minimized? Or should some other factor serve as the optimization criterion? Because different minimization criteria almost always provide different results, many signal timing strategies seek to minimize a combined function of more than one factor (e.g., vehicle stops and delay) (Vincent, Mitchell, and Robertson 1980). Recent theoretical work has departed from the more traditional delay/stop approach and has demonstrated optimal signal timings based on the maximization of travelers' economic welfare (Mannering, Abu-Eisheh, and Arnadottir 1991). Needless to say, the optimization-measurement problem makes this aspect of traffic analysis a fruitful area for future research.

To demonstrate one possible signal optimization strategy (putting aside the optimization issues just discussed), assume that the sole objective of signal timing is to minimize vehicle delay and that traffic can be represented by a simple $D/D/1$ queue. Such a strategy is illustrated by the following example.

EXAMPLE 6.5

A pretimed signal controls a four-way intersection with no turning permitted and zero lost time. The eastbound (eb) and westbound (wb) traffic volumes are 700 and 800 veh/h, respectively, and both movements share the same effective green and effective red portions of the cycle. The northbound (nb) and southbound (sb) directions also share cycle times with volumes of 400 and 250 veh/h, respectively. If the saturation flow of all approaches is 1800 veh/h, the cycle length is 60 seconds, and $D/D/1$ queuing applies, determine the effective red and effective green times that must be allocated to each directional combination (i.e., north-south, east-west) to minimize total vehicle delay, and compute the delay per cycle.

SOLUTION

Putting arrival and departure rates into common units of vehicles per second,

$$\lambda_{eb} = \frac{700 \text{ veh/h}}{3600 \text{ s/h}} = 0.194 \text{ veh/s}$$

$$\lambda_{wb} = \frac{800 \text{ veh/h}}{3600 \text{ s/h}} = 0.222 \text{ veh/s}$$

$$\lambda_{nb} = \frac{400 \text{ veh/h}}{3600 \text{ s/h}} = 0.111 \text{ veh/s}$$

$$\lambda_{sb} = \frac{250 \text{ veh/h}}{3600 \text{ s/h}} = 0.069 \text{ veh/s}$$

$$\mu = \frac{1800 \text{ veh/h}}{3600 \text{ s/h}} = 0.5 \text{ veh/s}$$

Because the departure rate is the same for all approaches, the traffic intensities are $\rho_{eb} = 0.388$, $\rho_{wb} = 0.444$, $\rho_{nb} = 0.222$, and $\rho_{sb} = 0.138$. If it is assumed that approach capacity exceeds approach arrivals for all approaches (an assumption that will be tested later), the total vehicle delay at the intersection (using Eq. 6.5) is

$$D_t = \frac{\lambda_{eb} r_{eb}^2}{2(1 - \rho_{eb})} + \frac{\lambda_{wb} r_{wb}^2}{2(1 - \rho_{wb})} + \frac{\lambda_{nb} r_{nb}^2}{2(1 - \rho_{nb})} + \frac{\lambda_{sb} r_{sb}^2}{2(1 - \rho_{sb})}$$

or, by substituting the computed values of λ's and ρ's,

$$D_t = 0.1585 r_{eb}^2 + 0.1996 r_{wb}^2 + 0.07115 r_{nb}^2 + 0.04 r_{sb}^2$$

The problem states that east and west effective reds are equal and north and south effective reds are equal. So let r_{ew} be the effective red of eastbound and westbound directions (i.e., $r_{ew} = r_{eb} = r_{wb}$) and let r_{ns} be the effective red of northbound and southbound directions (i.e., $r_{ns} = r_{nb} = r_{sb}$). So, by definition, with a 60-second signal cycle, $r_{ns} = 60 - r_{ew}$. Substituting this into the total delay expression gives

$$D_t = 0.1585r_{ew}^2 + 0.1996r_{ew}^2 + 0.07115(60 - r_{ew})^2 + 0.04(60 - r_{ew})^2$$

$$= 0.46925r_{ew}^2 - 13.338r_{ew} + 400.14$$

At minimum total delay, $dD_t/dr_{ew} = 0$. Therefore, differentiating yields

$$\frac{dD_t}{dr_{ew}} = 0.9357r_{ew} - 13.338 = 0$$

This gives $r_{ew} = \underline{14.2\text{ s}}$ ($g_{ew} = \underline{45.8\text{ s}}$), and $r_{ns} = \underline{45.8\text{ s}}$ ($g_{ns} = \underline{14.2\text{ s}}$). For total delay,

$$D_t = 0.1585(14.2)^2 + 0.1996(14.2)^2 + 0.07115(45.8)^2 + 0.04(45.8)^2$$

$$= \underline{305.36\text{ veh-s}}$$

Finally, a check of the earlier assumption that approach capacity exceeds approach arrivals for all approaches must be undertaken. In the 60-second cycle, eastbound and westbound arrivals are 11.64 (0.194×60) and 13.32 (0.222×60), respectively. With 45.8 seconds of effective green, the capacity for both approaches is 22.9 (0.5×45.8), so the assumption is satisfied. The northbound and southbound arrivals are 6.67 (0.111×60) and 4.14 (0.069×60), respectively, and, with 14.2 seconds of effective green, the capacity for both approaches is 7.1 (0.5×14.2). Again, the assumption is satisfied and the method used in this example is valid.

6.6. TRAFFIC SIGNAL TIMING IN PRACTICE

In practice, the installation and operation of traffic signals to control conflicting traffic movements at intersections has advantages and disadvantages. The advantages include a potential reduction of some types of accidents (particularly angle accidents), interruption of traffic to provide pedestrians the opportunity to cross the street, interruption of traffic so that minor-street vehicles can enter the traffic stream, provision for the progressive flow of traffic in a signal-system corridor, possible improvements in capacity, and possible reductions in delay. On the other hand, a poorly timed signal or one that is not justified can have a negative impact on the operation of the intersection by increasing vehicle delay, increasing vehicle accidents (particularly rear-end accidents), causing a disruption of traffic progression (i.e., impacting the through movement of traffic), and encouraging the use of routes not intended for through traffic (i.e., routes through residential neighborhoods).

Traffic signals are costly to install, with even some basic pretimed signals costing in excess of $100,000. Because of the expense and the potential negative impacts of signalizing an intersection, guidelines and procedures are established by traffic departments to determine whether a signal is warranted. A good example of traffic-signal warrants are those provided in the "Manual of Uniform Traffic Control Devices for Streets and Highways" (U.S. Federal Highway Administration, 1988). Following is a summary of the 11 warrants presented in this manual (referred to as the MUTCD).

Warrant 1: Minimum Vehicle Volume When traffic volume is the principal consideration for signal installation, this warrant gives minimum volumes required on the major street (the street with the higher-volume approaches) and on the minor street (the street with the lower-volume approaches).

Warrant 2: Interruption of Continuous Traffic This warrant applies when traffic on the major street is so heavy that traffic on the minor street suffers excessive delay or hazard when entering or crossing the major street. This warrant gives required volumes for the major street and the higher-volume minor-street approach.

Warrant 3: Minimum Pedestrian Volume This warrant gives pedestrian volumes required to warrant signal installation. This warrant is used when pedestrian safety and delay are a principal consideration for signal installation.

Warrant 4: School Crossing Traffic signals are warranted at school crossings when traffic studies show that the frequency and adequacy of the time headways between vehicles does not meet specified requirements in relation to the number of children crossing.

Warrant 5: Progressive Movement Concerns about progressive movement of traffic can be used to justify a traffic signal when other warrants are not met. The idea is that installation of a traffic signal can maintain proper grouping of vehicles on a street and thus result in a reduction in vehicle delay.

Warrant 6: Accident Experience This warrant establishes criteria relating to the record of accident types and accident severities needed to justify signal installation.

Warrant 7: System Warrants In some instances, traffic signal installation may be warranted to encourage concentration and progression of traffic flow on the traffic network. This warrant establishes the traffic network characteristics needed to justify signal installation on signal-system grounds.

Warrant 8: Combination of Warrants In some cases, signals may be justified when no single warrant is met but where two or more of warrants 1, 2, and 3 are satisfied to 80% or more of their stated values.

Warrant 9: Four-Hour Volumes This warrant applies when the traffic volume in each of any four hours in an average day exceeds specified values. This warrant attempts to account for the effects of traffic flow variations over the course of a day.

Warrant 10: Peak-Hour Delay The peak-hour delay warrant applies when traffic conditions are such that for one hour of the day the minor-street traffic suffers undue delay when entering or crossing the major street. This warrant is satisfied by delay estimates and approach volumes exceeding specified amounts.

Warrant 11: Peak-Hour Volume This warrant is similar to warrant 10 in that it applies when traffic conditions are such that for one hour of the day the minor-street traffic suffers undue delay when entering or crossing the major street. However, the criteria for assessing delay are different in that standardized graphs are used to determine if the warrant is met.

Even when a warrant is met, some caution is usually applied before proceeding with signal installation. Sound judgment must be exercised due to the potential downsides of signal installation (e.g., the potential increase in certain types of accidents).

6.6.1   Traffic Signal Timing Procedure

The development of a traffic signal timing plan can be complex, particularly if the intersection has multi-lane approaches and requires protected turning movements (e.g., a left-turn arrow, with associated phasing). However, the timing plan analysis can be simplified by dealing with each approach separately. This section provides the fundamentals needed to time a fixed traffic signal (i.e., a pretimed signal). As timing plans become more complex, they simply build upon these fundamental principles. The reader is encouraged to review the materials in other references to see how actuated signals and progressive timing plans are developed (Kell and Fullerton, 1991).

There are eight steps in the development of a traffic signal timing plan: select signal phasing, calculate equivalent straight-through passenger cars, select critical lane volumes, calculate change interval, calculate minimum cycle length, allocate green time, check pedestrian crossing time, and prepare a signal indication summary. Following is a discussion of these eight steps.

Select Signal Phasing

Recall that a cycle is the sum of individual phases. The most basic traffic signal cycle will be made up of two phases as shown in Fig. 6.3. In this case, phase one is the traffic movement for the north-southbound vehicles, and phase two is the movement of the east-westbound vehicles. These phases will alternate during the continuous operation of the signal. It is important, however, to note that each time the phase changes there will be lost time associated with the traffic movement. Although the lost time may be only 3 to 5 seconds per phase change, the accumulated total lost time throughout the day can be significant. During this lost time there is no movement of vehicles, consequently, delay is increased.

Each time a phase is added (such as the protected left-turn phase added in the three-phase operation diagram shown in Fig. 6.3) the lost time increases. Therefore a primary concern in signal timing is to keep the number of phases

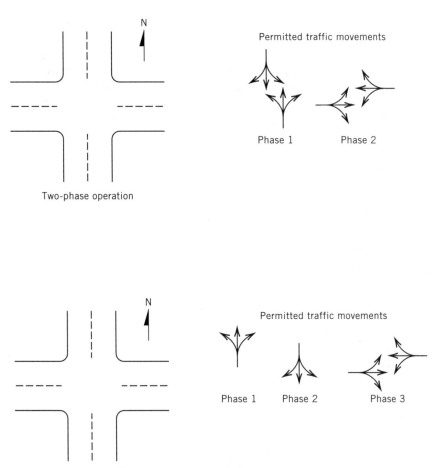

Figure 6.3 Illustration of two-phase and three-phase signal operation.

to a minimum. Because protected-turn phases add to lost time, they should be used only when warranted. As a rule, a separate left-turning phase should be considered for the signal timing plan when

1. The product of left-turning vehicles and opposing traffic volume exceeds 50,000 during the peak hour on a two-lane highway, or 100,000 on a four-lane highway.
2. Two or more vehicles are still waiting to turn left at the end of the phase.
3. There are more than 50 vehicles turning left during the peak hour and the approach speeds are greater than 72 km/h.
4. There are five or more accidents associated with turning movements during a 12-month period.

The signal phasing selection process is presented in the following example.

EXAMPLE 6.6

The intersection shown in Fig. 6.4 does not satisfy left-turn phase requirements for waiting vehicles, approach speeds greater than 72 km/h, or turning-movement accidents. Determine if the product of left-turning vehicles and opposing vehicles suggests the use of a left-turn phase (Note in Fig. 6.4 that the traffic volume numbers in parentheses are heavy vehicles, such as buses and large trucks.)

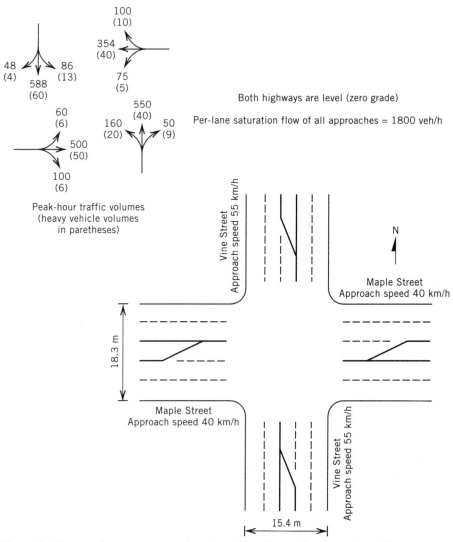

Figure 6.4 Intersection geometry and peak traffic volumes for example problems.

SOLUTION

There are 180 northbound vehicles that turn left during the peak hour. The product of the northbound left-turning vehicles and the opposing traffic (right-turn and straight-through vehicles) is 126,000 (180 × [648 + 52]). Because the product is greater than 100,000 (the requirement for a four-lane highway), a left-turn phase is suggested. The other approaches do not require a left-turn phase using this criterion because the products for all other approaches are less than 100,000 (eastbound left-turn/opposing traffic product is 33,264, westbound left-turn/opposing traffic product is 52,480, southbound left-turn/opposing traffic product is 64,900). Therefore, a three-phase traffic signal plan should be used.

Calculate Equivalent Straight-Through Passenger Cars

Traffic arrives at an intersection in mixed-vehicle form with respect to vehicle type (i.e., car, buses, and trucks) and turning movement (left-turning, right-turning, and straight-through). Heavy vehicles are slower to accelerate and decelerate and they have larger turning radii than passenger cars. Also, all types of vehicles require additional time to execute both left and right turns. Because turning movements and heavy vehicles require additional green time to complete their maneuvers, a method has been developed to account for this. Mixed-traffic volumes and non-through movements are converted to equivalent straight-through passenger cars for the signal timing analysis, using the adjustment factors in Table 6.1 (The factors in this table are derived from turning movement times in the *Highway Capacity Manual* [Transportation Research Board 1994].) The equivalent straight-through passenger car determination is demonstrated by the following example.

EXAMPLE 6.7

Calculate the equivalent straight-through passenger cars for the northbound, southbound, eastbound, and westbound traffic streams shown in Fig. 6.4. (Note in Fig. 6.4 that the traffic volume numbers in parentheses are heavy vehicles, such as buses and large trucks.)

Table 6.1 Adjustment Factors for Equivalent Straight-Through Passenger Cars

Vehicle Type and Movement	Adjustment Factor
Passenger car (straight-through)	1.0
Heavy vehicle	1.5
Left-turning	1.6
Right-turning	1.4

SOLUTION

Using Table 6.1, calculations for the total northbound equivalent straight-through passenger cars for each movement are (rounding up to the nearest 1.0 vehicle)

$$\textit{Left turns} \quad = 160(1.6) + 20(1.6)(1.5) = 304 \text{ veh/h}$$

$$\textit{Right turns} \quad = 50(1.4) + 9(1.4)(1.5) \quad = 89 \text{ veh/h}$$

$$\textit{Straight through} = 550(1.0) + 40(1.0)(1.5) = 610 \text{ veh/h}$$

The summation gives 1003 veh/h total equivalent straight-through passenger cars. The equivalent straight-through passenger car traffic for the other directions is calculated in the same manner. The results give 923 veh/h, 839 veh/h, and 707 veh/h for southbound, eastbound, and westbound, respectively.

Select Critical-Lane Volumes

After the equivalent straight-through passenger car volumes have been calculated, the critical-lane volumes for each approach can be determined. The left-turn volumes are considered a separate movement if a left-turn bay is provided (as shown in Fig. 6.4). The equivalent straight-through and right-turn movements are combined and distributed over the number of approach lanes. The distribution of traffic over the lanes is best done by a traffic survey that determines actual lane volumes by observing vehicles. If such a survey is not available, the amount of traffic in the most heavily used lane in the approach (i.e., critical lane volume) can be approximated as the total approach volume multiplied by a critical-lane factor. For through lanes or shared through and turning lanes, the factor is 0.525 and 0.367 for two-and three-lane approaches, respectively. For two-lane exclusive left-turn approaches, the factor is 0.515, and for two-lane exclusive right-turn approaches, the factor is 0.565. For through or shared-lane approaches with more than three lanes, and exclusive-turn approaches with more than two lanes, a traffic survey is recommended (Transportation Research Board 1994).

EXAMPLE 6.8

A traffic survey shows that the through volumes of all approaches are distributed equally among the through lanes. Using the equivalent straight-through passenger cars calculated in Example 6.7, determine the critical-lane volumes for east-west and north-south approaches to the Vine and Maple streest intersection.

SOLUTION

In Example 6.7 the northbound left-turning equivalent straight-through passenger car volume was determined to be 304 vehicles. The straight-through equiva-

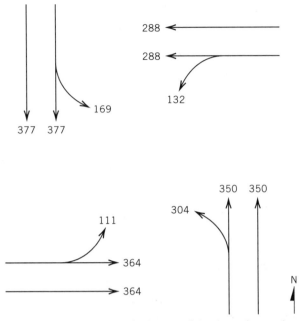

Figure 6.5 Sketch of equivalent straight-through cars for each approach.

lent was 610 vehicles with 89 right-turn equivalents. Combining the right-turn and straight-through movements gives 699 vehicles (610 + 89). As shown by the traffic survey (and stated in the example), the straight-through traffic is distributed equally over the lanes as 350 vehicles per lane (699/2, and rounded up to the nearest vehicle for practical reasons). The other approaches are handled in the same way with the straight-through passenger cars distributed as shown in Fig. 6.5.

The critical-lane volumes are then determined for each phase of the traffic-signal timing plan. As discussed earlier in Example 6.6, a three-phase timing plan is to be used. Consequently, critical lanes must be selected for the north-south movement, the east-west movement, and the left-turning phase. Observing the east-west traffic volumes shown in Figure 6.5, it can be seen that the eastbound movement of 364 vehicles per hour per lane is greater than the 288 vehicles per hour per lane for the westbound movement. Therefore, the critical-lane volume for the east-west phase is 364 vehicles per hour per lane. Likewise, the 377 vehicles per lane for the southbound approach represents the critical volumes for the north-south phase. The left-turn phase will use 304 vehicles per hour as the critical-lane volume.

Calculate Change Interval

Recall that the change interval is the yellow plus all-red times (the short period of time in which all approaches have a red signal). This interval provides for clearance of the intersection before conflicting traffic movements are given a signal indication that allows them to enter the intersection (expressed in seconds). In the past, a yellow indication was used to warn the driver of a signal change and provide clearance time. Today, however, there is routine red-indication abuse and frequent running of red indications after the yellow time. As a result, the all-red indication has become commonly used.

Accepted formulas for calculating yellow and all-red times are

$$YT = t_p + \frac{V}{2a + 2g_r G} \tag{6.10}$$

$$AR = \frac{w + l}{V} \tag{6.11}$$

where YT is the yellow time (usually rounded to the nearest 0.5 second), t_p is the driver perception/reaction time taken as 1.0 second, V is the speed of the vehicle in m/s, a is the deceleration rate for the vehicle taken as 3.05 m/s^2, G is the percent grade divided by 100, g_r is the acceleration due to gravity (i.e., 9.807 m/s^2), AR is the all-red time, w is the width of the cross street in meters, and l is the length of the vehicle (taken as a conservative 6 m).

Typically the yellow time is in the range of 3 to 5 seconds. Yellow times that are shorter than 3 seconds and longer than 5 seconds are not practical because long yellow times encourage motorists to continue to enter the intersection, and short times can place drivers in a dilemma zone. A dilemma zone is created for a driver if a safe stop before the intersection cannot be accomplished but continuing through the intersection at a constant speed (i.e., without accelerating) will result in a red indication. If a dilemma zone exists, drivers always make the wrong decision whether they decide to stop or continue through the intersection. Figure 6.6 shows a sketch of the dilemma zone. Referring to this figure, suppose a

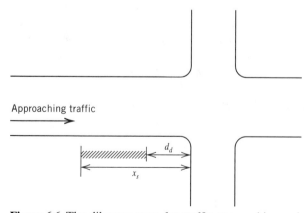

Figure 6.6 The dilemma zone for traffic approaching a signalized intersection.

vehicle traveling at a constant speed requires distance x_s to stop. If the vehicle is closer to the intersection than distance d_a, then it can enter before the all-red indication. If the vehicle is in the shaded area (i.e., $x_s - d_a$ from the intersection) when the yellow light is displayed, the driver is in the dilemma zone and can neither stop in time nor continue through the intersection at a constant speed without passing through a red indication. In order to avoid a dilemma zone the total change interval (yellow plus all-red times) should always be equal to or greater than the sum of Eqs. 6.10 and 6.11.

EXAMPLE 6.9

Determine the yellow and all-red times for vehicles traveling on Vine and Maple Streets as shown in Fig. 6.4.

SOLUTION

For the Vine Street phasing (applying Eqs. 6.10 and 6.11),

$$YT = 1.0 + \frac{(55 \times 1000/3600)}{2(3.05)}$$

$$= \underline{\underline{3.5 \text{ s}}}$$

$$AR = \frac{18.3 + 6}{55 \times 1000/3600}$$

$$= \underline{\underline{1.6 \text{ s}}} \text{ (rounded to the nearest 0.1)}$$

For the Maple Street phasing (applying Eq. 6.11),

$$YT = 1.0 + \frac{(40 \times 1000/3600)}{2(3.05)}$$

$$= \underline{\underline{2.8 \text{ s}}} \text{ (rounded to the nearest 0.1)}$$

$$AR = \frac{15.4 + 6}{40 \times 1000/3600}$$

$$= \underline{\underline{1.9 \text{ s}}} \text{ (rounded to the nearest 0.1)}$$

The yellow time for Maple Street (2.8 s) is rounded to 3.0 s because many traffic signal controllers handle timing only to the nearest 0.5 s. The yellow time for the left-turn phase on Vine Street is calculated by using the same values as the Vine Street phasing. Thus a yellow time of 3.5 s is obtained.

For the all-red times, Maple's 1.9 s and Vine's 1.6 s are both rounded up to 2.0 s (the nearest 0.5 s interval, which can be handled by most signal controllers).

The all-red time on Vine Street for the left-turn phase will be the same as for the through movements. A turning vehicle will travel a circular path to clear the approaching lanes. The travel distance for the turning vehicle will be approximately 10 m, which is shorter than the width of Maple Street; however, because a turning maneuver requires more time, the all-red time is also set as a conservative 2.0 s.

Calculate Minimum Cycle Length

The cycle length is simply the summation of the individual phases. In practice, cycle lengths are kept as short as possible, typically within the 40- to 60-second range. However, complex intersections with five or more phases can have cycle lengths of 120 seconds or more. It should be noted that cycle lengths greater than 120 seconds are unusual and should be used only in exceptional circumstances.

A practical equation for the calculation of optimum cycle length was developed by Webster (1958), who assumed that vehicles arrive at an intersection on a random basis. He developed a practical expression for cycle length that sought to minimize vehicle delay. Webster's optimum cycle-length formula is

$$c = \frac{1.5LT + 5}{1.0 - \sum_{i=1}^{n} y_i} \tag{6.12}$$

where c is the cycle length (usually rounded up to the nearest 5-second increment), LT is the lost time approximated as the total yellow and all-red times per cycle, and y_i is the ratio of the critical-lane volume to the per-lane saturation flow for signal phase i.

EXAMPLE 6.10

Using Webster's delay formula, calculate the optimum cycle length using the information provided in the preceding examples for the Maple Street–Vine Street intersection.

SOLUTION

In Example 6.9 it was determined that the yellow times for Vine Street and Maple Street phases are 3.5 seconds for the Vine Street phase, 3.5 seconds for the Vine Street left-turn phase, and 3.0 seconds for the Maple Street phase. The summation of these yellow times gives a total yellow time of 10 seconds. Also, in Example 6.9, it was determined that the all-red time for each phase was 2.0 seconds, giving a total all-red time of 6 seconds. Thus the total lost time is approximated as 16 seconds (10 + 6). From Example 6.8 it was determined that the critical lane volumes are 377 veh/h per lane for the Vine Street phase, 304 veh/h for the Vine Street left-turn phase, and 364 veh/h per lane for the Maple Street phase. Using these values in Equation 6.12 gives

$$c = \frac{1.5(16) + 5}{1.0 - \left[\left(\dfrac{377}{1800}\right) + \left(\dfrac{304}{1800}\right) + \left(\dfrac{364}{1800}\right)\right]}$$

$$= 69.14 \text{ s}$$

Rounding this up to the nearest 5-second increment gives a cycle length of 70 seconds.

Allocate Green Time

After a cycle length has been calculated, the next step in the traffic signal timing process is to determine how much green time should be allocated to each phase. The cycle length is the sum of all green times plus all lost time. In other words, the cycle length minus the sum of yellow and all-red times (i.e., the approximated lost time) will leave the total green time available for all phases. This total green time is divided among the phases that comprise the cycle. The green time is usually distributed to each phase in proportion to the ratio of critical-lane volume of the phase to total critical-lane traffic (summation of the critical-lane volumes of all phases). This green time allocation procedure is demonstrated by the following example.

EXAMPLE 6.11

Determine the green time allocations for the 70-second cycle length found in Example 6.10.

SOLUTION

Because there are 16 seconds of lost time, the total available green time is 54 seconds (70 − 16). Summing the critical-lane volumes gives a total critical-lane volume of 1045 veh/h (377 + 304 + 364). Using the ratio of critical-lane volumes to total critical-lane volume gives 19.5 seconds [(377/1045) × 54] of green time (GT) allocated to the Vine Street phase, 15.7 seconds [(304/1045) × 54] for the Vine Street left-turn phase, and 18.8 seconds [(364/1045) × 54] for the Maple Street phase. Rounding values (which always involves some compromise), we allocate 20 seconds, 15 seconds, and 19 seconds to the Vine Street phase, Vine Street left-turn phase, and Maple Street phase, respectively.

Check Pedestrian Crossing Time

In urban areas and other locations where pedestrians are present, the signal timing plan should be checked for its ability to provide adequate pedestrian crossing time. At locations where streets are wide and green times are short, it is possible that pedestrians can be caught in the middle of the intersection when the phase changes. To avoid this problem, the minimum green time required for pedestrian crossing should be checked against the apportioned green time for the phase. If there is not enough green time for pedestrians to safely cross the street, the apportioned green time should be increased to meet pedestrian needs.

To calculate the minimum green time required for pedestrian movements, the time required for a pedestrian to walk across the street must be determined. For signal timing purposes, it is assumed that pedestrians walk at a rate of 1.2 m/s. In areas where there are a large number of elderly pedestrians, a walking speed of 0.9 m/s is appropriate. In addition to the crossing time, a pedestrian reaction time of 7.0 seconds is added to the required green time. In equation form the minimum pedestrian green time is

$$PGT = 7 + \frac{w}{PWS} - YT - AR \qquad (6.13)$$

where PGT is the pedestrian green time in seconds, w is the width of the street in meters, PWS is the pedestrian-walking speed in m/s, YT is the yellow time in seconds, and AR is the all-red time in seconds.

EXAMPLE 6.12

Determine the minimum amount of pedestrian green time required for the intersection of Vine and Maple streets. (Assume pedestrian walking speed is 1.2 m/s.)

SOLUTION

As shown in Fig. 6.4, a pedestrian who crosses Maple Street will cross while Vine Street has a green interval. The minimum green time needed on Vine Street is (using Eq. 6.13) is

$$PGT = 7 + \frac{18.3}{1.2} - 3.5 - 2.0$$

$$= \underline{\underline{16.75 \text{ s}}}$$

From Example 6.11, Vine Street is assigned 20 seconds of green time, thus there is enough green time for pedestrians to cross the street. The minimum green

Table 6.2 Signal-Indication Summary for the Intersection of Vine and Maple Streets (Total Cycle Length = 70 Seconds)

Street	Phase 1	Phase 2	Phase 3
Vine Street (through)	$GT = 20$ s $YT = 3.5$ s $AR = 2.0$ s	$RT = 20.5$ s	$RT = 24$ s
Vine Street (left turn)	$RT = 25.5$ s	$GT = 15$ s $YT = 3.5$ s $AR = 2.0$ s	$RT = 24$ s
Maple Street (through)	$RT = 25.5$ s	$RT = 20.5$ s	$GT = 19$ s $YT = 3.0$ s $AR = 2.0$ s

time needed on Maple Street (using Eq. 6.13) is

$$PGT = 7 + \frac{15.4}{1.2} - 3.0 - 2.0$$

$$= \underline{\underline{14.83 \text{ s}}}$$

From Example 6.11, Maple Street is assigned 19 seconds of green time, so the minimum pedestrian green time is again met.

Prepare Signal Indication Summary

After the pedestrian green time is evaluated, the calculation of interval times for each phase is complete, and a signal indication summary can be prepared. The signal indication summary shows the phases for each movement and, at this time, the phases and cycle length can be finalized and conflict checks can be made. The signal indication summary for the Vine and Maple streets intersection is given in Table 6.2. Note that the red times (RTs) presented in this table are simply equal to the summation of green time (GT), yellow time (YT), and all-red time (AR) for the particular phase.

NOMENCLATURE FOR CHAPTER 6

a deceleration rate for vehicle at an intersection

AR all-red time

c time for a signal to complete one cycle

C	approach capacity
d	average vehicle delay per cycle assuming $D/D/1$ queuing
d'	average vehicle delay per cycle assuming random arrivals
d_d	distance from the intersection that the dilemma zone is avoided
d_m	maximum delay of any vehicle
D	deterministic arrivals or departures
D_t	total vehicle delay
g	effective green
g_r	acceleration due to gravity
G	grade
GT	green time
l	vehicle length
LT	lost time
n	number of vehicles
P_q	proportion of the signal cycle with a queue
P_s	proportion of stopped vehicles
PGT	pedestrian green time
PWS	pedestrian walking speed
Q	number of vehicles in the queue
Q_m	maximum number of vehicles in the queue
r	effective red
RT	red time
s	saturation flow
t	time
t_0	time after the start of effective green until queue dissipation
t_p	driver perception/reaction time
V	travel speed of vehicle
w	width of street
x	ratio of approach arrivals to approach departures
x_s	distance required to stop
y_i	ratio of traffic volume to saturation flow

YT yellow time

λ arrival rate

μ departure rate

ρ traffic intensity

REFERENCES

Allsop, R. E. "Delay at a Fixed Time Traffic Signal, I: Theoretical Analysis." *Transportation Science,* vol. 6, no. 3, 1972.

Kell, James H., and Iris J. Fullerton. *Manual of Traffic Signal Design.* 2d ed. Institute of Transportation Engineers, Prentice Hall, 1991.

Mannering, F., S. Abu-Eisheh, and A. Arnadottir. "Dynamic Traffic Equilibrium with Discrete/Continuous Econometric Models." *Transportation Science*, vol. 24, no. 2, 1990.

Transportation Research Board. *Traffic Flow Theory: A Monograph.* Special Report 165. Washington, DC: National Research Council, 1975.

Transportation Research Board. *Highway Capacity Manual.* Special Report 209. Washington, DC: National Research Council, 1994.

U.S. Federal Highway Administration. *Manual of Uniform Traffic Control Devices for Streets and Highways.* Washington, DC: U.S. Government Printing Office, 1988.

Vincent, R. A., A. I. Mitchell, and D. I. Robertson. "User Guide to TRANSYT Version 8." Transport and Road Research Laboratory Report 888. Crowthorne, England, 1980.

Webster, F. V. "Traffic Signal Settings." Road Research Technical Paper No. 39. London: Great Britain Road Research Laboratory, 1958.

PROBLEMS

6.1. An intersection approach has a saturation flow of 1500 veh/h and vehicles arrive at the approach at the rate of 800 veh/h. The approach is controlled by a pretimed signal with a cycle length of 60 seconds and $D/D/1$ queuing holds. Local standards dictate that signals should be set so that all approach queues dissipate 10 seconds before the end of the effective green portion of the cycle. Assuming that approach capacity exceeds arrivals, determine the maximum length of effective red that will satisfy the local standards.

6.2. An approach to a pretimed signal has 30 seconds of effective red and $D/D/1$ queuing holds. The total delay at the approach is 83.33 veh-s/cycle and the saturation flow is 1000 veh/h. If the capacity of the approach equals the number of arrivals per cycle, determine the approach flow and cycle length.

6.3. An approach to a pretimed signal has 25 seconds of effective green in a 60-second cycle. The approach volume is 500 veh/h and the saturation flow is 1400 veh/h. Calculate the average vehicle delay using $D/D/1$ queuing, Webster's formula, and Allsop's formula.

6.4. An approach to a pretimed signal has a maximum of 8 vehicles in a queue in a given cycle. If the saturation flow is 1440 veh/h and the effective red time is 40 seconds, how much time will it take this queue to dissipate after the start of the effective green? (Assume that approach capacity exceeds arrivals and $D/D/1$ queuing applies.)

6.5. Recent computations at an approach to a pretimed signalized intersection indicate that the average delay per vehicle is 16.6 seconds (as calculated by Allsop's equation). The volume-to-capacity ratio (i.e., $\lambda c/\mu g$) is 0.8, the saturation flow is 1600 veh/h, and the effective green time is 50 seconds. If the uniform delay (assuming $D/D/1$ queuing) is 11.25 seconds per vehicle, determine the arrival flow (in veh/h) and the cycle length.

6.6. An approach to a pretimed signal has 25 seconds of effective green, a saturation flow of 1300 veh/h, and a volume-to-capacity ratio less than 1. If the cycle length is 60 seconds and Allsop's delay formula estimates a delay that is 34 s/veh higher than that estimated by using the $D/D/1$ delay formula, determine the vehicle arrival rate.

6.7. An approach to a pretimed signal with a 60-second cycle has 8.9 vehicles in the queue at the beginning of the effective green. Four of the 8.9 vehicles in the queue are left over from the previous cycle (i.e., at the end of the previous cycle's effective green). The saturation flow of the approach is 1500 veh/h, total delay for the cycle is 5.78 vehicle-minutes, and at the end of the effective green there are 2 vehicles left in the queue. Determine the arrival rate assuming that it is unchanged over the duration of the observation period (i.e., from the beginning to the end of this 5.78 vehicle-minute delay cycle). (Assume $D/D/1$ queuing.)

6.8. At the beginning of an effective red, vehicles are arriving at an approach at the rate of 500 veh/h, and 16 vehicles are left in the queue from the previous cycle (i.e., at the end of the previous cycle's effective green). However, due to the end of a major sporting event, the arrival rate is continuously increasing at a constant rate of 200 veh/h/min (i.e., after 1 minute the arrival rate will be 700 veh/h; after 2 minutes, 900 veh/h, etc.). The saturation flow of the approach is 1800 veh/h, the cycle length is 60 seconds, and the effective green time is 40 seconds. Determine the total vehicle delay until complete queue dissipation. (Assume $D/D/1$ queuing.)

6.9. The saturation flow for an intersection approach is 3600 veh/h. At the beginning of a cycle (effective red) no vehicles are queued. The signal is timed so that when the queue (from the continuously arriving vehicles) is

13 vehicles long the effective green begins. If the queue dissipates 8 seconds before the end of the cycle and the cycle length is 60 seconds, what is the arrival rate assuming $D/D/1$ queuing?

6.10. The saturation flow for a pretimed signalized intersection approach is 1800 veh/h. The cycle length is 80 seconds. Its is known that the arrival rate during the effective green is twice the arrival rate during the effective red. During one cycle, there are 2 vehicles in the queue at the beginning of the cycle (the beginning of the effective red) and 7.9 vehicles in the queue at the end of the effective red (i.e., the beginning of the effective green). If the queue clears exactly at the end of the effective green, and $D/D/1$ queuing applies, determine the total vehicle delay in the cycle (in veh-s).

6.11. Vehicles arrive at an approach to a pretimed signalized intersection. The arrival rate over the cycle is given by the function $\lambda(t) = 0.22 + 0.012t$ ($\lambda(t)$ is in veh/s and t is in seconds). There are no vehicles in the queue when the cycle (effective red) begins. The cycle length is 60 seconds and the saturation flow is 3600 veh/h. Determine the effective green and red times that will allow the queue to clear exactly at the end of the cycle (i.e., the end of the effective green), and determine the total vehicle delay over the cycle (assuming $D/D/1$ queuing).

6.12. At the start of the effective red at an intersection approach with a pretimed signal, vehicles begin to arrive at a rate of 800 veh/h for the first 40 seconds and then 500 veh/h from then on. The approach has a saturation flow of 1200 veh/h, an effective green of 20 seconds, and the cycle length is 40 seconds. What is the total vehicle delay two full cycles after the 800 veh/h arrival rate begins? (Assume $D/D/1$ queuing.)

6.13. Consider the intersection of Vine and Maple streets as shown in Fig. 6.4. Suppose Vine Street's northbound and southbound approaches are both on an 8% upgrade. How would the pretimed signal indication summary (Table 6.2) change?

6.14. A new shopping center opens near the intersection of Vine and Maple streets (i.e., the intersection shown in Fig. 6.4). The effect is to increase Vine Street northbound left turns (during the peak hour) from 160 to 200 for passenger cars and from 20 to 25 for heavy vehicles. All other traffic volumes remain the same. What would the revised green times, yellow times, and all-red times be for each phase of the pretimed signal?

6.15. Two additional 3.1-m lanes are added to Vine Street (i.e., the street in the intersection shown in Fig. 6.4), one lane in each direction. If the peak-hour traffic volumes are unchanged, what would the revised green times, yellow times, and all-red times be for each phase of the pretimed signal?

6.16. A northbound approach of a north-south highway crosses an east-west highway (at the same elevation) at right angles. The intersection of the highways is controlled by a pretimed signal, and the northbound traffic is allocated 20 seconds of green time, 3 seconds of yellow time, and 2 seconds of all-red time. The approach speed is 45 km/h and the intersection is near a retirement community. How wide can the east-west highway be with this signal timing?

6.17. A pretimed four-phase signal has critical-lane volumes for the first three phases of 200, 187, and 210 (saturation flows are 1800 veh/h per lane for all phases). The lost time is known to be 4 seconds for each phase. If the cycle length is 60 seconds, what is the estimated green time of the fourth phase?

Chapter 7

Level of Service Analysis

7.1 INTRODUCTION

The underlying objective of level of service analysis is to quantify a roadway's performance with regard to specified traffic volumes (i.e., its ability to efficiently handle a specified volume of traffic). This performance can be measured in terms of travel delay (as the roadway becomes increasingly congested) as well as other factors. The comparative performance of various roadway segments (which is determined from an analysis of traffic) is important because it can be used as a basis to allocate scarce roadway construction and improvement funds. Although the material presented in Chapter 5 covered the basic elements of traffic analysis, a number of important practical issues must be addressed before this material can be applied to traffic analysis in a meaningful way. In particular, a number of terms used in Chapter 5 must be more carefully examined to allow a field analysis of traffic that will ultimately lead to an assessment of the traffic-related performance of a roadway (i.e., the level of traffic congestion on the facility). Capacity and flow (typically given in units of vehicles per hour) are two such terms. In Chapter 5, capacity, q_m, is simply defined as the highest traffic flow that a roadway is capable of supporting. For level of service analysis, a consistent and reasonably precise method of determining capacity must be developed within this definition. Because it can readily be shown that the capacity of a roadway section is a function of factors such as roadway type (e.g., freeway, multilane highway without full access control, or rural road), free-flow speed, number of lanes, and widths of lanes and shoulders, the method of capacity determination clearly must account for a wide variety of physical and operational roadway characteristics.

With regard to traffic flow, recall that Chapter 5 defines traffic flow with units of *vehicles* per hour. Two practical issues arise when using this unit of measure. First, in many cases vehicular traffic consists of a variety of vehicle types with substantially different performance characteristics. These performance differentials are likely to be magnified by changing roadway geometrics, such as upgrades or downgrades, which have a differential effect on the acceleration and deceleration capabilities of the various types of vehicles (e.g., grades have a larger impact on the performance of large trucks relative to automobiles). As a result, traffic must not only be defined in terms of vehicles per unit time but also in terms of vehicle composition, because it is clear that a 1500-veh/h traffic flow, consisting of 100% automobiles, will differ sigificantly with regard to operating speed and traffic density when compared to a 1500-veh/h traffic flow that consists of 50% automobiles and 50% heavy trucks.

The other flow-related concern is the temporal distribution of traffic. In practice, the analysis of roadway traffic usually focuses on the most critical condition, which is the most congested hour within a 24-hour daily period (the temporal distribution of traffic will be discussed in more detail in Section 7.7). However, within this most congested *peak hour,* traffic flow is likely to be nonuniform (as illustrated in Fig. 7.1). It is therefore necessary to arrive at some method of defining and measuring the nonuniformity of flow within the peak hour.

To summarize, the objective of level of service analysis is to provide a practical method of quantifying the degree of traffic congestion and being able to relate this to the overall traffic-related performance of the roadway. The following sections of this chapter discuss and demonstrate accepted standards for level of service analysis.

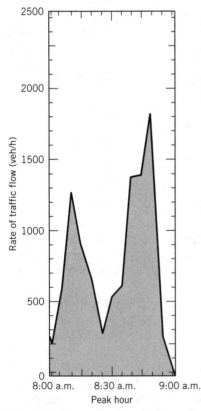

Figure 7.1 Example of nonuniform flow over a peak hour.

7.2   LEVEL OF SERVICE

To begin to quantify a roadway's degree of congestion, a qualitative measure describing traffic operational conditions and their perception by drivers is first needed. Such a measure is referred to as *level of service* and is intended to capture factors such as speed and travel time, freedom to maneuver, and safety. Current practice designates six levels of service ranging from A to F, with level of service A representing the best operating conditions and level of service F the worst. The *Highway Capacity Manual* (Transportation Research Board 1994) defines level of service (LOS) for freeways (divided highways with two or more lanes in each direction and full access control) as follows:

Level of Service A LOS A represents free-flow conditions (i.e., traffic operates at free-flow speeds as defined in Chapter 5). Individual users are virtually unaffected by the presence of others in the traffic stream. Freedom to select desired speeds and to maneuver within the traffic stream is extremely high. The general level of comfort and convenience provided to drivers is excellent.

Level of Service B LOS B also allows speeds at or near free-flow speeds, but the presence of other users in the traffic stream begins to be noticeable. Freedom to select desired speeds is relatively unaffected, but there is a slight decline in the freedom to maneuver within the traffic stream relative to LOS A.

Level of Service C LOS C has speeds at or near free-flow speeds, but the freedom to maneuver is noticeably restricted (e.g., lane changes require careful attention on the part of drivers). The general level of comfort and convenience declines significantly at this level. Disruptions in the traffic stream, such as an incident (e.g., vehicular accident or disablement), can result in significant queue formation and vehicular delay. In contrast, the effects of incidents at LOS A or LOS B are minimal, and cause only minor delay in the immediate vicinity of the event.

Level of Service D LOS D represents the conditions where speeds begin to decline slightly with increasing flow. The freedom to maneuver becomes more restricted and drivers experience reductions in physical and psychological comfort. Incidents can generate lengthy queues because the higher density associated with this LOS provides little space to absorb disruptions in the traffic flow.

Level of Service E LOS E represents operating conditions at or near the roadway's capacity. Even minor disruptions to the traffic stream, such as vehicles entering from a ramp or changing lanes, can cause delays as other vehicles give way to allow such maneuvers. In general, maneuverability is extremely limited and drivers experience considerable physical and psychological discomfort.

Level of Service F LOS F describes a breakdown in vehicular flow. Queues form quickly behind points in the roadway where the arrival flow rate temporarily exceeds the departure rate, as determined by the roadway's capacity (see Chapter 5). Such points occur at minor incidents and on- and off-ramps where incoming traffic results in capacity being exceeded. Vehicles often proceed at reasonable speeds and then are required to stop in a cyclic fashion. The cyclic formation and dissipation of queues is a key characterization of LOS F.

LOS A

LOS D

LOS B

LOS E

LOS C

LOS F

Figure 7.2 Illustration of freeway level of service (A to F). (Reproduced by permission from Transportation Research Board, *Highway Capacity Manual,* Special Report 209, National Research Council, Washington, DC, 1994).

A visual perspective of level of service for freeways is provided in Fig. 7.2. In dealing with level of service it is important to remember that roadway capacity (which will be shown as a function of the prevailing traffic and physical characteristics of the roadway) will always be reached when the roadway is operating at LOS E. This, however, is not a desirable condition because LOS E causes considerable driver discomfort that could increase the likelihood of vehicular accidents and overall delay. In roadway design, the possibility of a degradation in level of service to LOS E should be avoided, although this is not always possible due to financial and environmental constraints that may limit the design speed, number of lanes, and other factors that affect roadway capacity.

7.3 BASIC DEFINITIONS

In determining the level of service of a roadway segment, a few key definitions and associated notations must be well understood.

Hourly Volume Hourly volume is the actual traffic volume on a roadway in vehicles per hour, given the symbol V. Generally, the highest volume in a 24-hour period (i.e., the peak-hour volume) is used for V in traffic analysis computations.

Peak-Hour Factor The peak-hour factor accounts for the nonuniformity of traffic flow over the peak hour (as shown in Fig. 7.1). It is denoted PHF and is defined as the ratio of the hourly volume (V) to the maximum 15-min rate of flow (V_{15}) expanded to an hourly volume. Therefore,

$$\text{PHF} = \frac{V}{V_{15} \times 4} \tag{7.1}$$

Equation 7.1 indicates that the further the PHF is from unity, the more *peaked* or nonuniform the traffic flow during the hour. For example, consider two roads both of which have a peak-hour volume, V, of 2000 veh/h. The first road has 1000 vehicles arriving in the highest 15-min interval, and the second road has 600 vehicles arriving in the highest 15-min interval. The first road has a more nonuniform flow, and this is substantiated by the fact that its PHF of 0.5 [i.e., $2000/(1000 \times 4)$] is further from unity than the second road's PHF of 0.83 [i.e., $2000/(600 \times 4)$].

Service Flow Service flow is the actual rate of flow for the peak 15-min period expanded to an hourly volume and expressed in vehicles per hour. Service flow is denoted SF and is defined as

$$\text{SF} = \frac{V}{\text{PHF}} \tag{7.2}$$

or

$$\text{SF} = V_{15} \times 4 \tag{7.3}$$

Returning to the previous peak-hour factor example, note that the road with PHF = 0.5 has a service flow of 4000 veh/h and the road with PHF = 0.83 has a service flow of 2400 veh/h. In determining level of service, service flow (not peak-hour volume) will be used. Thus the road with a service flow of 4000 veh/h will have a considerably worse level of service even though both roads have the same hourly volume, V.

These definitions apply to all basic roadway types: freeways, multilane highways, and two-lane highways (one lane in each direction). However, there are a number of additional terms that must be introduced before a roadway level of service analysis can be undertaken. These additional terms are best defined within specific roadway types as presented in the following sections.

7.4   BASIC FREEWAY SEGMENTS

A basic freeway segment is defined as a section of a divided roadway having two or more lanes in one direction, full access control, and traffic that is unaffected by merging or diverging movements near ramps or lane additions or lane deletions. It is important to note that capacity analysis for divided roadways focuses on the traffic in one direction only. This is a logical approach because the concern is to measure the highest level of congestion and, due to directional imbalance of traffic flows (i.e., typically during morning rush hours the high volumes are going toward the central city and during evening rush hours the high volumes are going away from the central city), consideration of traffic volumes in both directions is likely to seriously understate the true level of traffic congestion.

Determination of a roadway's level of service begins by specifying *ideal* roadway conditions. Recall that in the introduction of this chapter the effect of vehicle performance and roadway design characteristics on traffic flow was discussed qualitatively. In practice, the effect of such factors on traffic flow is measured quantitatively, relative to traffic and roadway design conditions that are considered ideal. For freeways, ideal conditions can be categorized as those relating to lane widths and/or lateral clearances, the effects of heavy vehicles (such as large trucks and buses), and driver population characteristics. Studies have shown that the ideal lane width is 12 ft (3.6 m) and objects (e.g., telephone poles or retaining walls) should be no closer than 6 ft (1.8 m) from the edge of the traveled pavement (at the roadside or median). Also, under ideal conditions there should be passenger cars only in the traffic stream with no heavy vehicles such as buses or large trucks, and the driver population should be weekday drivers or commuters (i.e., regular users) who, due to their presumed familiarity with traffic and roadway conditions, will behave so as to enhance the efficient flow of traffic.

With the concept of ideal conditions established, the term *maximum service flow*, MSF_i, can be defined for a given level of service i as the highest service flow that can be achieved while maintaining the specified level of service i, assuming ideal roadway conditions. Because ideal conditions specify the presence of passenger cars only, and because it is desirable to have the maximum service flow rate independent of the number of lanes, MSF_i is in units of passenger cars per hour per lane (pcphpl). Accepted level of service criteria for a given maximum service flow rate have been found to be a function of the freeway's free-flow speed. The free-flow speed is a term that was first introduced in Chapter 5 as the speed of traffic as the traffic density approaches zero. In practice, free-flow speed is determined by the design speed of the roadway (i.e., the design speed of the horizontal and vertical curves as discussed in Chapter 3), the frequency of on-ramps and off-ramps and number of vehicles entering and exiting the traffic stream, the general density of the surrounding development, the complexity of the driving environment (e.g., possible distractions from roadway signs and so on), and speed limits. For freeways, free-flow speeds are determined directly from the field by measuring the mean speed of passenger cars when flow rates are 1300 pcphpl or less. Table 7.1 provides the level of service criteria corresponding to maximum service flows, traffic densities, and speeds. Also note that each level of service has a maximum volume-to-capacity ratio that corresponds to the maximum service flow rate. Within this context, one of the basic relationships underlying Table 7.1 can be expressed as

$$MSF_i = c_j \times (v/c)_i \tag{7.4}$$

where MSF_i is the maximum service flow rate per lane for level of service i under ideal conditions in pcphpl, $(v/c)_i$ is the maximum volume-to-capacity ratio associated with level of service i for a specified number of freeway lanes (see Table 7.1), and c_j is the per-lane capacity under ideal conditions for a freeway with a specified number of lanes j. The per-lane capacity c_j has been determined to be 2200 pcphpl for four-lane freeways (two lanes in each direction) and 2300 pcphpl for freeways with six or more lanes. Note that the value of c_j equals the maximum service flow rate at LOS E in Table 7.1 because the maximum volume-to-capacity ratio at LOS E is equal to one [i.e., $(v/c)_E = 1$].

A graphical representation of Table 7.1 is provided in Fig. 7.3.

7.4.1 Service Flow Rates and Level of Service

The concept of a maximum service flow provides an important benchmark for determining a roadway's level of service, but, because ideal conditions are seldom realized in practice, a method of converting the maximum service flow rate into an equivalent service flow rate (which accounts for actual prevailing conditions) is needed. Once this is achieved, the highest service flow rate at prevailing conditions for a given level of service (SF_i) can be related to the service flow rate obtained from actual vehicle counts (i.e., SF in Eqs. 7.2 and 7.3) to determine the roadway's level of service (as will soon be demonstrated by example). In calculating service flow rates under prevailing conditions, correction factors are used along with the number of lanes (in each direction) such that

$$SF_i = MSF_i \times N \times f_w \times f_{HV} \times f_p \tag{7.5}$$

Table 7.1 Level of Service Criteria for Freeways

Level of Service	Maximum Density (pc/mi/ln)	Minimum Speed (mph)	Maximum Service Flow Rate (pcphpl)	Maximum v/c Ratio
		Free-Flow Speed = 70 mph		
A	10.0	70.0	700	0.318/0.304
B	16.0	70.0	1,120	0.509/0.487
C	24.0	68.5	1,644	0.747/0.715
D	32.0	63.0	2,015	0.916/0.876
E	36.7/39.7	60.0/58.0	2,200/2,300	1.000
F	var	var	var	var
		Free-Flow Speed = 65 mph		
A	10.0	65.0	650	0.295/0.283
B	16.0	65.0	1,040	0.473/0.452
C	24.0	64.5	1,548	0.704/0.673
D	32.0	61.0	1,952	0.887/0.849
E	39.3/43.4	56.0/53.0	2,200/2,300	1.000
F	var	var	var	var
		Free-Flow Speed = 60 mph		
A	10.0	60.0	600	0.272/0.261
B	16.0	60.0	960	0.436/0.417
C	24.0	60.0	1,440	0.655/0.626
D	32.0	57.0	1,824	0.829/0.793
E	41.5/46.0	53.0/50.0	2,200/2,300	1.000
F	var	var	var	var
		Free-Flow Speed = 55 mph		
A	10.0	55.0	550	0.250/0.239
B	16.0	55.0	880	0.400/0.383
C	24.0	55.0	1,320	0.600/0.574
D	32.0	54.8	1,760	0.800/0.765
E	44.0/47.9	50.0/48.0	2,200/2,300	1.000
F	var	var	var	var

Note: In table entries with split values, the first value is for four-lane freeways, and the second is for six- and eight-lane freeways.

Source: Transportation Research Board, *Highway Capacity Manual,* Special Report 209, National Research Council, Washington, DC, 1994.

where SF_i is the service flow rate (in veh/h) for level of service i under prevailing conditions for N lanes (in one direction) in vehicles per hour, f_w is a factor to adjust for the effects of less than ideal lane widths and/or lateral clearances (distances from the roadway edge to objects on the side of the roadway), f_{HV} is a factor to adjust for the effect of vehicles other than passenger cars in the traffic stream (i.e., heavy vehicles such as large trucks, buses, and recreational vehicles), and f_p is a factor to adjust for the effect of nonideal driver populations (e.g.,

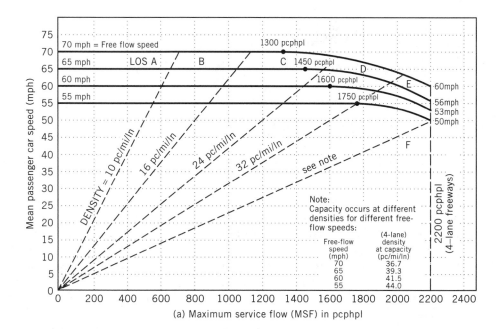

(a) Maximum service flow (MSF) in pcphpl

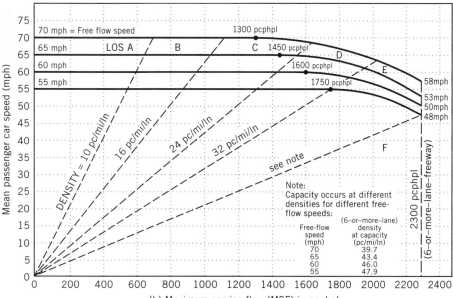

(b) Maximum service flow (MSF) in pcphpl

Figure 7.3 Speed-flow curves and level of service criteria: (a) four-lane freeways, (b) six-or-more-lane freeways. (Reproduced by permission from Transportation Research Board, *Highway Capacity Manual,* Special Report 209, National Research Council, Washington, DC, 1994).

drivers who are not regular users). The combination of Eqs. 7.4 and 7.5 provides another equation that will prove useful in forthcoming level of service computations:

$$\text{SF}_i = c_j \times (v/c)_i \times N \times f_w \times f_{HV} \times f_p \tag{7.6}$$

Eqs. 7.4, 7.5, and 7.6 form the basis for freeway level of service analysis.

7.4.2 Lane Width and/or Lateral Clearance Adjustment

When lane widths are narrower than the ideal 12 ft (3.6 m) and/or obstructions (e.g., retaining walls or utility poles) are closer than 6 ft (1.8 m) from the traveled pavement (at the roadside or at the median), the adjustment factor f_w is used to reflect the impact on level of service. Such an adjustment is needed because narrow lanes and obstructions close to the traveled lane cause traffic to slow as a result of reduced psychological comfort and limits on driver maneuvering and accident avoidance options. This, in turn, leads to an effective reduction in roadway capacity relative to the capacity that would be available if ideal lane widths and lateral clearances were provided.

The adjustment factors used in current practice are presented in Table 7.2. Although the definition of lane width is unambiguous, some elaboration of what is meant by an obstruction is needed. An obstruction is a right-side (road-side) or left-side (median-side) object that can either be continuous (e.g., a retaining wall or barrier) or periodic (e.g., light posts or utility poles). Table 7.2 provides corrections for obstructions on one side of the roadway (either median or road-side) and for both sides (both median and roadside). For the case where obstructions are on both sides of the roadway and distances from the traveled pavement edge to objects are unequal (e.g., 2 ft [0.6 m] to right-side obstructions and 4 ft

Table 7.2 Adjustment Factor for Restricted Lane Width and Lateral Clearance (for Freeways)

Distance from Traveled Way to Obstruction[a] (ft)	Adjustment Factor					
	Obstructions on One Side			Obstructions on Two Sides		
	Lane Width[a] (ft)					
	≥12	11	10	≥12	11	10
≥6	1.00	0.95	0.90	1.00	0.95	0.90
4	0.99	0.94	0.89	0.98	0.93	0.88
2	0.97	0.92	0.88	0.95	0.90	0.86
0	0.92	0.88	0.84	0.86	0.82	0.78

[a] Interpolation may be used for lane width or distance from traveled way to obstruction.

Source: Transportation Research Board, *Highway Capacity Manual,* Special Report 209, National Research Council, Washington, DC, 1994.

[1.2 m] to left-side obstructions), the average distance is used to arrive at the f_w. In this case, the values in Table 7.2 would have to be interpolated.

As an example, suppose we have a freeway with 11-ft (3.4-m) lanes with the unequal obstructions (on both sides of the roadway) at 2 ft (0.6 m) and 4 ft (1.2 m) as previously discussed. With an average of 3-ft (0.9-m) obstructions on both sides, f_w is 0.915 $[(0.93 + 0.90)/2]$ for 11-ft (3.4-m) lanes. This implies that 8.5% of the capacity is lost due to nonideal lane widths and lateral clearances.

7.4.3 Heavy Vehicle Adjustment

Large trucks, buses, and recreational vehicles have performance characteristics (slow acceleration and inferior braking) and dimensions (length, height, and width) that have an adverse effect on roadway capacity. Recall that ideal conditions stipulate that no heavy vehicles are present in the traffic stream, and when prevailing conditions indicate that presence of such vehicles, the adjustment factor f_{HV} is used to translate ideal to prevailing conditions. The f_{HV} correction term is found using a two-step process. The first step is to determine the passenger-car equivalent (pce) for each large truck, bus, and/or recreational vehicle in the traffic stream. These values represent the number of passenger cars that would consume the same amount of roadway capacity as a single large truck, bus, or recreational vehicle. These passenger-car equivalents are denoted E_T for large trucks and buses, and E_R for recreational vehicles. They are a function of roadway grades because steep grades will tend to magnify the poor performance of heavy vehicles as well as the sight distance problems caused by their larger dimensions (i.e., the visibility afforded to drivers in vehicles following heavy vehicles). For roadway segments where no single section has a grade of more than 3% for more than 0.25 mi (0.4 km), or is longer than 0.5 mi (0.8 km) if the grade is less than 3%, the passenger car equivalency factors can be obtained from Table 7.3 with terrain types defined as follows (Transportation Research Board 1994):

> *Level Terrain* Any combination of grades and horizontal and vertical alignments permitting heavy vehicles to maintain approximately the same speed as passenger cars. This generally includes short grades of no more than 2%.

**TABLE 7.3 Passenger Car Equivalents on
Extended Roadway Sections (for Freeways and
Multilane Highways)**

	Type of Terrain		
Category	Level	Rolling	Mountainous
E_T for trucks and buses	1.5	3.0	6.0
E_R for recreational vehicles	1.2	2.0	4.0

Source: Transportation Research Board, *Highway Capacity Manual*, Special Report 209, National Research Council, Washington, DC, 1994.

Table 7.4 Passenger Car Equivalents for Trucks and Buses on Specific Upgrades (for Freeways and Multilane Highways)

Grade (%)	Length (mi)	E_T Percent Trucks and Buses								
		2	4	5	6	8	10	15	20	25
<2	All	1.5	1.5	1.5	1.5	1.5	1.5	1.5	1.5	1.5
2	0–¼	1.5	1.5	1.5	1.5	1.5	1.5	1.5	1.5	1.5
	¼–½	1.5	1.5	1.5	1.5	1.5	1.5	1.5	1.5	1.5
	½–¾	1.5	1.5	1.5	1.5	1.5	1.5	1.5	1.5	1.5
	¾–1	2.5	2.0	2.0	2.0	1.5	1.5	1.5	1.5	1.5
	1–1½	4.0	3.0	3.0	3.0	2.5	2.5	2.0	2.0	2.0
	>1½	4.5	3.5	3.0	3.0	2.5	2.5	2.0	2.0	2.0
3	0–¼	1.5	1.5	1.5	1.5	1.5	1.5	1.5	1.5	1.5
	¼–½	3.0	2.5	2.5	2.0	2.0	2.0	2.0	1.5	1.5
	½–¾	6.0	4.0	4.0	3.5	3.5	3.0	2.5	2.5	2.0
	¾–1	7.5	5.5	5.0	4.5	4.0	4.0	3.5	3.0	3.0
	1–1½	8.0	6.0	5.5	5.0	4.5	4.0	4.0	3.5	3.0
	>1½	8.5	6.0	5.5	5.0	4.5	4.5	4.0	3.5	3.0
4	0–¼	1.5	1.5	1.5	1.5	1.5	1.5	1.5	1.5	1.5
	¼–½	5.5	4.0	4.0	3.5	3.0	3.0	3.0	2.5	2.5
	½–¾	9.5	7.0	6.5	6.0	5.5	5.0	4.5	4.0	3.5
	¾–1	10.5	8.0	7.0	6.5	6.0	5.5	5.0	4.5	4.0
	>1	11.0	8.0	7.5	7.0	6.0	6.0	5.0	5.0	4.5
5	0–¼	2.0	2.0	1.5	1.5	1.5	1.5	1.5	1.5	1.5
	¼–⅓	6.0	4.5	4.0	4.0	3.5	3.0	3.0	2.5	2.0
	⅓–½	9.0	7.0	6.0	6.0	5.5	5.0	4.5	4.0	3.5
	½–¾	12.5	9.0	8.5	8.0	7.0	7.0	6.0	6.0	5.0
	¾–1	13.0	9.5	9.0	8.0	7.5	7.0	6.5	6.0	5.5
	>1	13.0	9.5	9.0	8.0	7.5	7.0	6.5	6.0	5.5
6	0–¼	4.5	3.5	3.0	3.0	3.0	2.5	2.5	2.0	2.0
	¼–⅓	9.0	6.5	6.0	6.0	5.0	5.0	4.0	3.5	3.0
	⅓–½	12.5	9.5	8.5	8.0	7.0	6.5	6.0	6.0	5.5
	½–¾	15.0	11.0	10.0	9.5	9.0	8.0	8.0	7.5	6.5
	¾	15.0	11.0	10.0	9.5	9.0	8.5	8.0	7.5	6.5
	>1	15.0	11.0	10.0	9.5	9.0	8.5	8.0	7.5	6.5

Note: If the length of grade falls on a boundary, apply the longer category; interpolation may be used to find equivalents for intermediate percent grades.

Source: Transportation Research Board, *Highway Capacity Manual,* Special Report 209, National Research Council, Washington, DC, 1994.

Rolling Terrain Any combination of grades and horizontal and vertical alignment that causes heavy vehicles to reduce their speed substantially below those of passenger cars, but does not cause heavy vehicles to operate at their limiting speed on the given grade for any significant length of time (i.e., not having $F_{net}(V) = 0$ due to high grade resistance as illustrated in Fig. 2.6).

Mountainous Terrain Any combination of grades and horizontal and vertical alignments that causes heavy vehicles to operate at their limiting speed on the given grade for significant distances or at frequent intervals.

If a roadway has grades greater than 3% that are longer than 0.25 mi (0.4 km) or has grades less than 3% but longer than 0.5 mi (0.8 km), the values in Table 7.3 are no longer valid. In these cases, more detailed tables are used. Tables 7.4 and 7.5 are the tables used for positive grades (upgrades). These tables assume typical large trucks (with average weight-to-horsepower between 125 and 150 lb/hp [746 and 895 N/kW]) and recreational vehicles (with average weight-to-horsepower ratios between 30 and 60 lb/hp [179 and 358 N/kW]). Note that the equivalency factors presented in these tables increase with increasing grade and length of grade, but decrease with increasing heavy vehicle percentages. This decrease with increasing percentages is because heavy vehicles tend to group together as their percentages increase on steep, extended grades, thus decreasing their adverse impact on the traffic stream.

When two or more grades are present, a distance-weighted average may be used if all grades are less than 4% and the total combined length of the grades is less than 4000 ft (1220 m). For example, a 2% upgrade for 1000 ft (305 m) followed immediately by a 3% upgrade for 2000 ft (610 m) would use the equivalency factor for a 2.67% upgrade $[(2 \times 1000 + 3 \times 2000)/3000]$ for 3000 ft (914 m) or 0.568 mi. For additional information on combining two grades within

Table 7.5 Passenger Car Equivalents for Recreational Vehicles on Specific Upgrades (for Freeways and Multilane Highways)

Grade (%)	Length (mi)	E_R Percent Recreational Vehicles								
		2	4	5	6	8	10	15	20	25
≤2	All	1.2	1.2	1.2	1.2	1.2	1.2	1.2	1.2	1.2
3	0–½	1.2	1.2	1.2	1.2	1.2	1.2	1.2	1.2	1.2
	>½	2.0	1.5	1.5	1.5	1.5	1.5	1.2	1.2	1.2
4	0–¼	1.2	1.2	1.2	1.2	1.2	1.2	1.2	1.2	1.2
	¼–½	2.5	2.5	2.0	2.0	2.0	2.0	1.5	1.5	1.5
	>½	3.0	2.5	2.5	2.0	2.0	2.0	2.0	1.5	1.5
5	0–¼	2.5	2.0	2.0	2.0	1.5	1.5	1.5	1.5	1.5
	¼–½	4.0	3.0	3.0	3.0	2.5	2.5	2.0	2.0	2.0
	>½	4.5	3.5	3.0	3.0	3.0	2.5	2.5	2.0	2.0
6	0–¼	4.0	3.0	2.5	2.5	2.5	2.0	2.0	2.0	1.5
	¼–½	6.0	4.0	4.0	3.5	3.0	3.0	2.5	2.5	2.0
	>½	6.0	4.5	4.0	4.0	3.5	3.0	3.0	2.5	2.0

Note: If the length of grade falls on a boundary, apply the longer category; interpolation may be used to find equivalents for intermediate percent grades.

Source: Transportation Research Board, *Highway Capacity Manual,* Special Report 209, National Research Council, Washington, DC, 1994.

the same section when grades exceed 4% or the combined lengths are greater than 4000 ft (1220 m), determining the length of a grade that starts or ends on a vertical curve, and the critical part of a grade when more than one grade exists in the roadway segment (e.g., if a 4% grade were immediately followed by a 2% grade, the 4% grade would be used because the vehicle could be assumed to accelerate on the 2% portion), see the *Highway Capacity Manual* (Transportation Research Board 1994).

Negative grades (downgrades) also have an impact on equivalency factors because the comparatively poor braking characteristics of heavy vehicles have a more deleterious effect on the traffic stream on downgrades than on level terrain. Table 7.6 gives the passenger car equivalents for trucks and buses on downgrades. It is assumed that recreational vehicles are not significantly impacted by downgrades; thus downgrade values for E_R are drawn from the level terrain column in Table 7.3.

Once the appropriate equivalency factors have been obtained, the following equation is applied to arrive at the heavy vehicle correction factor f_{HV}:

$$f_{HV} = \frac{1}{1 + P_T(E_T - 1) + P_R(E_R - 1)} \tag{7.7}$$

where P's are the proportions of heavy vehicles in the traffic stream and E's are the equivalency factors from Tables 7.3, 7.4, 7.5, and/or 7.6.

As an example of how a heavy vehicle correction factor is computed, consider a freeway with a 0.75-mi- (1.2-km-) long 4% upgrade with a traffic stream having 8% trucks, 2% buses, and 2% recreational vehicles. Tables 7.4 and 7.5 must be used

Table 7.6 Passenger Car Equivalents for Trucks and Buses on Specific Downgrades (for Freeways and Multilane Highways)

Downgrade (%)	Length of Grade (mi)	Passenger Car Equivalent, E_T Percent Trucks/Buses			
		5	10	15	20
<4	All	1.5[a]	1.5[a]	1.5[a]	1.5[a]
4	≤4	1.5[a]	1.5[a]	1.5[a]	1.5[a]
4	>4	2.0	2.0	2.0	1.5[a]
5	≤4	1.5[a]	1.5[a]	1.5[a]	1.5[a]
5	>4	5.5	4.0	4.0	3.0
≥6	≤4	1.5[a]	1.5[a]	1.5[a]	1.5[a]
≥6	>4	7.5	6.0	5.5	4.5

[a] Value for level terrain.

Source: Transportation Research Board, *Highway Capacity Manual*, Special Report 209, National Research Council, Washington, DC, 1994.

Table 7.7 Adjustment Factor for Driver Population (for Freeways)

Traffic Stream Type	Adjustment Factor (f_p)
Weekday, commuter (familiar users)	1.00
Recreational or other	0.75–0.99

Source: Transportation Research Board, *Highway Capacity Manual,* Special Report 209, National Research Council, Washington, DC, 1994.

because the grade is too steep and long for Table 7.3 to apply. The corresponding equivalency factors for this roadway are $E_T = 5.5$ (applying the longer cagetory because 0.75 mi [1.2 km] is on the boundary between two categories, and using a combined truck and bus percentage of 10) and $E_R = 3.0$ as obtained from Tables 7.4 and 7.5 respectively. Also, from the given percentages of heavy vehicles in the traffic stream, $P_T = 0.1$ and $P_R = 0.02$. Substituting these values into Eq. 7.7 gives $f_{HV} = 0.67$ or a 33% reduction in effective roadway capacity relative to the ideal condition of having no heavy vehicles in the traffic stream.

7.4.4 Driver Population Adjustment

Under ideal conditions, a traffic stream is assumed to consist of regular weekday drivers and commuters. Such drivers have a high familiarity with the roadway and generally maneuver and respond to the maneuvers of other drivers in a safe and predictable fashion. There are times, however, when the traffic stream has a driver population that is less familiar with the roadway in question (e.g., weekend drivers or recreational drivers). Such drivers can cause a significant reduction in roadway capacity relative to the ideal condition of having only familiar drivers.

To account for the composition of the driver population, the f_p adjustment factor is used; its recommended values are given in Table 7.7. Note that for nonideal driver populations (i.e., "recreational or other" in Table 7.7), the loss in roadway capacity can vary from 1% to 25%. The exact value of the nonideal driver correction is dependent on local conditions such as roadway characteristics and the surrounding environment (e.g., possible driver distractions such as scenic views, and so on). When nonideal driver populations are present, judgment is necessary to determine the exact value of this term. This usually involves collection of data on local conditions (for further information see the *Highway Capacity Manual* [Transportation Research Board 1994]).

7.4.5 Freeway Traffic Analysis

With all the terms in Eqs. 7.4, 7.5, and 7.6 defined, these equations can now be applied to determine freeway level of service and freeway capacity. The manner in which this is done is best demonstrated by example.

EXAMPLE 7.1

A six-lane freeway (three lanes in each direction) is on rolling terrain with a 70 mph (113 km/h) free-flow speed, 10-ft (3-m) lanes, with obstructions 2 ft (0.6 m) from both the right and left edges of the traveled pavement. The traffic stream consists of urban commuters. A directional weekday peak-hour volume of 2200 vehicles is observed with 700 vehicles arriving in the most congested 15-min period. If the traffic stream has 15% large trucks and buses and no recreational vehicles, determine the level of service.

SOLUTION

The approach to take to determine the level of service is to compute the volume-to-capacity ratio (v/c) of the freeway and compare it with the maximum volume-to-capacity ratios for specified levels of service as given in Table 7.1. To arrive at the freeway's volume-to-capacity ratio, Eq. 7.6 is rearranged giving

$$v/c = \frac{SF}{c_j \times N \times f_w \times f_{HV} \times f_p}$$

where, from Eq. 7.3,

$$SF = V_{15} \times 4 = 700 \times 4 = 2800 \text{ veh/h}$$

and

$c_j = 2300$ pcphpl (for six-lane freeway)
$N = 3$ (given)
$f_p = 1.0$ (commuters, Table 7.7)
$f_w = 0.86$ (10-ft [3-m] lanes, obstructions 2 ft [0.6 m] on both sides, Table 7.2)
$E_T = 3.0$ (rolling terrain, Table 7.3)

From Eq. 7.7 we obtain

$$f_{HV} = \frac{1}{1 + 0.15(3 - 1)} = 0.769$$

Substituting, we find that

$$v/c = \frac{2800}{2300 \times 3 \times 0.86 \times 0.769 \times 1.0} = \underline{\underline{0.614}}$$

which gives LOS C from Table 7.1, because the maximum v/c for LOS B (with 70-mph [113-km/h] free-flow speed and a six-lane freeway) is 0.487, and the maximum v/c for LOS C is 0.715 (i.e., $0.487 < 0.614 < 0.715$).

This problem can also be solved using maximum service flow. To do so, Eq. 7.5 is applied:

$$MSF = \frac{2800}{3 \times 0.86 \times 0.769 \times 1.0} = \underline{\underline{1411.28 \text{ pcphpl}}}$$

From Table 7.1 we see that the freeway operates at LOS C because the maximum MSF for LOS B (with 70-mph [113-km/h] free-flow speed and a six-lane freeway) is 1120 pcphpl, and the maximum MSF for LOS C is 1644 pcphpl (i.e., 1120 < 1411.28 < 1644). Using MSF, this problem can also be solved graphically by applying Fig. 7.3. Using part (*a*) of this figure (four-lane freeway), we draw a vertical line up from 1411.28 pcphpl (on the figure's *x*-axis) and find that this line intersects the 70-mph (113-km/h) free-flow speed curve in the LOS C region.

EXAMPLE 7.2

Consider the freeway and traffic conditions in Example 7.1. At some point farther along the roadway there is a 5% upgrade that is 0.5 mi (0.8 km) long. All other characteristics are the same as in Example 7.1. What is the level of service of this portion of the roadway, and how many vehicles can be added before the roadway reaches capacity (assuming that the proportion of vehicle types and the peak-hour factor remain constant)?

SOLUTION

To determine the LOS of this section of the freeway, we note that all adjustment factors are the same as those in Example 7.1 except f_{HV}, which must now be determined using an equivalency factor, E_T, drawn from the specific upgrade tables (in this case Table 7.4). From Table 7.4, $E_T = 6.0$, which gives

$$f_{HV} = \frac{1}{1 + 0.15(6 - 1)} = 0.571$$

and

$$v/c = \frac{2800}{2300 \times 3 \times 0.86 \times 0.571 \times 1.0} = \underline{\underline{0.826}}$$

which gives LOS D from Table 7.1, because the maximum v/c for LOS C (with 70-mph [113-km/h] free-flow speed and a six-lane freeway) is 0.715, and the maximum v/c for LOS D is 0.876 (i.e., 0.715 < 0.826 < 0.876).

To determine how many vehicles can be added before capacity is reached, the service flow at capacity must be computed. Because roadway capacity occurs at LOS E, and the highest volume-to-capacity ratio under LOS E is 1.0 (Table 7.1, with volume equal to capacity), the service flow at capacity can be calculated from Eq. 7.6 as

$$SF_E = c_j \times (v/c)_E \times N \times f_w \times f_{HV} \times f_p$$

$$SF_E = 2300 \times 1.0 \times 3 \times 0.86 \times 0.571 \times 1.0 = 3388.3 \text{ veh/h}$$

Recall that service flow is based on the highest 15-min volume in the peak hour. To determine the number of vehicles that can be added to the entire peak hour, service flow must be converted to an equivalent hourly volume. By rearranging Eq. 7.2, we get

$$V = SF \times PHF$$

where

$$PHF = \frac{V}{V_{15} \times 4}$$

or, because initial $V = 2200$ veh/h (given) and $V_{15} = 700$ vehicles (given),

$$PHF = \frac{2200}{700 \times 4} = 0.786$$

so that

$$V = 3388.3 \times 0.786 = \underline{2663.2 \text{ veh/h}}$$

This means that about 463 vehicles (2663 − 2200) can be added to the peak hour before capacity is reached. It should be noted that the assumption that the peak-hour factor would remain constant as the roadway approaches capacity is not very realistic. In practice, it is observed that as a roadway approaches capacity, the PHF gets closer to one. This implies that the flow rate over the peak hour becomes more uniform. This uniformity is the result of, among other factors, motorists adjusting their departure and arrival times to avoid congested periods within the peak hour. Graphically this means that the "valleys" shown in Fig. 7.1 will tend to fill in.

7.5 MULTILANE RURAL AND SUBURBAN HIGHWAYS

Multilane highways in suburban and rural settings are highways that do not meet freeway standards because (1) vehicles may enter or leave the roadway at at-grade intersections and driveways and may cross the median at certain points (i.e., multilane highways do not have full access control), (2) traffic signals may be present, (3) design standards (e.g., design speeds) are typically lower than those for freeways, and (4) the visual setting and development along multilane highways is more distracting to drivers than along freeways. Multilane highways are usually four or six lanes (total, both directions), have posted speed limits between 40 mph (64.4 km/h) and 55 mph (88.5 km/h), and can have physical medians, medians that are two-way left-turn lanes (TWLTLs), or opposing directional volumes that may not be divided by a median at all. Some examples of rural and suburban multilanes are given in Fig. 7.4.

Determining level of service on rural and suburban multilane highways differs from the procedure previously discussed for freeways. As we shall soon see, while freeway and multilane analyses have some elements in common, there are a number of important differences. The procedure that we will present is valid only for sections of highway that are not significantly influenced by large queue formations and dissipations resulting from traffic signals (this is generally taken as having traffic signals spaced 2.0 mi [3.2 km] or more apart), do not have significant on-street parking, do not have bus stops with high usage, and do not have significant pedestrian activity.

Ideal conditions for multilane highways are defined as having level terrain (no grades greater than 2%), 12-ft (3.6-m) lanes, objects (e.g., utility poles or retaining

(a) Divided multilane highway in a rural environment

(b) Divided multilane highway in a suburban environment

(c) Undivided multilane highway in a rural environment

(d) Undivided multilane highway in a suburban environment

Figure 7.4 Illustration of rural and suburban multilane highways. (Reproduced by permission from Transportation Research Board, *Highway Capacity Manual,* Special Report 209, National Research Council, Washington, DC, 1994).

walls) no closer than 6 ft (1.8 m) from the edge of the traveled pavement (at the roadside or median), no direct access points along the roadway, a divided highway, passenger cars only in the traffic stream, and a free-flow speed of 60 mph (97 km/h) or more. As was the case in the freeway level of service analysis, adjustments must be made when nonideal conditions are encountered.

Level of service estimation for multilane highways is best done by using the speed-flow relationships shown in Fig. 7.5, which is a graphical representation of the values given in Table 7.8. The level of service estimation procedure will be to estimate the free-flow speed and the service flow rate, and to use the intersection of a vertical line from the flow with the corresponding free-flow curve to determine level of service. For example, a multilane highway with a free-flow speed of 55 mph (89 km/h) and a service flow rate of 800 pcphpl would be operating at LOS B. The LOS problem for multilane highways is simply one of determining the free-flow speed, which can be obtained directly from field studies or by other means, and the service flow rate, which will be a function of the traffic volume, peak-hour factor, number of lanes, and a heavy vehicle adjustment factor.

7.5.1 Free-Flow Speed Determination

As previously defined, free-flow speed is the mean speed of passenger cars under low to moderate flow rates, which is usually taken as up to 1400 passenger cars per hour per lane (pcphpl) for multilane highways, as shown graphically in

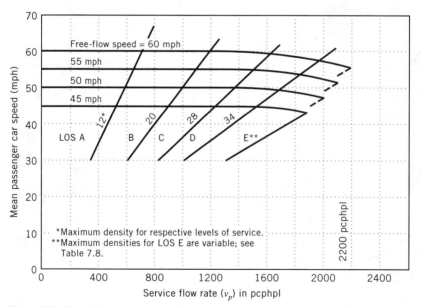

Figure 7.5 Speed-flow curves and level of service criteria for multilane highways. (Reproduced by permission from Transportation Research Board, *Highway Capacity Manual,* Special Report 209, National Research Council, Washington, DC, 1994).

Table 7.8 Level of Service Criteria for Multilane Highways

					Free-Flow Speed											
	60 mph				55 mph				50 mph				45 mph			
Level of Service	Max Density (pc/mi/ln)	Average Speed (mph)	Max v/c	Max Service Flow Rate (pcphpl)	Max Density (pc/mi/ln)	Average Speed (mph)	Max v/c	Max Service Flow Rate (pcphpl)	Max Density (pc/mi/ln)	Average Speed (mph)	Max v/c	Max Service Flow Rate (pcphpl)	Max Density (pc/mi/ln)	Average Speed (mph)	Max v/c	Max Service Flow Rate (pcphpl)
A	12	60	0.33	720	12	55	0.31	660	12	50	0.30	600	12	45	0.28	540
B	20	60	0.55	1,200	20	55	0.52	1,100	20	50	0.50	1,000	20	45	0.47	900
C	28	59	0.75	1,650	28	54	0.72	1,510	28	50	0.70	1,400	28	45	0.66	1,260
D	34	57	0.89	1,940	34	53	0.86	1,800	34	49	0.84	1,670	34	44	0.79	1,500
E	40	55	1.00	2,200	41	51	1.00	2,100	43	47	1.00	2,000	45	42	1.00	1,900

Note: The exact mathematical relationship between density and v/c has not always been maintained at LOS boundaries because of the use of rounded values. Density is the primary determinant of LOS. LOS F is characterized by highly unstable and variable traffic flow. Prediction of accurate flow rate, density, and speed at LOS F is difficult.

Source: Transportation Research Board, *Highway Capacity Manual*, Special Report 209, National Research Council, Washington, DC, 1994.

Fig. 7.5. Unlike freeways (which require direct free-flow speed measurement), multilane highway free-flow speeds can be determined in a number of ways. The direct measurement approach, which is to simply measure the mean speed of passenger cars at flow rates at or below 1400 pcphpl, is still the preferred option. If this approach is used, the resultant mean speed (the free-flow speed) accounts for possible nonideal roadway characteristics, including lane width, lateral clearance, type of median, and number of access points. However, conducting the necessary speed study may not always be possible due to financial constraints and/or other reasons. If this is the case, two other empirical approaches to determine free-flow speed for multilane highways can be used. The first is to use the 85th-percentile speed from existing speed data (this speed is often used to determine the roadway's speed limit and is the speed that has 85% of all traffic traveling slower than the speed and 15% greater than the speed). In determining this 85th-percentile speed, the speeds of heavy vehicles may be included if the terrain is level, but the speeds of passenger cars only should be used for rolling or mountainous terrain (see earlier terrain definitions in section 7.4.3). The second approach is to simply use the speed limit of the facility.

If either the 85th-percentile speed or speed limit is used to determine free-flow speed, free-flow speed corrections must be applied. To do this, we first estimate the free-flow speed for ideal roadway conditions (i.e., 12-ft [3.6-m] lanes, etc.). We denote this speed FFS_I and it is determined from the 85th-percentile and speed limits as shown in Table 7.9. Then free-flow speed for existing conditions (which may not be ideal) is determined from

$$FFS = FFS_I - F_M - F_{LW} - F_{LC} - F_A \qquad (7.8)$$

where FFS is the estimated free-flow speed in mph, FFS_I is the estimated free-flow speed in mph for ideal conditions, F_M is an adjustment for median type, F_{LW} is an adjustment for lane width, F_{LC} is an adjustment for lateral clearance, and F_A is an adjustment for the number of access points along the roadway. Note that this equation is used only when speeds are obtained from the 85th-percentile speed or speed limit. If the mean speed of passenger cars is obtained at flow levels less than 1400 pcphpl, this equation is not used because FFS is obtained directly.

Values for the adjustment factor for median type F_M are provided in Table 7.10. This table shows that undivided highways have a free-flow speed that is

Table 7.9 Determination of Free-Flow Speed with Ideal Roadway Conditions (FFS_I) for Multilane Highways

Data Basis for Free-Flow Speed	Equation for FFS_I
85th percentile of speed	FFS_I = 85th-percentile speed in mph + [3 − 0.1 × (85th-percentile speed in mph)]
Speed limit	For 40- and 45-mph speed limits: FFS_I = 7 + (speed limit in mph) For 50- and 55-mph speed limits: FFS_I = 5 + (speed limit in mph)

**Table 7.10 Adjustment for Median Type
(for Multilane Highways)**

Median Type	Reduction in Free-Flow Speed (mph)
Undivided Highways	1.6
Divided Highways (including TWLTLs)	0.0

Source: Transportation Research Board, *Highway Capacity Manual,* Special Report 209, National Research Council, Washington, DC, 1994.

1.6 mph (2.6 km/h) lower than divided highways (which include those with two-way left-turn lanes). Table 7.11 gives free-flow speed reductions resulting from lane widths that are less than the ideal 12 ft (3.6 m) (the values in this table are for the adjustment factor F_{LW}). In this table, lane widths greater than 12 ft (3.6 m) are assumed to be 12 ft (3.6 km), and no data exist for lane widths less than 10 ft (3.0 m).

The adjustment factor for possible nonideal lateral clearances (F_{LC}) is determined first by computing the total lateral clearance, which is defined as

$$TLC = LC_R + LC_L \tag{7.9}$$

where TLC is the total lateral clearance in feet, LC_R is the lateral clearance on the right side of the traveled lanes to obstructions (e.g., retaining walls, signs, trees, utility poles, and so on), and LC_L is the lateral clearance on the left side of the traveled lanes to obstructions. For undivided highways, there is no adjustment for left-side lateral clearance because this is already taken into account in the F_M term (i.e., $LC_L = 6$ ft [1.8 m] in Eq. 7.9). If an individual lateral clearance (either left or right side) exceeds 6 ft (1.8 m), 6 ft (1.8 m) is used in Eq. 7.9. Finally, highways with TWLTLs are considered to have an LC_L equal to 6 ft (1.8 m). Once Eq. 7.9 is applied, the value for F_{LC} can be determined directly from Table 7.12.

**Table 7.11 Adjustment for Lane Width
(for Multilane Highways)**

Lane Width (ft)	Reduction in Free-Flow Speed (mph)
10	6.6
11	1.9
12	0.0

Source: Transportation Research Board, *Highway Capacity Manual,* Special Report 209, National Research Council, Washington, DC, 1994.

**Table 7.12 Adjustment for Lateral Clearance
(for Multilane Highways)**

Four-Lane Highways		Six-Lane Highways	
Total Lateral Clearance[a] (ft)	Reduction in Free-Flow Speed (mph)	Total Lateral Clearance[a] (ft)	Reduction in Free-Flow Speed (mph)
12	0.0	12	0.0
10	0.4	10	0.4
8	0.9	8	0.9
6	1.3	6	1.3
4	1.8	4	1.7
2	3.6	2	2.8
0	5.4	0	3.9

[a] Total lateral clearance is the sum of the lateral clearances of the median (if greater than 6 ft, use 6 ft) and shoulder (if greater than 6 ft, use 6 ft). Therefore, for analysis purposes, total lateral clearance cannot exceed 12 ft.

Source: Transportation Research Board, *Highway Capacity Manual,* Special Report 209, National Research Council, Washington, DC, 1994.

The final adjustment factor in Eq. 7.8 is the adjustment factor for the number of access points, F_A. An access point is defined to include intersections and driveways (on the right side of the highway in the direction being considered) that significantly influence traffic flow and, as such, do not generally include driveways to individual residences or service driveways at commercial sites. For up to 40 access points per mile (25 access points per kilometer), studies show that every access point per mile reduces the free-flow speed by approximately 0.25 mph (0.4 km/h). If the number of access points per mile exceeds 40 (access points per kilometer exceeds 25), a constant 10 mph (16.1 km/h) reduction in free-flow speed is used. With *NAPM* equal to the number of access points per

**Table 7.13 Number of Access Points
for General Development Environments
(for Multilane Highways)**

Type of Development	Access Points per Mile (One Side of Roadway)
Rural	0–10
Low-Density Suburban	11–20
High-Density Suburban	21 or more

Source: Transportation Research Board, *Highway Capacity Manual,* Special Report 209, National Research Council, Washington, DC, 1994.

mile, we have

$$
\begin{aligned}
\text{if } NAPM &\le 40 \quad F_A = 0.25 \times NAPM \\
\text{if } NAPM &> 40 \quad F_A = 10
\end{aligned}
\tag{7.10}
$$

To get some idea of the typical number of access points per mile in different development environments, see Table 7.13. This provides some important background information on access point densities in rural and suburban environs.

EXAMPLE 7.3

A four-lane undivided highway has 11-ft (3.4-m) lanes, with 4-ft (1.2-m) shoulders on the right side. There are seven access points per mile (four access points per kilometer) and the 85th-percentile is 51 mph (82 km/h). What is the estimated free-flow speed?

SOLUTION

The problem is solved with a direct application of Eq. 7.8:

$$
FFS = FFS_I - F_M - F_{LW} - F_{LC} - F_A
$$

where

$FFS_I = 48.9$ mph (78.7 km/h) (from Table 7.9, $FFS_I = 51 + [3.0 - 0.1 \times 51]$)
$F_M = 1.6$ mph (2.6 km/h) (from Table 7.10, undivided highways)
$F_{LW} = 1.9$ mph (3.1 km/h) (from Table 7.11, 11 ft [3.4 m] lane width)
$F_{LC} = 0.4$ mph (0.6 km/h) (from Table 7.12, with $TLC = 4 + 6 = 10$, from
    Eq. 7.9, with $LC_L = 6$ ft because the highway is undivided)
$F_A = 1.75$ mph (2.8 km/h) (from Eq. 7.10, $F_A = 0.25 \times 7 = 1.75$)

Substitution gives

$$
FFS = 48.9 - 1.6 - 1.9 - 0.4 - 1.75 = \underline{\underline{43.25 \text{ mph } (69.6 \text{ km/h})}}
$$

which means that the nonideal conditions reduced the ideal free-flow speed by 5.65 mph (9.1 km/h).

7.5.2   Service Flow Rate Determination

The service flow rate used in Fig. 7.5 is determined by making two adjustments to the hourly traffic volume: one for the peak-hour factor and one for heavy vehicles. The appropriate equation is

$$v_p = \frac{V}{(N)(\text{PHF})(f_{HV})} \qquad (7.11)$$

where v_p is the service flow rate in passenger cars per hour per lane (pcphpl), V is the hourly volume, N is the number of lanes, PHF is the peak-hour factor as defined in Eq. 7.1, and f_{HV} is the heavy vehicle adjustment as defined in Eq. 7.7. The determination of f_{HV} for multilane highways is exactly the same as that for freeways, with Tables 7.3, 7.4, and 7.5 used to arrive at E_T's and E_R's.

Using v_p from Eq. 7.11 and *FFS* from Eq. 7.8, level of service can be readily determined. This is demonstrated by the following example.

EXAMPLE 7.4

A six-lane rural multilane divided highway is on rolling terrain with two access points per mile (one access point per kilometer) and has 10-ft (3-m) lanes, with a 5-ft (1.5-m) shoulder on the right side and a 3-ft (0.9-m) shoulder on the left side. The peak-hour factor is 0.80 and the directional peak-hour volume is 3000 veh/h. There are 6% large trucks, 2% buses, and 2% recreational vehicles. No speed studies are available, but the speed limit is 55 mph (89 km/h). Determine the level of service.

SOLUTION

We begin by determining the *FFS* by applying Eq. 7.8:

$$FFS = FFS_I - F_M - F_{LW} - F_{LC} - F_A$$

where

$\quad FFS_I = 60$ mph (97 km/h) (from Table 7.9, $FFS_I = 5 + 55$)
$\quad\quad F_M = 0.0$ mph (0 km/h) (from Table 7.10, divided highways)
$\quad\quad F_{LW} = 6.6$ mph (10.6 km/h) (from Table 7.11, 10 ft [3 m] lane width)
$\quad\quad F_{LC} = 0.9$ mph (1.4 km/h) (from Table 7.12, with $TLC = 5 + 3 = 8$,
$\quad\quad\quad\quad$ from Eq. 7.9)
$\quad\quad F_A = 0.5$ mph (0.8 km/h) (from Eq. 7.10, $F_A = 0.25 \times 2 = 0.5$)

Substitution gives

$$FFS = 60.0 - 0.0 - 6.6 - 0.9 - 0.5 = 52.0 \text{ mph (83.7 km/h)}$$

Next, we determine the service flow rate using Eq. 7.11:

$$v_p = \frac{V}{(N)(\text{PHF})(f_{HV})}$$

where

$$V = 3000 \text{ veh/h (given)}$$
$$N = 3 \text{ (given)}$$
$$\text{PHF} = 0.8 \text{ (given)}$$
$$E_T = 3.0 \text{ (rolling terrain, Table 7.3)}$$
$$E_R = 2.0 \text{ (rolling terrain, Table 7.3)}$$

From Eq. 7.8 we obtain

$$f_{HV} = \frac{1}{1 + 0.08(3 - 1) + 0.02(2 - 1)} = 0.847$$

Substitution gives

$$v_p = \frac{3000}{(3)(0.8)(0.847)} = \underline{\underline{1475.8 \text{ pcphpl}}}$$

As shown in Fig. 7.6, we draw a speed-flow curve for 52 mph (84.7 km/h) *FFS* (using the same shape of the curves for 60, 55, 50, and 45 mph as shown in Fig. 7.5) and note that the 1475.8 pcphpl service flow rate intersects this curve in LOS D. Therefore, this highway is operating at LOS D.

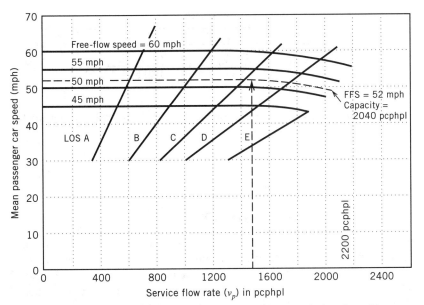

Figure 7.6 Solution to Example 7.4. (Reproduced by permission from Transportation Research Board, *Highway Capacity Manual,* Special Report 209, National Research Council, Washington, DC, 1994).

EXAMPLE 7.5

A local manufacturer wishes to open a factory near the section of highway described in Example 7.4. How many large trucks can be added to the peak-hour directional volume before capacity is reached? (Assume only trucks and buses are added and that the PHF remains constant).

SOLUTION

Note that the *FFS* will remain unchanged at 52 mph (83.7 km/h). Table 7.8 shows that the capacity with *FFS* = 55 mph (88.6 km/h) is 2100 pcphpl, and with *FFS* = 50 mph (80.5 km/h) is 2000 pcphpl, so a linear interpolation gives us a capacity of 2040 pcphpl at *FFS* = 52 mph (83.7 km/h). The current number of large trucks and buses in the peak-hour traffic stream is 240 (0.08 × 3000), and the current number of recreational vehicles is 60 (0.02 × 3000). Let us denote the number of new trucks added as V_{nt}. The combination of Eq. 7.7 and 7.11 gives

$$v_p = \frac{V + V_{nt}}{(N)(\text{PHF})\left[\dfrac{1}{1 + \left(\dfrac{240 + V_{nt}}{V + V_{nt}}\right)(E_T - 1) + \left(\dfrac{60}{V + V_{nt}}\right)(E_R - 1)}\right]}$$

With V = 3000 veh/h, E_T = 3, E_R = 2, N = 3, PHF = 0.8, and v_p = 2040 pcphpl, we have

$$2040 = \frac{3000 + V_{nt}}{(3)(0.80)\left[\dfrac{1}{1 + \left(\dfrac{240 + V_{nt}}{3000 + V_{nt}}\right)(3 - 1) + \left(\dfrac{60}{3000 + V_{nt}}\right)(2 - 1)}\right]}$$

which gives V_{nt} = 452, which is the number of trucks that can be added to the peak hour before capacity is reached.

7.6 RURAL TWO-LANE HIGHWAYS

Two-lane highways are defined as roadways with one lane available in each direction. In terms of level of service determination, a key distinction between two-lane highways and the freeways and multilane highways previously discussed is that traffic in *both* directions must now be considered (previously we considered traffic in one direction only). This is because traffic in an opposing direction has a strong influence on level of service. For example, a high opposing traffic volume limits the opportunity to pass slow-moving vehicles (because such a pass requires the passing vehicle to occupy the opposing lane) and thus forces a lower traffic speed and, as a consequence, a lower level of service. It also follows that any geometric features that restrict passing sight distance (such as sight distance on horizontal and vertical curves) will have an adverse impact on the level of service. Finally, the type of terrain (i.e., level, rolling, or mountainous) plays a more critical role in level of service calculations relative to freeways and multilane highways, because of the sometimes limited ability to pass slower-moving vehicles on grades (i.e., in areas where passing is prohibited due to sight distance restrictions or opposing traffic does not permit safe passing).

With these points in mind, consider the following ideal conditions for rural two-lane highways (from the *Highway Capacity Manual,* Transportation Research Board 1994):

1. Design speed greater than or equal to 60 mph (97 km/h) (see Chapter 3)
2. Lane widths greater than or equal to 12 ft (3.6 m)
3. Clear shoulders wider than or equal to 6 ft (1.8 m)
4. No no-passing zones on the highway segment
5. All passenger cars in the traffic stream
6. A 50/50 directional split of traffic (e.g., 50% traveling northbound and 50% traveling southbound)
7. No impediments to through traffic due to traffic control or turning vehicles
8. Level terrain (as defined in section 7.4.3)

The capacity of a roadway under such conditions is 2800 passenger cars per hour (pcph), total, both directions. This leads to the basic service flow expression for two-lane, two-way rural highways:

$$SF_i = 2800 \times (v/c)_i \times f_d \times f_w \times f_{HV} \qquad (7.12)$$

where all terms are as defined for freeways (see Eq. 7.6) with the exception of f_d, which is an additional adjustment factor for the nonideal directional distribution of traffic (i.e., having more than 50% of the total traffic volume traveling in one of the two directions). The idea behind f_d is that, as the directional distribution of traffic deviates from 50/50, the 2800 pcph (total both directions) is adjusted downward toward the extreme case in which the directional distribution is 100/0. In such a case the total capacity in both directions becomes 2000 pcph or, because all of the flow is in one direction, 2000 pcphpl. The values for the directional distribution adjustment factor are provided in Table 7.14. Values for $(v/c)_i$ and

Table 7.14 Adjustment for Directional Distribution on Two-Lane Highways

Directional Distribution	100/0	90/10	80/20	70/30	60/40	50/50
Adjustment factor, f_d	0.71	0.75	0.83	0.89	0.94	1.00

Source: Transportation Research Board, *Highway Capacity Manual,* Special Report 209, National Research Council, Washington, DC, 1994.

Table 7.15 Level of Service Criteria for Two-Lane Highways

v/c Ratio[a]

LOS	Percent Time Delay	Level Terrain Avg[b] Speed	0	20	40	60	80	100	Rolling Terrain Avg[b] Speed	0	20	40	60	80	100	Mountainous Terrain Avg[b] Speed	0	20	40	60	80	100
				Percent No-Passing Zones							Percent No-Passing Zones							Percent No-Passing Zones				
A	≤ 30	≥ 58	0.15	0.12	0.09	0.07	0.05	0.04	≥ 57	0.15	0.10	0.07	0.05	0.04	0.03	≥ 56	0.14	0.09	0.07	0.04	0.02	0.01
B	≤ 45	≥ 55	0.27	0.24	0.21	0.19	0.17	0.16	≥ 54	0.26	0.23	0.19	0.17	0.15	0.13	≥ 54	0.25	0.20	0.16	0.13	0.12	0.10
C	≤ 60	≥ 52	0.43	0.39	0.36	0.34	0.33	0.32	≥ 51	0.42	0.39	0.35	0.32	0.30	0.28	≥ 49	0.39	0.33	0.28	0.23	0.20	0.16
D	≤ 75	≥ 50	0.64	0.62	0.60	0.59	0.58	0.57	≥ 49	0.62	0.57	0.52	0.48	0.46	0.43	≥ 45	0.58	0.50	0.45	0.40	0.37	0.33
E	> 75	≥ 45	1.00	1.00	1.00	1.00	1.00	1.00	≥ 40	0.97	0.94	0.92	0.91	0.90	0.90	≥ 35	0.91	0.87	0.84	0.82	0.80	0.78
F	100	< 45	—	—	—	—	—	—	< 40	—	—	—	—	—	—	< 35	—	—	—	—	—	—

[a] Ratio of flow rate to an ideal capacity of 2800 pcph in both directions.

[b] Average travel speed of all vehicles (in mph) for highways with design speed ≥ 60 mph; for highways with lower design speeds, reduce speed by 4 mph for each 10-mph reduction in design speed below 60 mph; assumes that speed is not restricted to lower values by regulation.

Source: Transportation Research Board, *Highway Capacity Manual*, Special Report 209, National Research Council, Washington, DC, 1994.

f_w are obtained from Tables 7.15 and 7.16, respectively. Determination of the heavy vehicle adjustment factor (f_{HV}) is slightly different from that used in the freeway and multilane highway cases in that trucks and buses are now considered separately. The equation thus becomes (compare to Eq. 7.7),

$$f_{HV} = \frac{1}{1 + P_T(E_T - 1) + P_B(E_B - 1) + P_R(E_R - 1)} \tag{7.13}$$

where P_T, P_B, and P_R are the proportions of large trucks, buses, and recreational vehicles in the traffic stream, and E_T, E_B, and E_R are their corresponding equivalency factors. The equivalency factors for the general terrain types of level, rolling, and mountainous (as defined in section 7.4.3) are shown Table 7.17. For details on the procedure used to evaluate two-lane rural highways on specific grades (e.g., a 5% grade 0.75 mi [1.2 km] long), the reader is referred to the *Highway Capacity Manual* (Transportation Research Board 1994).

Two points relating to Tables 7.15, 7.16, and 7.17 are worthy of note. First, the v/c terms shown in Table 7.15 differ from those used in similar tables for freeways and multilane highways (i.e., Tables 7.1 and 7.8). The v/c terms in Table 7.15 are implicitly adjusted to include reductions in level of service resulting from the combined effects of different terrain types and different percentages of no-passing zones. This explains why the maximum v/c for LOS E is sometimes less than one (e.g., see values of LOS E for mountainous terrain). The second point relates to the adjustment factor f_w (Table 7.16) and the passenger car equivalency factors (Table 7.17). Research has found that these factors vary by level of service (which was not the case for freeways, for example). As will be shown in forthcoming examples, this dependence on level of service will complicate the traffic analysis procedure.

As a final observation, note that Eq. 7.12 does not contain an adjustment factor for regular/nonregular users as was the case in Eq. 7.6 for freeways.

Table 7.16 Adjustment for Effects of Narrow Lanes and Restricted Shoulder Widths (for Two-Lane Highways)

Usable[a] Shoulder Width (ft)	12-ft		11-ft		10-ft		9-ft	
	LOS A–D	LOS[b] E	LOS A–D	LOS[b] E	LOS A–D	LOS[b] E	LOS A–D	LOS[b] E
≥ 6	1.00	1.00	0.93	0.94	0.84	0.87	0.70	0.76
4	0.92	0.97	0.85	0.92	0.77	0.85	0.65	0.74
2	0.81	0.93	0.75	0.88	0.68	0.81	0.57	0.70
0	0.70	0.88	0.65	0.82	0.58	0.75	0.49	0.66

[a] Where shoulder width is different on each side of the roadway, use the average shoulder width.

[b] Factor applies for all speeds less than 45 mph.

Source: Transportation Research Board, *Highway Capacity Manual,* Special Report 209, National Research Council, Washington, DC, 1994.

Table 7.17 Passenger Car Equivalents for Two-Lane Highways

Vehicle Type	Level of Service	Type of Terrain		
		Level	Rolling	Mountainous
Trucks, E_T	A	2.0	4.0	7.0
	B and C	2.2	5.0	10.0
	D and E	2.0	5.0	12.0
Recreational vehicles, E_R	A	2.2	3.2	5.0
	B and C	2.5	3.9	5.2
	D and E	1.6	3.3	5.2
Buses, E_B	A	1.8	3.0	5.7
	B and C	2.0	3.4	6.0
	D and E	1.6	2.9	6.5

Source: A. Werner and J. F. Morrall, "Passenger Car Equivalencies of Trucks, Buses, and Recreational Vehicles for Two-Lane Rural Highways," *Transportation Research Record 615,* 1976.

This is because the many other complexities of two-lane highways make the composition of drivers a less significant concern (Transportation Research Board 1994). This lack of significance was also observed in the multilane highway case.

EXAMPLE 7.6

A rural two-lane highway is on level terrain with 11-ft (3.4-m) lanes, 2-ft (0.6-m) paved shoulders, and 80% no-passing zones. The directional split is 80/20 and there are 5% large trucks, 2% buses, and 5% recreational vehicles. Determine the service flow of the roadway at capacity.

SOLUTION

The roadway reaches capacity at the maximum of LOS E, which, from Table 7.15, gives a v/c ratio of 1.0, on level terrain with 80% no-passing zones. Thus the service flow at capacity can be computed using Eq. 7.12:

$$SF_E = 2800 \times (v/c)_E \times f_d \times f_w \times f_{HV}$$

where

$v/c_E = 1.0$ (Table 7.15)
$f_d = 0.83$ (80/20 directional split, Table 7.14)
$f_w = 0.88$ (11-ft [3.4-m] lanes, 2-ft [0.6-m] shoulders, LOS E, Table 7.16)
$E_T = 2.0$ (level terrain, LOS E, Table 7.17)
$E_R = 1.6$ (level terrain, LOS E, Table 7.17)
$E_B = 1.6$ (level terrain, LOS E, Table 7.17)

From Eq. 7.13, we obtain

$$f_{HV} = \frac{1}{1 + 0.05(2 - 1) + 0.02(1.6 - 1) + 0.05(1.6 - 1)} = 0.916$$

Substituting these terms into the previous equation for SF_E,

$$SF_E = 2800 \times 1.0 \times 0.83 \times 0.88 \times 0.916 = \underline{\underline{1873.33 \text{ veh/h}}}$$

EXAMPLE 7.7

Consider the conditions described in Example 7.6. If the peak-hour vehicle count is 522 with a peak-hour factor of 0.90, determine the level of service.

SOLUTION

Note that both f_w and f_{HV} are dependent on the LOS, which is not yet known. Therefore, this problem must be approached by initially assuming a LOS, then computing a LOS based on this assumption, and then making certain that the computed LOS is consistent with the initially assumed LOS. To begin, assume LOS E. Under this assumption, f_d, f_w, and f_{HV} are all as determined in Example 7.6. Also from given information, the application of Eq. 7.2 gives

$$SF = \frac{V}{PHF} = \frac{522}{0.90} = 580 \text{ veh/h}$$

To compute LOS (using v/c), Eq. 7.12 is rearranged:

$$v/c = \frac{SF}{2800 \times f_d \times f_w \times f_{HV}}$$

Substituting, we obtain

$$v/c = \frac{580}{2800 \times 0.83 \times 0.88 \times 0.916} = \underline{\underline{0.31}}$$

From Table 7.15 (level terrain, 80% no-passing zones) LOS C is obtained (i.e., $0.17 < 0.31 < 0.33$), which is inconsistent with the earlier assumed LOS E. This means that the above computations must be reworked. If LOS D is assumed, only f_w will change from the assumed LOS E adjustment factors. In this case, $f_w = 0.75$, assuming LOS D with 11-ft (3.4-m) lanes and 2-ft (0.6-m) shoulders, as indicated in Table 7.16. The v/c is computed as

$$v/c = \frac{580}{2800 \times 0.83 \times 0.75 \times 0.916} = \underline{\underline{0.363}}$$

From Table 7.15, LOS D is obtained (i.e., 0.33 < 0.363 < 0.58), which is consistent with the assumed LOS D. Therefore, the highway's level of service is D.

As a final point, note that if LOS C had been assumed, $f_w = 0.75$ (as for the LOS D assumption), but a different f_{HV} would result because now $E_T = 2.2$, $E_B = 2.0$, and $E_R = 2.5$ (level terrain, LOS C, Table 7.17). So,

$$f_{HV} = \frac{1}{1 + 0.05(2.2 - 1) + 0.02(2.0 - 1) + 0.05(2.5 - 1)} = 0.866$$

Substituting,

$$v/c = \frac{580}{2800 \times 0.83 \times 0.75 \times 0.866} = \underline{\underline{0.384}}$$

From Table 7.15, LOS D is again indicated (i.e., 0.33 < 0.384 < 0.58), which is inconsistent with the assumed LOS C. Thus LOS D provides the only consistent answer.

7.7 DESIGN TRAFFIC VOLUMES

In the preceding sections of this chapter, consideration was given to the determination of level of service, given some hourly volume. However, a procedure for selecting an appropriate hourly volume is needed to compute level of service and to determine the number of lanes that need to be provided in a new roadway design to achieve some specified level of service. The selection of an appropriate hourly volume is complicated by two concerns. First, there is considerable variability in traffic volumes by time of day, day of week, time of year, and type of roadway. Figure 7.7 shows such variations in traffic volumes by hour of day and day of week for typical intercity and intracity routes. Figure 7.8 gives variations

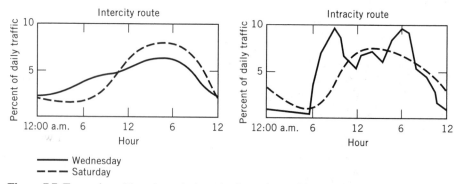

Figure 7.7 Examples of hourly variations for intercity and intracity routes.

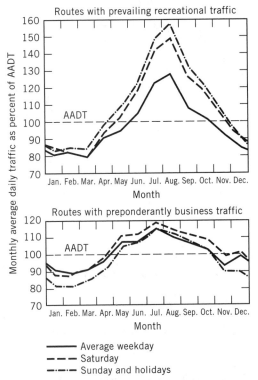

Figure 7.8 Examples of monthly traffic volume variations showing relative traffic trends by route type on rural roads (*Source:* T. Mutanyi, "A Method of Estimating Traffic Behavior on All Routes in a Metropolitan County," *Highway Research Record* 41, 1963).

by time of year by comparing monthly average traffic flows with the average annual daily traffic, AADT (in units of vehicles per day and computed as the total yearly traffic volume divided by the number of days in the year). The second concern is an outgrowth of the first in that, given the temporal variability in traffic flow, what hourly volume should be used for design and/or analysis? To answer this question, consider the example diagram shown in Figure 7.9. This figure plots hourly volume (as a percentage of AADT) against the cumulative number of hours that exceed this volume, per year. For example, the highest traffic flow in the year, on this sample roadway, would have an hourly volume of 0.148 × AADT (a volume that is exceeded by zero other hours). Sixty hours in the year would have a volume that exceeds 0.11 × AADT. In determining the number of lanes that should be provided on a new or redesigned roadway, it is obvious that using the worst single hour in a year (the hour with the highest traffic flow, which would be 0.148 × AADT from Fig. 7.9) would be a wasteful use of resources because additional lanes would be provided for a relative rare occurrence. In contrast, if the 100th highest volume is used for design, the design level of service will be exceeded 100 times a year, which will result in considerable

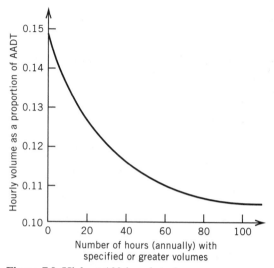

Figure 7.9 Highest 100 hourly volumes over a one-year period for a typical roadway.

driver delay. Clearly, some compromise between the expense of providing additional capacity (e.g., additional lanes) and the expense of incurring additional driver delay must be made.

A common design practice in the United States uses a design hourly volume (DHV) that is between the 10th and 50th highest volume hour of the year, depending on the type and location of the roadway (urban freeway, rural/suburban multilane highway, and so on), local traffic data, and engineering judgment. Perhaps the most common hourly volume used for roadway design is the 30th highest hourly volume of the year. In practice, the K-factor is used to convert average annual daily traffic (AADT) to the 30th highest hourly volume. K is defined as

$$K = \frac{\text{DHV}}{\text{AADT}} \tag{7.14}$$

where DHV is the design hourly volume (typically, the 30th highest annual hourly volume) and AADT is the roadway's average annual daily traffic. So, for example, Fig. 7.9 shows that the K-value corresponding to the 30th highest hourly volume is 0.12. More generally, K_i can be defined as the K-factor corresponding to the ith highest annual hourly volume. Again, for example, the 20th highest annual hourly volume would have a K-value, $K_{20} = 0.126$, from Fig. 7.9. If K is not subscripted, the 30th highest annual volume is assumed (i.e., $K = K_{30}$).

Finally, in the design and analysis of some highway types (e.g., freeways and multilane highways), the concern lies with directional traffic flows. Thus a factor is needed to reflect the proportion of peak-hour traffic volume traveling in the peak direction. This factor is denoted as D and is used to arrive at the directional design hour volume (DDHV) by application of

$$DDHV = K \times D \times AADT \qquad (7.15)$$

where all terms are as previously defined.

EXAMPLE 7.8

A freeway is to be designed as a passenger-car-only facility for an AADT of 35,000 vehicles per day. It is estimated that the freeway will have a free-flow speed of 70 mph (112.7 km/h). The design will be for commuters, and the peak-hour factor is estimated to be 0.85 with 65% of the peak-hour traffic traveling in the peak direction. Assuming that Fig 7.9 applies, determine the number of 12-ft (3.6-m) lanes required (assuming no lateral obstructions) to provide at least LOS C using the highest annual hourly volume and the 30th highest annual hourly volume.

SOLUTION

By inspection of Fig. 7.9, the highest annual hourly volume has $K_1 = 0.148$. Application of Eq. 7.15 gives

$$DDHV = K_1 \times D \times AADT$$

$$= 0.148 \times 0.65 \times 35,000 = 3367 \text{ veh/h}$$

With $V = DDHV$, Eq. 7.2 gives

$$SF = \frac{V}{PHF} = \frac{3367}{0.85} = 3961.18 \text{ veh/h}$$

To determine the number of lanes required, Eq. 7.6 will be rearranged as

$$N = \frac{SF_i}{c_j \times (v/c)_i \times f_w \times f_{HV} \times f_p}$$

where

$f_w = 1.0$ (12-ft [3.6-m] lanes, no obstructions)
$f_{HV} = 1.0$ (no heavy vehicles)
$f_p = 1.0$ (commuters, Table 7.7)

We will assume that three or more lanes will be needed (i.e., a six- or eight-lane freeway). This gives $c_j = 2300$ pcphpl. Also, Table 7.1 shows that the worst LOS C condition for a six- or eight-lane freeway has $(v/c)_C = 0.715$. Substituting, we obtain

$$N = \frac{3961.18}{2300 \times 0.715 \times 1.0 \times 1.0 \times 1.0} = \underline{\underline{2.41}}$$

Because 2.41 lanes are needed, 3 lanes must be provided to achieve at least a LOS C. Thus the assumption of a six- or eight-lane freeway (needed to get a value for c_j) was correct. For the 30th highest hourly annual volume, Fig. 7.9 gives $K_{30} = K = 0.12$, which, when used in Eq. 7.15, gives

$$DDHV = K \times D \times AADT$$

$$= 0.12 \times 0.65 \times 35,000 = 2730 \text{ veh/h}$$

so,

$$SF = \frac{V}{PHF} = \frac{2730}{0.85} = 3211.76 \text{ veh/h}$$

We will now assume that only two lanes will be required (i.e., a four-lane freeway). Therefore, c_j will be 2200 pcphpl instead of the previous 2300 pcphpl, and, from Table 7.1, the maximum allowable volume-to-capacity ratio at LOS C is 0.747 (for a four-lane freeway). All other terms are as before, so

$$N = \frac{3211.76}{2200 \times 0.747 \times 1.0 \times 1.0 \times 1.0} = \underline{\underline{1.95}}$$

This finding shows that only two lanes are needed to provide LOS C or better for the 30th highest annual hourly volume as opposed to the three lanes needed to satisfy the level of service conditions for the highest annual hourly volume.

NOMENCLATURE FOR CHAPTER 7

AADT	average annual daily traffic (in veh/day)
c	roadway capacity (in veh/h)
c_j	capacity per lane at free-flow speed j (in pcphpl), for freeways
D	directional factor
DHV	design hour volume
DDHV	directional design hour volume

E_B	passenger car equivalents for buses (two-lane rural highways only)
E_R	passenger car equivalents for recreational vehicles
E_T	passenger car equivalents for large trucks and buses, for freeways and multilane highways; passenger-car equivalents for large trucks only, for two-lane rural highways
f_d	adjustment factor for directional distribution of traffic (rural two-lane highways only)
f_{HV}	adjustment factor for heavy vehicles
f_p	adjustment factor for driver population (freeways only)
f_w	adjustment factor for lane widths and lateral clearances
FFS	estimated free-flow speed (multilane highways)
FFS_I	estimated free-flow speed for ideal conditions (multilane highways)
F_A	free-flow speed adjustment for access points (multilane highways)
F_{LC}	free-flow speed adjustment for lateral clearance (multilane highways)
F_{LW}	free-flow speed adjustment for lane width (multilane highways)
F_M	free-flow speed adjustment for median type (multilane highways)
K	factor used to convert AADT to 30th highest annual hourly volume
K_i	factor used to convert AADT to ith highest annual hourly volume
LC_L	left-side lateral clearance (multilane highways)
LC_R	right-side lateral clearance (multilane highways)
MSF_i	maximum service flow rate for level of service i (in pcphpl)
N	number of lanes in one direction
$NAPM$	number of access points per mile (multilane highways)
PHF	peak-hour factor
SF	service flow (in veh/h)
TLC	total lateral clearance (multilane highways)
v_p	service flow rate for multilane highways (in pcphpl)
V	hourly volume (in veh/h)
V_{15}	highest 15-minute volume
v/c	volume-to-capacity ratio

REFERENCES

Transportation Research Board. *Highway Capacity Manual.* Special Report 209. Washington, DC: National Research Council, 1994.

PROBLEMS

7.1. A six-lane freeway has regular weekday users and commuters and currently operates at maximum LOS C conditions. The free-flow speed is 65 mph (104.7 km/h), lanes are 11 ft (3.4 m) wide, and there is a 6-ft (1.8-m) shoulder on the right side and a 4-ft (1.2-m) shoulder on the left side. The freeway is on rolling terrain with 10% large trucks and buses (no recreational vehicles) and the peak hour factor is 0.90. If 20% of all of the daily directional traffic occurs during the peak hour, determine the total daily directional traffic volume.

7.2. Consider the freeway in Problem 7.1. At one point along this freeway there is a 4% upgrade with a directional hourly traffic volume of 3500 vehicles. If all other conditions are as described in Problem 7.1, how long can this grade be before the freeway level of service drops to E?

7.3. A four-lane freeway is located on rolling terrain with a free-flow speed of 70 mph, 12-ft (3.6-m) lanes, and no lateral obstructions within 8 ft (2.4 m) of the pavement edges. The traffic stream consists of cars, buses, and large trucks (no recreational vehicles). A weekday directional peak-hour volume of 1800 vehicles (regular users) is observed, with 700 arriving in the most congested 15-min period. If a level of service no worse than C is desired, determine the maximum number of large trucks and buses that can be present in the peak-hour traffic stream.

7.4. Consider the freeway and traffic conditions described in Problem 7.3. If 180 of the 1800 vehicles observed in the peak hour were large trucks and buses, what would the level of service of this freeway be on a 4.5-mi (7.2-km), 5% downgrade?

7.5. A six-lane freeway in a scenic area has 10-ft (3-m) lanes with 2-ft (0.6-m) shoulders on both sides of the roadway. The peak-hour factor is 0.80 and there are 8% large trucks and buses and 6% recreational vehicles in the traffic stream. One upgrade is 5% and 0.5 mi (0.8 km) long. An analyst has determined that the freeway is operating at capacity on this upgrade during the peak hour. If the peak-hour traffic volume is 2470 vehicles, what value for the driver population adjustment was used?

7.6. A four-lane freeway is on mountainous terrain with 11-ft (3.4-m) lanes, a 5-ft (1.5-m) right-side shoulder, a 3-ft (0.9-m) left-side shoulder, and a 60-mph (96.6-km/h) free-flow speed. During the peak hour there are 12% large trucks and buses and 6% recreational vehicles. The driver population adjustment is determined to be 0.90. The freeway currently operates at capacity during the peak hour, in the direction in question. If an additional 11-ft (3.4-m) lane is added, and all other factors stay the same, what will the new level of service be?

7.7. A section of freeway has a 3% upgrade that is 1500 ft (457 m) long followed by a 1000-ft (304.8-m) long 4% upgrade. It is a four-lane freeway with 12-ft (3.6-m) lanes and 3-ft (0.9-m) shoulders on both sides. The free-flow speed is 60 mph (96.6 km/h) and the directional hourly traffic flow is 2000 vehicles with 5% large trucks and buses (no recreational vehicles). If the peak-hour factor is 0.90 and all of the drivers are regular users, what is the level of service of this compound-grade freeway section?

7.8. Consider Example 7.2 in which it was determined that 463 vehicles could be added to the peak hour before capacity is reached. Assuming rolling terrain as in Example 7.1, how many passenger cars could be added to the original traffic mix before peak-hour capacity is reached? (Assume that only passenger cars are added and that the number of large trucks and buses originally in the traffic stream remains constant.)

7.9. An eight-lane freeway is on rolling terrain and has 11-ft (3.4-m) lanes with 4-ft (1.2-m) shoulders on both sides. The free-flow speed is 60 mph (96.6 km/h). The directional peak-hour traffic volume is 5400 vehicles with 6% large trucks and 5% buses (no recreational vehicles). The traffic stream consists of regular users and the peak hour factor is 0.95. It has been decided that large trucks will be banned from the freeway during the peak hour. What will be the freeway's volume-to-capacity ratio and level of service before and after the ban? (Assume that the trucks are removed and all other traffic attributes are unchanged.)

7.10. A 5% upgrade on a six-lane freeway is 1.2 mi (1.9 km) long. On this section of freeway, the directional peak-hour volume is 3800 vehicles with 2% large trucks and 4% buses (no recreational vehicles), the peak-hour factor is 0.90, and all drivers are regular users. The free-flow speed is 65 mph, the lanes are 12 ft (3.6 m) wide, and there are no lateral obstructions within 10 ft (3 m) of the roadway. A bus strike will eliminate all bus traffic, but it is estimated that for each bus removed from the roadway, six additional passenger cars will be added as travelers seek other means of travel. What is the volume-to-capacity ratio and level of service of the upgrade section before and after the bus strike?

7.11. A suburban multilane highway (two lanes in each direction) is on level terrain. The free-flow speed is determined to be 45 mph (72.5 km/h). The peak-hour directional traffic flow is 1300 vehicles with 6% large trucks and buses and 2% recreational vehicles. If the peak-hour factor is 0.85, determine the highway's level of service.

7.12. Consider the multilane highway in Problem 7.11. If the proportion of vehicle types and peak-hour factor remain constant, how many vehicles can be added to the directional traffic flow before capacity is reached?

7.13. An undivided rural multilane highway has four lanes. The lanes are 10 ft (3 m) wide and the right shoulder is 2 ft (0.6 m) wide. There are four access points per mile. If the free-flow speed is measured at 49.9 mph (80.3 km/h), what would you estimate the speed limit and the 85th-percentile speed on this highway to be?

7.14. A suburban, six-lane multilane highway has a peak-hour factor of 0.90, 11-ft (3.3-m) lanes with 4-ft (1.2-m) shoulders on the right side and two-way left-turn lanes in the median. The directional peak-hour traffic flow is 3900 vehicles with 8% large trucks and buses and 2% recreational vehicles. What would the level of service of this highway be on a 4% upgrade that is 1.2 mi (1.9 km) long if the speed limit is 55 mph (89 km/h) and there are 13 access points per mile (8 access points per kilometer)?

7.15. A divided rural multilane highway has four lanes and is on rolling terrain. The highway has 10-ft (3-m) lanes with 6-ft (1.8-m) right-side shoulders and 3-ft (0.9-m) left-side shoulders. The 85th-percentile speed is measured at 58 mph (93.4 km/h). Previously there were 4 access points per mile (2.5 access points per kilometer) but recent development has increased the number of access points to an average of 12 per mile (7.5 per kilometer). Before the development, the peak-hour factor was 0.95 and the directional hourly volume 2300 vehicles with 10% large trucks and buses and 3% recreational vehicles. After the development, the peak-hour directional flow is 2700 vehicles with the same vehicle percentages and peak-hour factor. What is the level of service before and after the development?

7.16. A suburban multilane highway has four lanes and a free-flow speed of 55 mph (88.6 km/h). One upgrade is 5% and 0.62 mi (1 km) long. Currently, trucks are not permitted on the highway but there are 2% buses (no recreational vehicles) in the directional peak-hour volume of 1900 vehicles (the peak-hour factor is 0.80). Local authorities are considering allowing trucks on this upgrade. If so, they estimate that 150 large trucks will use the highway during the peak hour. What would the level of service be before and after the trucks are allowed?

7.17. A four-lane undivided suburban multilane highway has 11-ft (3.4-m) lanes, a 4-ft (1.2-m) shoulder, and 10 access points per mile (6 access points per kilometer). It is determined that the roadway currently operates at capacity with a peak-hour factor of 0.80. If the highway is on level terrain with 8% trucks and buses (no recreational vehicles) and the speed limit is 55 mph (89 km/h), what is the directional hourly volume?

7.18. A new four-lane divided suburban multilane highway is being planned with 12-ft (3.6-m) lanes, 6-ft (1.8-m) shoulders on both sides, and a 50 mph (80 km/h) speed limit. One 3% downgrade is 4.5 mi (7.2 km) long and there will be 4 access points per mile (2.5 access points per kilometer). The peak-hour directional volume along this grade is estimated to consist of 1800 passenger cars, 140 large trucks, 40 buses, and 10 recreational vehicles. If the peak-hour factor is estimated to be 0.85, what level of service will this section of highway operate under?

7.19. A six-lane divided suburban multilane highway has a free-flow speed of 50 mph (80.5 km/h). It is on mountainous terrain with a traffic stream consisting of 6% large trucks and buses and 2% recreational vehicles. One direction of the highway currently operates at maximum LOS C conditions, and the highway has a peak-hour factor of 0.90. How many vehicles can be added to this highway before capacity is reached, assuming the proportion of vehicle types remains the same but that the peak-hour factor increases to 0.95?

7.20. A four-lane undivided rural multilane highway has 11-ft (3.4-m) lanes and 5-ft (1.5-m) shoulders. At one point along the highway there is a 4% upgrade that is 0.59 mi (0.95 km) long. There are 13 access points along this grade. The peak-hour traffic volume has 2000 passenger cars and 212 trucks and buses (no recreational vehicles) and 600 of these vehicles arrive in the most congested 15-min period. The 85th-percentile speed is 60 mph (96.6 km/h). To improve the level of service, the local transportation agency is considering reducing the number of access points by blocking some roads and rerouting their traffic. How many of the 13 access points must be blocked to achieve LOS D?

7.21. A rural two-lane highway is currently operating at capacity in mountainous terrain. Passing is not permitted and the highway has 11-ft (3.4-m) lanes and 2-ft (0.6-m) shoulders. The traffic stream consists of cars and trucks only and there is a 70/30 directional split. A recent traffic count revealed 256 vehicles (total, both directions) arriving in the most congested 15-min interval. What is the percentage of trucks in the traffic stream?

7.22. A two-lane rural highway carries a peak-hour volume of 180 vehicles and has a peak-hour factor of 0.87. The highway has 11-ft (3.4-m) lanes, 2-ft (0.6-m) shoulders, and is on mountainous terrain with 80% no-passing zones. The traffic stream has a 60/40 directional split with 5% large trucks, 10% recreational vehicles, and no buses. Determine the level of service.

7.23. An engineer designs a rural two-lane highway on rolling terrain with 10-ft (3-m) lanes and 2-ft (0.6-m) shoulders. The peak-hour traffic count is 280 vehicles with a peak-hour factor of 0.80. The traffic stream has 10% large trucks, 2% buses, and 5% recreational vehicles. There are 40% no-passing zones. What directional split is necessary to ensure that the roadway operates at LOS C or better?

7.24. A rural two-lane highway is on mountainous terrain with passing permitted throughout. The highway has 11-ft (3.4-m) lanes with 4-ft (1.2-m) shoulders. The highway is oriented north and south and, during the peak hour, 480 vehicles are going northbound and 320 vehicles are going southbound. If the peak-hour factor is 0.75 and there are 4% large trucks, 3% buses, and 1% recreational vehicles, what is the level of service?

7.25. A four-lane freeway has 12-ft (3.6-m) lanes, no lateral obstructions, passenger cars only, a driving population of regular users, a peak-hour-directional distribution of 0.70, a peak-hour factor of 0.80, and a free-flow speed of 70 mph (112.7 km/h). Assuming Fig. 6.9 applies, if the AADT is 30,000 veh/day, determine the level of service for the 10th, 50th, and 100th highest annual hourly volumes.

7.26. A four-lane freeway operates at capacity during the peak hour. It has 11-ft (3.4-m) lanes and 4-ft (1.2-m) shoulders on both sides. The freeway has regular users only and 8% large trucks and buses (no recreational vehicles), and is on rolling terrain with a peak-hour factor of 0.85. It is known that 12% of the AADT occurs in the peak hour and that the directional factor is 0.6. What is the freeway's AADT?

Chapter **8**

Principles of Traffic Forecasting

8.1 INTRODUCTION

The modification of a highway network either by new road construction or operating improvements on existing roads (e.g., use or retiming of traffic signalization) must be predicated upon some expectation or forecast of traffic volumes. For new road construction, forecasts of traffic are needed to determine an appropriate pavement (number of equivalent axle loads as discussed in Chapter 4) and geometric design (number of lanes, shoulder widths, and so on) that will provide an acceptable level of service. For operational improvements, traffic forecasts are needed to estimate the effectiveness of alternate improvement options.

In forecasting vehicle traffic, two interrelated elements must be considered: (1) overall regional traffic growth/decline and (2) traffic diversion. The long-term trend of traffic growth/decline is clearly an important concern, because projects such as highway construction and operational improvements must be undertaken with some idea of what future traffic conditions will be. In the design of these projects, engineers must seek to provide a sufficient highway level of service and an acceptable pavement ride quality for future traffic volumes. One would expect that factors affecting long-term regional traffic growth/decline trends are primarily economic and, to a historically lesser extent, social in nature. The economics of the region in which a highway action is being undertaken determine the amount of traffic-generating activities (e.g., work and shopping) and the spatial distribution of residential, industrial, and commercial areas. The social aspects of the population determine attitudes and behavioral tendencies with regard to traffic-generating decisions. For example, some regional populations may have social characteristics that make them more likely than other regional populations to make fewer trips, to carpool, to vanpool, or to take public transportation (buses or subways), all of which significantly impact the amount of highway traffic.

In addition to overall regional traffic growth/decline, there is the more microscopic, short-term phenomenon of traffic diversion. As new roads are constructed, as operational improvements are made, and/or as roads gradually become more congested, traffic will divert as drivers change routes or trip-departure times in an effort to avoid congestion and improve the level of service that they experience. Thus the highway network must be viewed as a system, and, consequently, with the realization that a capacity or level-of-service change on any one segment of the highway network will impact traffic flows on many of the surrounding highway segments.

From this discussion, it is obvious that traffic forecasting is a formidable problem, because it requires accurate regional economic forecasts as well as accurate forecasts of highway users' social and behavioral attitudes regarding trip-oriented decisions, in order to predict growth/decline trends and traffic diversion. Virtually everyone is aware of how inaccurate economic forecasts can be, which is testament to the complexity and uncertainty associated with such forecasts. Similarly, one can readily imagine the difficulty associated with forecasting individuals' trip-oriented decisions.

In spite of the enormous obstacles facing the accurate forecasting of traffic, over the years highway engineers and analysts have persisted in the development and refinement of a wide variety of traffic forecasting techniques. An overriding consideration has always been the ease with which a technique can be implemented in terms of input data requirements and the ability of users to comprehend the underlying methodological approach. The field has evolved such that many traffic analysts can legitimately argue that the more recent developments in traffic forecasting are largely beyond the reach of practice-oriented implementation. In many respects, this is an expected evolution since, due to the complexity of the traffic forecasting problem, there will always be a tendency for theoretical work to exceed the limits of practical implementability. Unfortunately, the outgrowth of the methodological gap between theory and practice has resulted in the use of a wide variety of traffic forecasting techniques, the selection of which is a function of the technical expertise of forecasting agencies' personnel as well as time and financial concerns.

In the past, textbooks have attempted to cover the full range of traffic forecasting techniques from the readily implementable simplistic approaches to the more theoretically refined methods. In so doing, such textbooks often sacrificed depth of coverage and, as a consequence, traffic forecasting frequently gave the appearance of being confusing and disjointed. This book attempts to convey the basic principles underlying traffic forecasting as opposed to reviewing the many techniques available to forecast traffic. This is achieved by focusing on an approach that is fairly advanced, technically, and one that, we believe, effectively and efficiently conveys the fundamental concepts of traffic forecasting. For more information on traffic forecasting techniques that are more implementable or more theoretically advanced than the concepts provided in this chapter, the reader is referred to other sources (Meyer and Miller 1984; Sheffi 1985; Ben-Akiva and Lerman 1985; Hamed and Mannering 1993; Mannering, Abu-Eisheh, and Arnadottir 1990).

8.2   TRAVELER DECISIONS

Forecasts of highway traffic should, at least in theory, be predicated upon some understanding of traveler decisions, because the various decisions that travelers make regarding trips will ultimately determine the quantity, spatial distribution (by route), and temporal distribution (by time) of vehicles on a highway network. Within this context, travelers can be viewed as making four distinct but interrelated decisions regarding trips: (1) temporal decisions, (2) destination decisions, (3) modal decisions, and (4) spatial or route decisions. The temporal decision includes the decision to travel and, more important, when to travel. The destination decision is concerned with the selection of a specific destination (e.g., shopping center, etc.), and the modal decision relates to how the trip is to be made (e.g., by automobile, bus, walking, or bicycling). Finally, spatial decisions focus on which route is to be taken from the traveler's origin (i.e., the traveler's initial location) to the traveler's desired destination. Being able to understand, let alone predict, such decisions is a monumental task. The remaining sections of this chapter seek to define the dimensions of this decision-prediction task and, through examples and illustrations, to demonstrate methods of forecasting traveler decisions and, ultimately, traffic volumes.

8.3   SCOPE OF THE TRAFFIC FORECASTING PROBLEM

Because traffic forecasting is predicated upon the accurate forecasting of traveler decisions, two factors must be addressed in the development of an effective traffic forecasting methodology: (1) the complexity of the traveler decision-making process and (2) system equilibration.

To begin the development of a fuller understanding of the complexity of traveler decisions (and traffic forecasting), consider the schematic presented in Fig. 8.1. This figure indicates that traveler socioeconomics and activity patterns constitute a major driving force in the decision-making process. Socioeconomics, including factors such as household income, number of household members, and traveler age, affect the types of activities that the traveler is likely to be involved in (e.g., work, yoga classes, shopping, children's day care, dancing lessons, community meetings, and so on), which in turn act as a primary factor in many travel decisions. Socioeconomics can also have a direct effect on travel-related decisions by, for example, limiting modal availability (e.g., travelers in low-income households may be forced to take a bus due to the nonavailability of a household automobile).

If we look more directly at the decision to travel, mode/destination choice, and highway route choice, Fig. 8.1 indicates that both long-term and short-term factors affect these decisions. For the decision to travel, as well as mode/destination choice, the long-term factors of modal availability, residential and commercial distributions, and modal infrastructure play a significant role. These factors are considered long-term because they change relatively slowly over time.

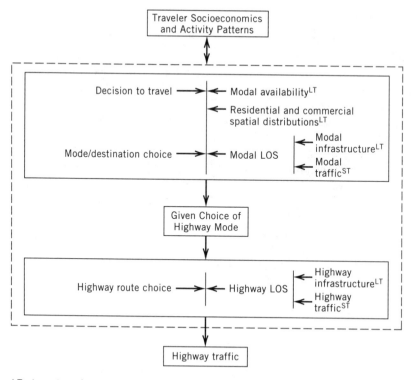

LT: Long-term factors.
ST: Short-term factors.

Figure 8.1 Overview of the process by which highway traffic is determined.

For example, the development and/or relocation of residential neighborhoods and commercial centers is a process that may take years. Similarly, changes in modal infrastructure (e.g., construction/relocation of highways, subways, commuter rail systems) and modal availability (e.g., changes in automobile ownership, bus routing/scheduling) are also factors that evolve over relatively long periods of time. In contrast, a short-term factor, such as modal traffic, is one that can vary in short periods of time as shown in Chapter 7 (see Fig. 7.1).

Moving further down the illustration presented in Fig. 8.1, we see that the choice of a traveler's highway route is also determined by both long-term (highway infrastructure) and short-term (highway traffic) factors. The outcome of the combination of these traveler decisions is, of course, highway traffic, the prediction of which is the objective of traffic forecasting.

Aside from the complexities involved in the traveler decision-making process, the issue of system equilibration (mentioned in the beginning of this section) must also be considered. Note that Fig. 8.1 not only indicates that long- and short-term factors affect traveler decisions and choices, but also that these decisions and choices in turn affect the long and short-term factors. Such a simultaneous relationship is most apparent when considering the relationship between traveler choices and short-term factors. For example, consider a traveler's choice of highway route. One would expect that a traveler would be more likely to select a route between an origin and a destination that provides a shorter travel time. The travel time on various routes will be a function of route distance (i.e., highway infrastructure, long-term) and route traffic (i.e., higher traffic volumes reduce travel speed and increase travel times as discussed in Chapter 5). But travelers' decisions to take specific routes ultimately determine the route traffic upon which their route decisions are based. This interdependence between traveler decisions and modal traffic is schematically presented in Fig. 8.2. In addition to these short-term effects, persistently high traffic volumes may lead to a change in the highway infrastructure (e.g., construction of additional lanes and/or new highways to reduce congestion), again resulting in an interdependence. This interdependence creates the problem of equilibration that is common to many modeling applications. Perhaps the most recognizable equilibration problem is determination of price in a classic model of economic supply and demand for a product. From a modeling perspective, as will be shown, equilibration adds yet another dimension of difficulty to an already complex traffic forecasting problem.

It is safe to say that no existing traffic forecasting methodology has come close to accurately capturing the complexities involved in traveler decisions or fully addressing the issue of equilibration. However, within rather obvious limitations, the field of traffic forecasting has, over the years, made progress toward more accurately modeling traveler decision complexities and equilibration concerns. This evolution of traffic forecasting methodology has led to the popular approach of viewing traveler decisions as a sequence of three distinct decisions, as shown in Fig. 8.3, the result of which is forecasted traffic flow (a direct outgrowth of the highway route choice decision). Clearly, the sequential structure of traveler decisions is a considerable simplification of the actual decision-making process in which all trip-related decisions are considered simultaneously by the traveler. However, this sequential simplification permits the development of a sequence of mathematical models of traveler behavior that can be applied to forecast traffic flow. The following sections of this chapter present and discuss typical

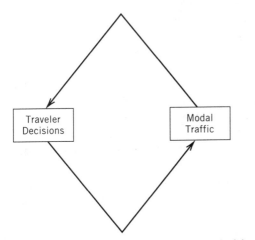

Figure 8.2 Interdependence of traveler decisions and traffic flow.

Sequence of Models	Cumulative Model Results (relevant to highway mode)
Trip generation ⟶	• Origin of trips • Number of trips • Departure time of trips
Mode/destination choice ⟶	• Origin of highway trips • Number of highway trips • Departure time of highway trips • Destination of highway trips
Highway route choice ⟶	• Traffic flow

Figure 8.3 Overview of the sequential approach to traffic estimation.

functional forms of the mathematical models used to forecast the three sequential traveler decisions shown in Fig. 8.3.

8.4 TRIP GENERATION

The first traveler decision to be modeled in the sequential approach to traffic forecasting is trip generation. The objective of trip generation modeling is to develop an expression that predicts exactly when a trip is to be made. This is an inherently difficult task due to the wide variety of trip types (e.g., working or shopping) and activities (e.g., eating lunch or visiting friends) undertaken by a traveler in a sample day, as is schematically shown in Fig. 8.4. To address the complexity of the trip generation decision, the following approach is typically taken:

1. **Aggregation of Decision-Making Units** Predicting trip generation behavior is simplified by considering the trip generation behavior of a household (i.e., a group of travelers sharing the same domicile) as opposed to the behavior of individual travelers. Such an aggregation of traveler decisions is justified on the basis of the comparatively homogeneous nature of household members (socially and economically) and their often intertwined trip generating activities (e.g., joint shopping trips, and so on).

2. **Segmentation of Trips by Type** Different types of trips have different characteristics that make them more or less likely to be taken at various times of the day. For example, work trips are more likely to be taken in the morning hours than are shopping trips, which are more likely to be taken during the evening hours. Also, it is more likely that a traveler will

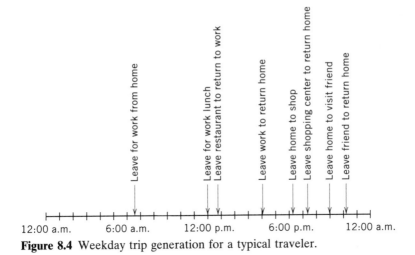

Figure 8.4 Weekday trip generation for a typical traveler.

take multiple shopping trips during the course of a day as opposed to multiple work trips. To account for this, three distinct trip types are used: (1) work trips, including trips to and from work, (2) shopping trips, and (3) social/recreational trips, which include vacations, visiting friends, church meetings, sporting events, and so on.

3. **Temporal Aggregation** Although research has been undertaken to develop mathematical expressions that predict when a traveler is likely to make a trip (Hamed and Mannering 1993), more often, trip generation focuses on the number of trips made over some period of time. Thus trips are aggregated temporally, and trip generation models seek to predict the number of trips per hour or per day.

8.4.1 Typical Trip Generation Models

Trip generation models generally assume a linear form, in which the number of vehicle-based (automobile, bus, or subway) trips is a function of various socio-, economic and/or distributional (residential and commercial) characteristics. An example of such a model is

$$T_{ij} = b_0 + b_1 z_{1j} + b_2 z_{2j} + \ldots + b_n z_{nj} \tag{8.1}$$

where T_{ij} is the number of vehicle-based trips of type i (i.e., shopping or social/recreational) in some specified time period made by household j, z_{nj} is characteristic n (e.g., income, employment in neighborhood, number of household members) of household j, and b_n is a coefficient estimated from traveler survey data and corresponding to characteristic n.

The estimated coefficients (b's) are most often estimated by the method of least squares regression using data collected from traveler surveys. A brief description and example of this method are presented in Appendix 8A.

EXAMPLE 8.1

A simple linear regression model is estimated for shopping-trip generation during a shopping-trip peak hour (e.g., Saturday afternoon). The model is

number of peak-hour vehicle-based shopping trips per household

= 0.12 + 0.09 (household size)

+ 0.011 (annual household income [in thousands of dollars])

− 0.15 (employment in the household's neighborhood [in hundreds])

A particular household has six members and an annual income of $50,000. They live in a neighborhood with 450 retail employees. They move to a new home in a neighborhood with 150 retail employees. Calculate the predicted number of vehicle-based peak-hour shopping trips the household makes before and after the move.

SOLUTION

Note that the signs of the model coefficients (b's, +0.09, and +0.011) indicate that as household size and income increase, the number of shopping trips also increases. This is reasonable because wealthier, larger households can be expected to make more vehicle-based shopping trips. The negative sign of the employment coefficient (−0.15) indicates that as retail employment in a household's neighborhood increases, fewer vehicle-based shopping trips will be generated. This reflects the fact that larger retail employment in a neighborhood implies more shopping opportunities nearer to the household, thereby increasing the possibility that a shopping trip can be conducted without the use of a vehicle (i.e., a nonvehicle-based trip, such as walking).

Turning to the problem solution, before the household moves, we find that

$$\text{number of vehicle trips} = 0.12 + 0.09(6) + 0.011(50) - 0.15(4.5)$$
$$= 0.535 \text{ vehicle trips}$$

After the household moves, we obtain

$$\text{number of vehicle trips} = 0.12 + 0.09(6) + 0.011(50) - 0.15(1.5)$$
$$= 0.985 \text{ vehicle trips}$$

Thus the model predicts that the move will result in 0.45 additional peak-hour vehicle-based shopping trips due to the decline in neighborhood shopping opportunities as reflected by the decline in neighborhood retail employment.

EXAMPLE 8.2

A model for social/recreational trip generation is estimated, with data collected during a major holiday, as

number of peak-hour vehicle-based social/recreational trips per household

= 0.04 + 0.018 (household size)

+ 0.009 (annual household income [in thousands of dollars])

+ 0.16 (number of nonworking household members)

If the family described in Example 8.1 has one working member, how many peak-hour social/recreational trips are predicted?

SOLUTION

The positive signs of the model coefficients indicate that increasing household size, income, and number of nonworking household members result in more social/recreational trips. Again, wealthier and larger households can be expected to be involved in more vehicle-based trip generating activities, and the larger the number of nonworking household members, the larger the number of people available, at home, to make peak-hour social/recreational trips.

The solution to this problem simply is

$$\text{number of vehicle trips} = 0.04 + 0.018(6) + 0.009(50) + 0.16(5)$$

$$= \underline{\underline{1.398}} \text{ vehicle trips}$$

EXAMPLE 8.3

A neighborhood has a retail employment of 205, and 700 households that can be categorized into four types, each type having identical characteristics as follows:

Type	Household Size	Annual Income	Number of Nonworkers in the Peak Hour	Workers Departing
1	2	$40,000	1	1
2	3	$50,000	2	1
3	3	$55,000	1	2
4	4	$40,000	3	1

There are 100 type 1, 200 type 2, 350 type 3, and 50 type 4 households. Assuming that shopping, social/recreational, and work vehicle-based trips all peak at the same time (for exposition purposes), determine the total number of peak hour trips (work, shopping, social/recreational) using the generation models described in Examples 8.1 and 8.2.

SOLUTION

For vehicle-based shopping trips, we obtain

$$\text{type 1: } 0.12 + 0.09(2) + 0.011(40) - 0.15(2.05) = 0.4325 \text{ trips/household}$$
$$\times 100 \text{ households}$$
$$= 43.25 \text{ trips}$$

$$\text{type 2: } 0.12 + 0.09(3) + 0.011(50) - 0.15(2.05) = 0.6325 \text{ trips/household}$$
$$\times 200 \text{ households}$$
$$= 126.5 \text{ trips}$$

$$\text{type 3: } 0.12 + 0.09(3) + 0.011(55) - 0.15(2.05) = 0.6875 \text{ trips/household}$$
$$\times 350 \text{ households}$$
$$= 240.625 \text{ trips}$$

$$\text{type 4: } 0.12 + 0.09(4) + 0.011(40) - 0.15(2.05) = 0.6125 \text{ trips/household}$$
$$\times 50 \text{ households}$$
$$= 30.625 \text{ trips}$$

Therefore, there will be a total of 441 vehicle-based shopping trips.
For vehicle-based social/recreational trips, we obtain

$$\text{type 1: } 0.04 + 0.018(2) + 0.009(40) + 0.16(1) = 0.596 \text{ trips/household}$$
$$\times 100 \text{ households}$$
$$= 59.6 \text{ trips}$$

$$\text{type 2: } 0.04 + 0.018(3) + 0.009(50) + 0.16(2) = 0.864 \text{ trips/household}$$
$$\times 200 \text{ households}$$
$$= 172.8 \text{ trips}$$

$$\text{type 3: } 0.04 + 0.018(3) + 0.009(55) + 0.16(1) = 0.749 \text{ trips/household}$$
$$\times 350 \text{ households}$$
$$= 262.15 \text{ trips}$$

$$\text{type 4: } 0.04 + 0.018(4) + 0.009(40) + 0.16(3) = 0.952 \text{ trips/household}$$
$$\times 50 \text{ households}$$
$$= 47.6 \text{ trips}$$

Therefore, there will be a total of 542.15 vehicle-based shopping trips.
For vehicle-based work trips, there will be 100 generated from type 1 household (1 × 100), 200 from type 2 (1 × 200), 700 from type 3 (2 × 350), and 50 from type 4 (1 × 50) for a total of 1050 vehicle-based work trips. Summing the totals of all three trip types gives 2033 peak-hour vehicle-based trips.

It should be noted that the trip generation models used in Examples 8.1, 8.2, and 8.3 are a simplified representation of the actual trip generation decision-making process. First, there are many more traveler and household characteristics that affect trip generating behavior (age, lifestyles, and so on), and, second, the models have no variables to capture the equilibration concept discussed earlier. The equilibration concern is important, because if the highway system is heavily congested, travelers are likely to make fewer peak-hour trips either as a result of canceling trips or postponing them until a less congested time period. Unfortunately, such obvious model defects must often be accepted due to data and/or resource limitations.

8.5 MODE AND DESTINATION CHOICE

After the number of trips generated per unit time is known, the next step in the sequential approach to traffic forecasting is to determine traveler mode and destination. As was the case with trip generation, trips are classified as work, shopping, and social/recreational. For both shopping and social/recreational trips, a traveler will have the option to choose a mode of travel (e.g., automobile, vanpool, or bus) as well as a destination (e.g., different shopping centers). In contrast, work trips offer only the mode option, because the choice of work location (destination) is usually a long-term decision that is beyond the time range of most traffic forecasts.

8.5.1 Theoretical Approach

Following recent advances in the traffic forecasting field, development of a model for mode/destination choice necessitates the use of some consistent theory of travelers' decision-making processes. Of the decision-making theories available, one that is based on the microeconomic concept of utility maximization has enjoyed comparatively widespread acceptance in mode/destination choice modeling. The basic assumption is that a traveler will select the combination of a mode and destination that provides the most utility in the economic sense. The problem, then, becomes one of developing an expression for the utility provided by various mode/destination alternatives. Because it is unlikely that individual travelers' utility functions can ever be specified with certainty, the unspecifiable portion is assumed to be random. To illustrate this approach, consider a utility function of the following form:

$$V_{mk} = \sum_{n} b_{mn} z_{kmn} + \varepsilon_{mk} \qquad (8.2)$$

where V_{mk} is the total utility (specifiable and unspecifiable) provided by mode/destination alternative m to a traveler k; b_{mn} is the coefficient estimated from traveler survey data for mode/destination alternative m corresponding to mode/destination or traveler characteristic n; z_{kmn} is the traveler or mode/destination characteristic n (e.g., income, travel time of mode, commercial floor space at destination, or population at destination) for mode/destination alternative m for traveler k; and ε_{mk} is the unspecifiable portion of the utility of mode/destination alternative m for traveler k, which will be assumed to be random. For notational convenience, define the specifiable nonrandom portion of utility V_{mk} as

$$U_{mk} = \sum_n b_{mn} z_{kmn} \tag{8.3}$$

With these definitions of utility, the probability that a traveler will choose some alternative, say m, is equal to the probability that the given alternative's utility is greater than the utility of all other possible alternatives. The probabilistic component arises from the fact that the unspecifiable portion of the utility expression is not known and is assumed to be a random variable. The basic probability statement is

$$P_{mk} = \text{prob}[U_{mk} + \varepsilon_{mk} > U_{sk} + \varepsilon_{sk}] \quad \text{for all } s \neq m \tag{8.4}$$

where P_{mk} is the probability that traveler k will select alternative m, prob[·] is the notation for probability, and s is the notation for available alternatives. With this basic probability and utility expression and an assumed random distribution of the unspecifiable components of utility (ε_{mk}), a probabilistic choice model can be derived and the coefficients in the utility function (b_{mn}'s in Eqs. 8.2 and 8.3) can be estimated with data collected from traveler surveys, along the same lines as was done for the coefficients in the trip generation models. A popular approach to deriving such a probabilistic choice model is to assume that the random, unspecifiable component of utility (ε_{mk} in Eq. 8.2) is generalized extreme value distributed. With this assumption, a rather lengthy and involved derivation (see McFadden 1981) gives rise to the logit model formulation,

$$P_{mk} = \frac{e^{U_{mk}}}{\sum_s e^{U_{sk}}} \tag{8.5}$$

where e is the base of the natural logarithm (i.e., $e = 2.718$).

The coefficients that comprise the specifiable portion of utility (b_{mn}'s in Eq. 8.3) are estimated by the method of maximum likelihood, which essentially accomplishes the same objective as the least squares estimation procedure described in Appendix 8A. For further information on logit model coefficient estimation and maximum likelihood estimation techniques, the reader is referred to more specialized references (Ben-Akiva and Lerman 1985; McFadden 1981).

8.5.2   Logit Model Applications

With the total number of vehicle-based trips made in specific time periods known (from trip generation models), the allocation of trips to vehicle-based modes and likely destinations can be undertaken by applying appropriate logit models. This process is best demonstrated by example.

EXAMPLE 8.4

A simple work-mode-choice model is estimated from data in a small urban area to determine the probabilities of individual travelers selecting various modes. The mode choices include automobile drive-alone (DL), automobile shared-ride (SR) and bus (B), and the utility functions are estimated as

$$U_{DL} = 2.2 - 0.2(\text{cost}_{DL}) - 0.03(\text{travel time}_{DL})$$

$$U_{SR} = 0.8 - 0.2(\text{cost}_{SR}) - 0.03(\text{travel time}_{SR})$$

$$U_B = -0.2(\text{cost}_B) - 0.01(\text{travel time}_B)$$

where cost is in dollars and time is in minutes. Between a residential area and an industrial complex, 4000 workers (generating vehicle-based trips) depart for work during the peak hour. For all workers, the cost of driving an automobile is $4.00 with a travel time of 20 minutes, and the bus fare is 50 cents with a travel time of 25 minutes. If the shared-ride option always consists of two travelers sharing costs equally, how many workers will take each mode?

SOLUTION

Note that the utility function coefficients logically indicate that as modal costs and travel times increase, modal utilities decline and, consequently, so do modal selection probabilities (see Eq. 8.5). Substitution of cost and travel time values into the utility expressions gives

$$U_{DL} = 2.2 - 0.2(4) - 0.03(20)$$

$$= 0.8$$

$$U_{SR} = 0.8 - 0.2(2) - 0.03(20)$$

$$= -0.2$$

$$U_B = -0.2(0.5) - 0.01(25)$$

$$= -0.35$$

Substituting these values into Eq. 8.5 yields

$$P_{DL} = \frac{e^{0.8}}{e^{0.8} + e^{-0.2} + e^{-0.35}}$$

$$= \frac{2.226}{2.226 + 0.819 + 0.705}$$

$$= \frac{2.226}{3.749}$$

$$= 0.594$$

$$P_{SR} = \frac{0.819}{3.749}$$

$$= 0.218$$

$$P_B = \frac{0.705}{3.749}$$

$$= 0.188$$

Multiplying these probabilities by 4000 (the total number of workers departing in the peak hour) gives 2380 workers driving alone, 870 sharing a ride, and 750 using a bus.

EXAMPLE 8.5

A bus company is making costly efforts in an attempt to increase work-trip bus usage for the travel conditions described in Example 8.4. An exclusive bus lane is constructed that reduces bus travel time to 10 min.

(a) Determine the modal distribution of trips after the lane is constructed.
(b) If shared-ride vehicles are also permitted to use the facility and travel time for bus and shared-ride modes is 10 min, determine the modal distribution.
(c) Given the conditions described in part (b), determine the modal distribution if the bus company offers free bus service.

SOLUTION

(a) After the bus lane construction, the modal utilities of drive-alone and shared-ride are unchanged from those in Example 8.4. However, the bus modal utility becomes

$$U_B = -0.2(0.5) - 0.01(10)$$

$$= -0.2$$

From Eq. 8.5 with 4000 work trips, we find that

$$P_{DL} = \frac{e^{0.8}}{e^{0.8} + e^{-0.2} + e^{-0.2}}$$

$$= \frac{2.226}{3.8635}$$

$$= 0.576 \text{ and } 0.576(4000) = 2304 \text{ trips}$$

$$P_{SR} = \frac{0.819}{3.8635} = 0.212 \text{ and } 0.212(4000) = 848 \text{ trips}$$

$$P_B = \frac{0.819}{3.8635} = 0.212 \text{ and } 0.212(4000) = \underline{\underline{848}} \text{ trips}$$

or an increase of 98 bus patrons, from the prediction of Example 8.4.

(b) With the bus lane opened to shared-ride vehicles, only the modal utility of shared ride will change from those in part (a):

$$U_{SR} = 0.08 - 0.02(2) - 0.03(10)$$

$$= 0.1$$

From Eq. 8.5 with 4000 work trips, we obtain

$$P_{DL} = \frac{e^{0.8}}{e^{0.8} + e^{0.1} + e^{-0.2}}$$

$$= \frac{2.226}{4.15}$$

$$= 0.536; 0.536(4000) = \underline{\underline{2144}} \text{ trips}$$

$$P_{SR} = \frac{1.105}{4.15} = 0.267; 0.267(4000) = \underline{\underline{1068}} \text{ trips}$$

$$P_B = \frac{0.8187}{4.15} = 0.197; 0.197(4000) = \underline{\underline{788}} \text{ trips}$$

or a loss of 60 bus patrons and a gain of 220 shared-ride users relative to part (a).

(c) With free bus fare, the bus modal utility becomes (with other utilities unchanged from part [b]),

$$U_B = -0.2(0) - 0.01(10)$$

$$= -0.1$$

From Eq. 8.5 with 4000 work trips, we obtain

$$P_{DL} = \frac{e^{0.8}}{e^{0.8} + e^{0.1} + e^{-0.1}}$$

$$= \frac{2.226}{4.236}$$

$$= 0.525; 0.525(4000) = \underline{\underline{2102}} \text{ trips}$$

$$P_{SR} = \frac{1.105}{4.236} = 0.261; 0.261(4000) = \underline{\underline{1043}} \text{ trips}$$

$$P_B = \frac{0.905}{4.236} = 0.214; 0.214(4000) = \underline{\underline{855}} \text{ trips}$$

or 67 more bus patrons compared to part (b).

EXAMPLE 8.6

Consider a residential area and two shopping centers that are possible destinations. From 7:00 to 8:00 P.M. on Friday night, 900 vehicle-based shopping trips leave the residential area for the two shopping centers. A joint shopping trip destination/mode choice logit model (choice of either auto or bus) is estimated giving the following coefficients:

Variable	Auto Coefficient	Bus Coefficient
Auto constant	0.6	0.0
Travel time in minutes	−0.3	−0.3
Commercial floor space (in thousands of square meters)	0.12	0.12

Initial travel times to shopping centers 1 and 2 are as follows:

	By Auto	By Bus
Travel time to shopping center 1 (in minutes)	8	14
Travel time to shopping center 2 (in minutes)	15	22

If shopping center 2 has 40,000 m² of commercial floor space and shopping center 1 has 25,000 m², determine the distribution of Friday night shopping trips by destination and mode.

SOLUTION

The utility function coefficients indicate that as modal travel times increase, the likelihood of selecting the mode/destination combination declines. Also, as the destination's floor space increases, the probability of selecting that destination will increase as suggested by the positive coefficient (+0.12). This reflects the fact that bigger shopping centers will tend to have a greater variety of merchandise and hence be more attractive shopping destinations. Note that because this is a joint mode/destination choice model, there are four mode/destination combinations and four corresponding utility functions. Let U_{A1} be the utility of the auto mode to shopping center 1, U_{A2} be the utility of the auto mode to shopping center 2, and U_{B1} and U_{B2} be the utility of the bus mode to shopping centers 1 and 2, respectively. The utilities are

$$U_{A1} = 0.6 - 0.3(8) + 0.12(25)$$

$$= 1.2$$

$$U_{B1} = -0.3(14) + 0.12(25)$$

$$= -1.2$$

$$U_{A2} = 0.6 - 0.3(15) + 0.12(40)$$

$$= 0.9$$

$$U_{B2} = -0.3(22) + 0.12(40)$$

$$= -1.8$$

Substituting these values into Eq. 8.5 gives

$$P_{A1} = \frac{3.32}{6.246}$$

$$= 0.532$$

$$P_{B1} = \frac{0.301}{6.246}$$

$$= 0.048$$

$$P_{A2} = \frac{2.46}{6.246}$$

$$= 0.394$$

$$P_{B2} = \frac{0.165}{6.246}$$

$$= 0.026$$

Multiplying these probabilities by the 900 trips gives 479 trips by auto to shopping center 1, 43 trips by bus to shopping center 1, 355 trips by auto to shopping center 2, and 23 trips by bus to shopping center 2.

EXAMPLE 8.7

A joint destination/mode vehicle-based social/recreational trip logit model is estimated with the following coefficients:

Variable	Auto Coefficient	Bus Coefficient
Auto constant	0.9	0.0
Travel time in minutes	-0.22	-0.22
Population in thousands	0.16	0.16
Amusement floor space in thousands of square meters	1.1	1.1

It is known that 500 social/recreational trips will depart from a residential area during the peak hour. There are three possible trip destinations with the following characteristics:

	Travel Time (in minutes)		Population (in thousands)	Amusement Floor Space (in thousands m²)
	Auto	Bus		
Destination 1	14	17	12.4	1.3
Destination 2	5	8	8.2	0.92
Destination 3	18	24	5.8	2.1

Determine the distribution of trips by mode and destination.

SOLUTION

As was the case for the shopping mode/destination model presented in Example 8.6, the signs of the coefficient estimates indicate that increasing travel time decreases an alternative's selection probability. Also, increasing population (reflecting an increase in social opportunities) and increasing amusement floor space (reflecting more recreational opportunities) both increase the probability of an alternative being selected. With two modes and three destinations, there are six alternatives providing the following utilities (using the same subscripting notation as in Example 8.6),

$$U_{A1} = 0.9 - 0.22(14) + 0.16(12.4) + 1.1(1.3)$$
$$= 1.234$$
$$U_{B1} = -0.22(17) + 0.16(12.4) + 1.1(1.3)$$
$$= -0.326$$
$$U_{A2} = 0.9 - 0.22(5) + 0.16(8.2) + 1.1(0.92)$$
$$= 2.124$$
$$U_{B2} = -0.22(8) + 0.16(8.2) + 1.1(0.92)$$
$$= 0.564$$
$$U_{A3} = 0.9 - 0.22(18) + 0.16(5.8) + 1.1(2.1)$$
$$= 0.178$$
$$U_{B3} = -0.22(24) + 0.16(5.8) + 1.1(2.1)$$
$$= -2.042$$

Using Eq. 8.5 with 500 trips, the total number of trips to the six mode/destination alternatives are

$$P_{A1} = \frac{3.435}{15.607}$$

$$= 0.22 \text{ and } 0.22 \times 500 = \underline{\underline{110 \text{ trips}}}$$

$$P_{B1} = \frac{0.722}{15.607}$$

$$= 0.046 \text{ and } 0.046 \times 500 = \underline{\underline{23 \text{ trips}}}$$

$$P_{A2} = \frac{8.365}{15.607}$$

$$= 0.536 \text{ and } 0.536 \times 500 = \underline{\underline{268 \text{ trips}}}$$

$$P_{B2} = \frac{1.76}{15.607}$$

$$= 0.113 \text{ and } 0.113 \times 500 = \underline{\underline{57 \text{ trips}}}$$

$$P_{A3} = \frac{1.195}{15.607}$$

$$= 0.077 \text{ and } 0.077 \times 500 = \underline{\underline{38 \text{ trips}}}$$

$$P_{B3} = \frac{0.13}{15.607}$$

$$= 0.008 \text{ and } 0.008 \times 500 = \underline{\underline{4 \text{ trips}}}$$

EXAMPLE 8.8

Consider the situation described in Example 8.7. A labor dispute results in a bus union slowdown that increases travel times from the origin by 4, 2, and 8 min to destinations 1, 2, 3, respectively. If the total number of trips remains constant, determine the resulting distribution of trips by mode and destination.

SOLUTION

The mode/destination utilities are computed as

$$U_{A1} = 1.234 \text{ (as in Example 8.7)}$$

$$U_{B1} = -0.22(21) + 0.16(12.4) + 1.1(1.3)$$

$$= -1.206$$

$$U_{A2} = 2.124 \text{ (as in Example 8.7)}$$

$$U_{B2} = -0.22(10) + 0.16(8.2) + 1.1(0.92)$$

$$= -0.124$$

$$U_{A3} = 0.178 \text{ (as in Example 8.7)}$$

$$U_{B3} = -0.22(32) + 0.16(5.8) + 1.1(2.1)$$

$$= -3.802$$

Applying Eq. 8.5 with 500 trips gives the following distribution of trips among mode/destination alternatives:

$$P_{A1} = \frac{3.435}{14.45}$$

$$= 0.238 \text{ and } 0.238 \times 500 = \underline{\underline{119 \text{ trips}}}$$

$$P_{B1} = \frac{0.299}{14.45}$$

$$= 0.021 \text{ and } 0.021 \times 500 = \underline{\underline{10 \text{ trips}}}$$

$$P_{A2} = \frac{8.365}{14.45}$$

$$= 0.579 \text{ and } 0.579 \times 500 = \underline{\underline{290 \text{ trips}}}$$

$$P_{B2} = \frac{1.132}{14.45}$$

$$= 0.078 \text{ and } 0.078 \times 500 = \underline{\underline{39 \text{ trips}}}$$

$$P_{A3} = \frac{1.195}{14.45}$$

$$= 0.083 \text{ and } 0.083 \times 500 = \underline{\underline{41 \text{ trips}}}$$

$$P_{B3} = \frac{0.022}{14.45}$$

$$= 0.002 \text{ and } 0.002 \times 500 = \underline{\underline{1 \text{ trip}}}$$

8.6   HIGHWAY ROUTE CHOICE

To summarize, the trip generation and mode/destination choice models give total highway traffic demand between a specified origin (e.g., the neighborhood from which trips originate) and a destination (e.g., the geographic area to which trips are destined), in terms of vehicles per some time period (usually vehicles per hour). With this information in hand, the final step in the sequential approach to traffic forecasting—route choice—can be addressed. The result of the route choice decision will be traffic flow (generally in units of vehicles per hour) on specific highway routes, which is the desired output from the traffic forecasting process.

8.6.1 Highway Performance Functions

Route choice presents itself as a classic equilibrium problem, because travelers' route choice decisions are primarily a function of route travel times that are determined by traffic flow, which itself is a product of route choice decisions. This interrelationship between route choice decisions and traffic flow forms the basis of route choice theory and model development.

To begin modeling traveler route choice, a mathematical relationship between route travel time and route traffic flow is needed. Such a relationship is commonly referred to as a highway performance function. The most simplistic approach to formalizing this relationship is to assume a linear highway performance function in which travel time increases linearly with speed. An example of such a function is illustrated in Fig. 8.5. In this figure, the free-flow travel time refers to the travel time that a traveler would experience if no other vehicles were present to impede travel speed (as discussed in Chapter 5). This free-flow speed is generally computed assuming that a vehicle travels at the speed limit of the route.

Although the linear highway performance function has the appeal of simplicity, it is not a particularly realistic representation of the travel time/traffic flow relationship. Recall that Chapter 5 presented a relationship between traffic speed and flow that was parabolic in nature, with significant reductions in travel speed occurring as the traffic flow approached the roadway's capacity. This parabolic speed/flow relationship suggests a nonlinear highway performance function, such as that illustrated in Fig. 8.6. This figure shows route travel time increasing more quickly as traffic flow approaches capacity, which is consistent with the parabolic relationship presented in Chapter 5.

Both linear and nonlinear highway performance functions will be demonstrated, through example, using two theories of travel route choice: user equilibrium and system optimization. For other theories of route choice, the reader is referred to Sheffi (1985).

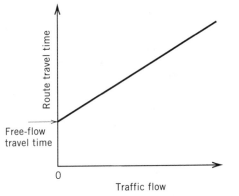

Figure 8.5 Linear travel time/flow relationship.

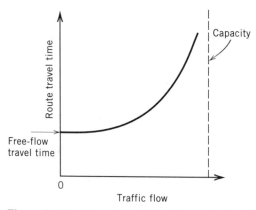

Figure 8.6 Nonlinear travel time/flow relationship.

8.6.2.   Theory of User Equilibrium

In developing theories of traveler route choice, two important assumptions are usually made. First, it is assumed that travelers will select routes between origins and destinations on the basis of route travel times only (i.e., they will tend to select the route with the shortest travel time). This assumption is not terribly restrictive, because travel time obviously plays the dominant role in route choice, but other more subtle factors that may influence route choice (e.g., scenery) are not accounted for. The second assumption is that travelers know the travel times that would be encountered on all available routes between their origin and destination. This is, potentially, a strong assumption, because a traveler may not have actually traveled on all available routes between an origin and destination and may repeatedly (day after day) choose one route based only on the perception that travel times on alternate routes are higher. However, in support of this assumption, studies have shown that travelers' perceptions of alternate route travel times are reasonably close to actual observed travel times (Mannering 1989).

With these assumptions, the theory of user equilibrium route choice can be made operational. The rule of choice underlying user equilibrium is that travelers will select a route so as to minimize their personal travel time between their origin and destination. User equilibrium is said to exist when individual travelers cannot improve their travel times by unilaterally changing routes. Stated differently (Wardrop 1952), user equilibrium can be defined as follows:

The travel time between a specified origin and destination on all used routes is equal, and less than or equal to the travel time that would be experienced by a traveler on any unused route.

EXAMPLE 8.9

Two routes connect a city and a suburb. During the peak-hour morning commute, a total of 4500 vehicles travel from the suburb to the city. Route 1 has a 60 km/h speed limit and is 6 km in length; route 2 is 3 km in length with a 45 km/h speed limit. Studies show that the total travel time on route 1 increases two minutes for every additional 500 vehicles added. Minutes of travel time on route 2 increase with the square of the number of vehicles expressed in thousands of vehicles per hour. Determine user equilibrium travel times.

SOLUTION

Determining free-flow travel times, in minutes, gives

Route 1:   6 km/60 km/h $\times$ 60 min/h = 6 min

Route 2:   3 km/45 km/h $\times$ 60 min/h = 4 min

With these data, the performance functions can be written as

$$t_1 = 6 + 4x_1$$
$$t_2 = 4 + x_2^2$$

where t_1 and t_2 are the average travel times on routes 1 and 2 in minutes, and x is the traffic flow in thousands of vehicles per hour. Also, we have the basic flow conservation identity,

$$q = x_1 + x_2 = 4.5$$

where q is the total traffic flow between the origin and destination in thousands of vehicles per hour. With Wardrop's definition of user equilibrium, it is known that the travel times on all used routes are equal. However, the first order of concern is to determine whether or not both routes are used. Figure 8.7 gives a graphic representation of the two performance functions. Note that because route 2 has a lower free-flow travel time, any total origin-to-destination traffic flow less than q' (in Fig. 8.7) will result in only route 2 being used, because the travel time on route 1 would be greater even if only one vehicle used it. At flows of q' and above, route 2 is sufficiently congested, and its travel time sufficiently high, so that route 1 becomes a viable alternative. To check if the problem's flow of 4500 vehicles per hour exceeds q', the following test is conducted:

(a) Assume that all traffic flow is on route 1. Substituting traffic flows of 4.5 and 0 into the performance functions gives $t_1(4.5) = 24$ min and $t_2(0) = 4$ min.

(b) Assume that all traffic flow is on route 2, giving $t_1(0) = 6$ min and $t_2(4.5) = 24.25$ min.

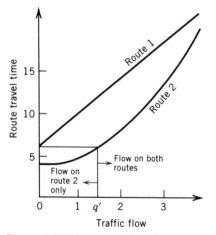

Figure 8.7 Illustration of performance curves for Example 8.9.

Thus, because $t_1(4.5) > t_2(0)$ and $t_2(4.5) > t_1(0)$, both routes will be used. If $t_1(0)$ would have been greater than $t_2(4.5)$, the 4500 vehicles would have been less than q' in Fig. 8.7 and only route 2 would have been used.

With both routes used, Wardrop's user equilibrium definition gives

$$t_1 = t_2$$

or

$$6 + 4x_1 = 4 + x_2^2$$

From flow conservation, $x_1 + x_2 = 4.5$. Substituting, we obtain

$$6 + 4(4.5 - x_2) = 4 + x_2^2$$

$$x_2 = 2.899 \text{ or } 2899 \text{ veh/h}$$

$$x_1 = 4.5 - x_2 = 4.5 - 2.899$$

$$= 1.601 \text{ or } 1601 \text{ veh/h}$$

which gives average route travel times of

$$t_1 = 6 + 4(1.601)$$

$$= 12.4 \text{ min}$$

$$t_2 = 4 + (2.899)^2$$

$$= \underline{\underline{12.4 \text{ min}}}$$

EXAMPLE 8.10

Peak-hour traffic demand between an origin and destination pair is initially 3500 vehicles. The two routes connecting the pair have performance functions $t_1 = 2 + 3(x_1/c_1)$ and $t_2 = 4 + 2(x_2/c_2)$, where t's are in minutes, and flows (x's) and capacities (c's) are in thousands of vehicles per hour. Initially, the capacities of routes 1 and 2 are 2500 and 4000 veh/h, respectively. A reconstruction project reduces capacity on route 2 to 2000 veh/h. Assuming user equilibrium before and during reconstruction, what reduction in total peak-hour origin–destination traffic flow is needed to ensure that total travel times (i.e., summation of all $x_a t_a$'s, where a denotes route) during reconstruction are equal to those before reconstruction?

SOLUTION

First, focusing on the roads before reconstruction, a check to see if both routes are used gives (using performance functions)

$$t_1(3.5) = 6.2 \text{ min}; t_2(0) = 4 \text{ min}$$

$$t_1(0) = 2 \text{ min}; t_2(3.5) = 5.75 \text{ min}$$

which, because $t_1(3.5) > t_2(0)$ and $t_2(3.5) > t_1(0)$, indicates that both routes are used. Setting route travel times equal and substituting performance functions gives

$$2 + \frac{3}{2.5}(x_1) = 4 + \frac{2}{4}(x_2)$$

and, from conservation of flow, $x_2 = 3.5 - x_1$, so that

$$2 + 1.2x_1 = 4 + 0.5(3.5 - x_1)$$

Solving gives $x_1 = 2.206$ and $x_2 = 3.5 - 2.206 = 1.294$. For travel times,

$$t_1 = 2 + 1.2(2.206)$$

$$= 4.647 \text{ min}$$

$$t_2 = 4 + 0.5(1.294)$$

$$= 4.647 \text{ min}$$

The total peak-hour travel time before reconstruction will simply be the average route travel time multiplied by the number of vehicles:

$$\text{total travel time} = 4.647(3500)$$

$$= 16,264.5 \text{ veh-min}$$

During reconstruction, the performance function of route 1 is unchanged, but the performance function of route 2 is altered because of the reduction in capacity to

$$t_2 = 4 + \frac{2}{2}x_2 = 4 + x_2$$

If we assume that both routes are used, $t_1 = t_2$. Also, it is known that the total travel time is

$$t_1(q) = t_2(q)$$
$$= 16{,}264.5 \text{ veh-min}$$

Using the performance function of route 2, we obtain

$$(4 + x_2)(q) = 16.2645 \text{ (using thousands of vehicles)}$$
$$q = \frac{16.2645}{4 + x_2}$$

From $t_1 = t_2$, and $x_1 = q - x_2$ (flow conservation),

$$2 + 1.2x_1 = 4 + x_2$$
$$2 + 1.2(q - x_2) = 4 + x_2$$
$$q = 1.67 + 1.83x_2$$

Equating the two expressions for q gives

$$1.67 + 1.83x_2 = \frac{16.2645}{4 + x_2}$$
$$1.83x_2^2 + 8.99x_2 - 9.5845 = 0$$

which gives $x_2 = 0.901$, $q = 1.67 + 1.83(0.901) = 3.319$, and $x_1 = 3.319 - 0.901 = 2.418$. Because flow exists on both routes, the earlier assumption that both routes would be used is valid, and a reduction of $181(3500 - 3319)$ vehicles in peak-hour flow is needed to ensure equality of total travel times.

EXAMPLE 8.11

Two highways serve a busy corridor with a traffic demand that is fixed at 6000 vehicles during the peak hour. The service functions for the two routes are $t_1 = 4 + 5(x_1/c_1)$ and $t_2 = 3 + 7(x_2/c_2)$, where the t's are travel times in minutes, the x's are the peak-hour traffic volumes expressed in thousands, and the c's are the peak-hour route capacities expressed in thousands of vehicles per hour. Initially, the capacities of routes 1 and 2 are 4400 veh/h and 5200 veh/h, respectively. If a highway reconstruction project cuts the capacity of route 2 to 2200 veh/h, how many additional vehicle hours will be added in the corridor assuming that user equilibrium conditions hold?

SOLUTION

To determine the initial number of vehicle hours, first check to see if both routes are used:

$$t_1(6) = 10.82 \text{ min}; \ t_2(0) = 3 \text{ min}$$

$$t_1(0) = 4 \text{ min}; \ t_2(6) = 11.08 \text{ min}$$

Both routes are used, because $t_2(6) > t_1(0)$ and $t_1(6) > t_2(0)$. At user equilibrium, $t_1 = t_2$, so that substituting performance functions gives

$$4 + \frac{5}{4.4}(x_1) = 3 + \frac{7}{5.2}(x_2)$$

With flow conservation, $x_2 = 6 - x_1$, so that

$$4 + 1.136(x_1) = 3 + 1.346(6 - x_1)$$

$$x_1 = 2.85$$

and

$$x_2 = 6 - 2.85$$

$$= 3.15$$

The total travel time in hours is $(t_1 x_1 + t_2 x_2)/60$ or, by substituting,

$$\frac{\{[(4 + 1.136(2.85)]2850 + [3 + 1.346(3.15)]3150\}}{60} = 723.88 \text{ veh-h}$$

For the reduced-capacity case, the route usage check is

$$t_1(6) = 10.82 \text{ min}; \ t_2(0) = 3 \text{ min}$$

$$t_1(0) = 4 \text{ min}; \ t_2(6) = 22.09 \text{ min}$$

Again, both routes are used $[t_2(6) > t_1(0) \text{ and } t_1(6) > t_2(0)]$. Equating performance functions (because travel times are equal) and using flow conservation $x_2 = 6 - x_1$,

$$4 + \frac{5}{4.4}(x_1) = 3 + \frac{7}{2.2}(x_2)$$

$$4 + 1.136x_1 = 3 + 3.182(6 - x_1)$$

$$x_1 = 4.19$$

and

$$x_2 = 6 - 4.19 = 1.81$$

which gives a total travel time of $(t_1 x_1 + t_2 x_2)/60$ or, by substituting,

$$\frac{\{[(4 + 1.136(4.19)]4190 + [3 + 3.182(1.81)]1810\}}{60} = 875.97 \text{ veh-h}$$

Thus the reduced capacity adds an additional 152.09 veh-h $(875.97 - 723.88)$.

8.6.3   Mathematical Programming Approach to User Equilibrium

Equating travel time on all used routes is a straightforward approach to user equilibrium, but one that can become cumbersome when many alternate routes are involved. The approach used to resolve this computational obstacle is to formulate the user equilibrium problem as a mathematical program. Specifically, user equilibrium route flows can be obtained by minimizing the following function (Sheffi 1985):

$$\min y(x) = \sum_n \int_0^{x_n} t_n(w) \, dw \qquad (8.6)$$

where n denotes a specific route and $t_n(w)$ is the performance function corresponding to route n (w denotes flow, x_n's). This function is subject to the constraints that the flow on all routes is greater than or equal to zero ($x_n \geq 0$) and that flow conservation holds (i.e., the flow on all routes between an origin and destination sums to the total number of vehicles, q, traveling between the origin and destination, $q = \sum_n x_n$).

Formulating the user equilibrium problem as a mathematical program permits an equilibrium solution to very complex highway networks (i.e., many origins and destinations) to be readily undertaken by computer. The reader is referred to Sheffi (1985) for an application of user equilibrium principles to such a network.

EXAMPLE 8.12

Solve Example 8.9 by formulating user equilibrium as a mathematical program.

SOLUTION

From Example 8.9, the performance functions are

$$t_1 = 6 + 4x_1$$
$$t_2 = 4 + x_2^2$$

Substituting these into Eq. 8.6 gives

$$\min y(x) = \int_0^{x_1} (6 + 4w)\, dw + \int_0^{x_2} (4 + w^2)\, dw$$

The problem can be viewed in terms of x_2 only by noting that flow conservation implies $x_1 = 4.5 - x_2$. Substituting, we obtain

$$y(x) = \int_0^{4.5-x_2} (6 + 4w)\, dw + \int_0^{x_2} (4 + w^2)\, dw$$

$$= 6w + 2w^2 \Big|_0^{4.5-x_2} + 4w + \frac{w^3}{3}\Big|_0^{x_2}$$

$$= 27 - 6x_2 + 40.5 - 18x_2 + 2x_2^2 + 4x_2 + \frac{x_2^3}{3}$$

To arrive at a minimum, the first derivative is set to zero, giving

$$\frac{dy(x)}{dx_2} = x_2^2 + 4x_2 - 20$$

$$= 0$$

which, solving for x_2, gives $x_2 = 2899$ veh/h, the same value as was found in Example 8.9. It can readily be shown that all other flows and travel times will also be the same as those computed in Example 8.9.

8.6.4 Theory of System Optimal Route Choice

From an idealistic point of view, one can visualize a single route choice strategy that results in the lowest possible number of total vehicle hours of travel for some specified origin/destination traffic flow. Such strategy is known as a system optimal route choice and is based on the choice rule that travelers will behave such that total system travel time will be minimized even though travelers may be able to decrease their own individual travel times by unilaterally changing routes. From this definition it is clear that system optimal flows are not stable, because there will always be a temptation for travelers to switch to nonsystem optimal routes in order to improve their own travel times. Thus system optimal flows are generally not a realistic representation of actual traffic. Nevertheless, system optimal flows often provide useful comparisons with the more realistic user equilibrium traffic forecasts.

The system optimal route choice rule is made operational by the following mathematical program:

$$\min y(x) = \sum_n x_n t_n(x_n) \qquad \qquad \textbf{(8.7)}$$

This program is subject to the constraint of flow conservation ($q = \sum_{n} x_n$) and nonnegativity ($x_n \geq 0$).

EXAMPLE 8.13

Determine the system optimal travel time for the situation described in Example 8.9.

SOLUTION

Using Eq. 8.7 and substituting the performance functions for routes 1 and 2, we obtain

$$y(x) = x_1(6 + 4x_1) + x_2(4 + x_2^2)$$
$$= 6x_1 + 4x_1^2 + 4x_2 + x_2^3$$

From flow conservation, $x_1 = 4.5 - x_2$; therefore,

$$y(x) = 6(4.5 - x_2) + 4(4.5 - x_2)^2 + 4x_2 + x_2^3$$
$$= x_2^3 + 4x_2^2 - 38x_2 + 108$$

To find the minimum, the first derivative is set to zero:

$$\frac{dy(x)}{dx_2} = 3x_2^2 + 8x_2 - 38 = 0$$

which gives $x_2 = 2.467$ and $x_1 = 4.5 - 2.467 = 2.033$. For system optimal travel times,

$$t_1 = 6 + 4(2.033)$$
$$= 14.13 \text{ min}$$
$$t_2 = 4 + (2.467)^2$$
$$= 10.08 \text{ min}$$

which are not user equilibrium travel times, because t_1 is not equal to t_2. In Example 8.9, the total user equilibrium travel time is computed as 930 veh-h [4500(12.4)/60]. For the system optimal total travel time [$(x_1t_1 + x_2t_2)/60$]),

$$\frac{[2033(14.13) + 2467(10.08)]}{60} = 893.2 \text{ veh-h}$$

Therefore, the system optimal solution results in a systemwide saving of 36.8 veh-h.

EXAMPLE 8.14

Two roads begin at a gate entrance to a park and both take different scenic routes to a single main attraction in the park. The park manager knows that 4000 vehicles arrive during the peak hour, and he distributes these vehicles among the two routes so that an equal number of vehicles take each route. The performance functions for the routes are $t_1 = 10 + x_1$ and $t_2 = 5 + 3x_2$ with the x's expressed in thousands of vehicles per hour and the t's in minutes. How many vehicle-hours would have been saved had the park manager distributed the vehicular traffic so as to achieve a system optimal solution?

SOLUTION

For the number of vehicle hours, assuming an equal distribution of traffic among the two routes,

$$\text{route 1:} \quad \frac{x_1 t_1}{60} = \frac{2000[10 + (2)]}{60}$$

$$= 400 \text{ veh-h}$$

$$\text{route 2:} \quad \frac{x_2 t_2}{60} = \frac{2000[5 + 3(2)]}{60}$$

$$= 366.67 \text{ veh-h}$$

for a total of 766.67 veh-h. With the system optimal traffic distribution, the performance functions are substituted into Eq. 8.7, giving

$$y(x) = (10 + x_1)x_1 + (5 + 3x_2)x_2$$

With flow conservation, $x_1 = 4.0 - x_2$ so that

$$y(x) = 4x_2^2 - 13x_2 + 56$$

Setting the first derivative to zero, we find that

$$\frac{dy(x)}{dx_2} = 8x_2 - 13$$

$$= 0$$

which gives $x_2 = 1.625$ and $x_1 = 4 - 1.625 = 2.375$. The total travel times are

$$\text{route 1:} \quad \frac{x_1 t_1}{60} = \frac{[2375(10 + 2.375)]}{60}$$

$$= 489.84 \text{ veh-h}$$

$$\text{route 2:} \quad \frac{x_2 t_2}{60} = \frac{\{1625(5 + 3[1.625])\}}{60}$$

$$= 267.45 \text{ veh-h}$$

which gives a total system travel time of 757.27 veh-h or a saving of 9.38 veh-h (766.67 − 757.29) over the equal distribution of traffic to the two routes.

8.7 THE STATE OF TRAFFIC FORECASTING IN PRACTICE

As mentioned earlier, the approach to traffic forecasting presented in this chapter provides an excellent exposition of the principles underlying the traffic forecasting problem. In practice, the most widely used approach to traffic forecasting is a four-step procedure: trip generation, mode choice, destination choice, and traffic assignment. This differs from the three-step procedure presented in this chapter in that mode and destination choice are modeled separately. Separating the mode and destination choice is theoretically incorrect, but it tends to simplify the modeling process considerably and makes model implementation easier. Although a logit formulation is often used for the mode choice decision in practice, the destination choice is often modeled by using the *gravity model* [see Meyer and Miller 1984]. The gravity model is based on the gravitational modeling principles covered in physics (i.e., the gravitational forces of planets) where the likelihood of a trip going to a destination is a function of the distance from the trip origin and some measure of attractiveness (i.e., the equivalent of mass in gravitational theory) of the destination. The use of the gravity model is a relic that remains from the earliest traffic forecasting efforts, before notions of utility maximization were fully implementable. Its continued use is not necessarily an indication of resistance in the traffic forecasting field to more theoretically consistent modeling approaches (although some resistance to more advanced approaches certainly exists), but more a reflection of the complexity of the underlying travel decision problem and the methodological and computational barriers that have made it difficult for the profession to evolve beyond the trip generation, mode choice, destination choice, and route choice stepped process.

Currently, the traffic forecasting profession is attempting to deal with some of the obvious limitations of the stepped process. These limitations include the difficulty in modeling trip chaining (i.e., the combining of trips such as shopping on the way to or from work), the temporal problem (i.e., the ability of trip makers to delay their trips in response to congestion), equilibrium concerns (e.g., the interaction among the models of the stepped modeling process), the use of the household as the decision-making unit instead of individual household members, the long-term effects of traveler decisions on residential and commercial locations (see Fig. 8.1), and the stability of model coefficients over time (i.e., as one forecasts into the future, it is a matter of some debate as to whether or not the estimated logit and regression model coefficients can be assumed to be the same as they were when the model was estimated). All of these limitations suggest that the stepped process is in need of considerable revision. However, the recent theoretical advances that have been made in the traffic forecasting field have proven difficult to implement, and, consequently, the stepped process remains as the most practical approach to the traffic forecasting problem.

APPENDIX 8A   METHOD OF LEAST SQUARES REGRESSION

Least squares regression is a popular method of developing mathematical relationships from empirical data. As mentioned in the text, it is a method that is well suited to the estimation of trip generation models. To illustrate the least squares approach, consider the hypothetical trip generation data presented in Table 8A.1, which could have been gathered from a typical survey of travelers.

To begin formalizing a mathematical expression, note that the objective is to predict the number of shopping trips made on a Saturday for each household, i; this number is referred to as the dependent variable (Y_i). This prediction is to be a function of the number of people in household i (z_i), which is referred to as the independent variable. A simple linear relationship between Y_i and z_i is

$$Y_i = b_0 + b_1 z_i \qquad \text{(8A.1)}$$

where the b's are coefficients to be determined and Y_i is the number of shopping trips predicted by the equation. Ideally, we want to determine the b's in Eq.

Table 8A.1 Example of Shopping Trip Generation Data

Household Number i	Number of Shopping Trips Made All Day Saturday Y_i	People in Household i z_i
1	3	4
2	1	2
3	1	3
4	5	4
5	3	2
6	2	4
7	6	8
8	4	6
9	5	6
10	2	2

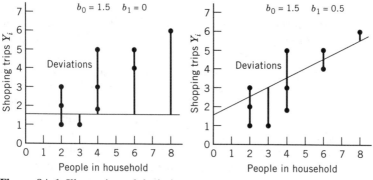

Figure 8A.1 Illustration of deviations.

8A.1 that will give predictions of the number of shopping trips (Y_i's) that are as close as possible to the actual observed number of shopping trips (Y_i's, as shown in Table 8A.1). The difference or deviation between the observed and predicted number of shopping trips can be expressed mathematically as

$$\text{deviation} = Y_i - (b_0 + b_1 z_i) \qquad \textbf{(8A.2)}$$

Graphically, such deviations are illustrated in Fig. 8A.1 for two groups of b_0 and b_1 values. In the first illustration of this figure, $b_0 = 1.5$ and $b_1 = 0$, which implies that the number of household members does not affect the number of shopping trips made. The second illustration has $b_0 = 1.5$ and $b_1 = 0.5$ and, as can readily be seen, the deviations (differences between the points representing the observed number of shopping trips and the line representing the equation, $b_0 + b_1 z_i$) are reduced relative to the first illustration. These two illustrations suggest the need for some method of determining the values of b_0 and b_1 that produce the smallest possible deviations relative to observed data. Such a method can be solved by a mathematical program whose objective is to minimize the sum of the square of deviations, or

$$\min y(b_0, b_1) = \sum_i (Y_i - b_0 - b_1 z_i)^2 \qquad \textbf{(8A.3)}$$

The minimization is accomplished by setting partial derivatives equal to zero:

$$\frac{\partial y}{\partial b_0} = -2 \sum_i (Y_i - b_0 - b_1 z_i)$$

$$= 0$$

$$\frac{\partial y}{\partial b_1} = -2 \sum_i z_i (Y_i - b_0 - b_1 z_i)$$

$$= 0$$

giving

$$\sum_i (Y_i - b_0 - b_1 z_i) = 0$$

$$\sum_i z_i(Y_i - b_0 - b_1 z_i) = 0$$

or

$$\sum_i Y_i - nb_0 - b_1 \sum_i z_i = 0$$

$$\sum_i z_i Y_i - b_0 \sum_i z_i - b_1 \sum_i z_i^2 = 0$$

where n is the total number of observations (or, in this case, households). Solving these equations simultaneously for b_0 and b_1 gives

$$b_1 = \frac{\sum_i (z_i - \bar{z})(Y_i - \bar{Y})}{\sum_i (z_i - \bar{z})^2} \qquad (8A.4)$$

$$b_0 = \bar{Y} - b_1 \bar{z} \qquad (8A.5)$$

This approach to determining the values of estimable coefficients (b's) is referred to as least squares regression, and it can be shown that for the data values given in Table 8A.1, the lowest deviations between the number of predicted and actual shopping trips will be given by the equation

$$Y_i = 0.33 + 0.7z_i$$

When many coefficient values (b's) must be determined, a matrix representation of the least squares solution is appropriate,

$$\mathbf{B} = (\mathbf{z'z})^{-1}\mathbf{z'Y} \qquad (8A.6)$$

where $\mathbf{B}$ is a vector of b_n's, $\mathbf{z}$ is a matrix of characteristics, and $\mathbf{Y}$ is the vector of dependent variables. For additional information on least squares regression, the reader is referred to Neter and Wasserman (1973).

NOMENCLATURE FOR CHAPTER 8

b_n estimated coefficients

P probability of an alternative being selected

q total origin-to-destination traffic flow

s notation for the set of available alternatives

T_{ij} number of household trips generated per unit time

t_a travel time on route a

U specifiable portion of an alternative's utility

V total alternative utility

w route flow operative for x_a

x_a traffic flow on route a

Y_i dependent variable

$y(\cdot)$ mathematical objective function

z household or alternative characteristic

ε unspecifiable portion of an alternative's utility (assumed to be a random variable)

REFERENCES

Ben-Akiva, M., and S. Lerman. *Discrete Choice Analysis: Theory and Application to Travel Demand.* Cambridge, MA: MIT Press, 1985.

Hamed, M., and F. Mannering. "Modeling Travelers' Post-Work Activity Involvement: Toward a New Methodology." *Transportation Science,* vol. 17, no. 4, 1993.

Mannering, F. "Poisson Analysis of Commuter Flexibility in Changing Route and Departure Times." *Transportation Research,* vol. 23B, no. 1, 1989.

Mannering, F., S. Abu-Eisheh, and A. Arnadottir. "Dynamic Traffic Equilibrium with Discrete/Continuous Econometric Models." *Transportation Science,* vol. 24, no. 2, 1990.

McFadden, D. "Econometric Models of Probabilistic Choice." In *Structural Analysis of Discrete Data with Econometric Applications,* edited by Manski and McFadden. Cambridge, MA: MIT Press, 1981.

Meyer, M., and E. Miller. *Urban Transportation Planning: A Decision-Oriented Approach.* New York: McGraw-Hill, 1984.

Neter J., and W. Wasserman. *Applied Linear Statistical Models.* Homewood, IL: Richard Irwin, 1973.

Sheffi, Y. *Urban Transportation Networks: Equilibrium Analysis with Mathematical Programming Models.* Englewood Cliffs, NJ: Prentice-Hall, 1985.

Wardrop, F. "Some Theoretical Aspects of Road Traffic Research." *Proceedings, Institution of Civil Engineers II,* vol. 1, 1952.

PROBLEMS

8.1. A large retirement village has a total retail employment of 100. All 1700 of the households residing in this village consist of two nonworking family members with household incomes of $20,000. Assuming that shopping and social/recreational trip rates both peak during the same hour (for exposition purposes), predict the total number of peak-hour trips generated by this village using the trip generation models of Examples 8.1 and 8.2.

8.2. Consider the retirement village described in Problem 8.1. Determine the amount of additional retail employment (in the village) necessary to reduce the total predicted number of peak-hour shopping trips to 100.

8.3. A large residential area has 1500 households with an average household income of $15,000, an average household size of 5.2, and, on the average, 1.2 working members. Using the model described in Example 8.2 (assuming that it was estimated using zonal averages instead of individual households), predict the change in the number of peak-hour social/recreational trips if employment in the area increased by 20% and household income by 10%.

8.4. If small express buses leave the origin described in Example 8.4 and all are filled to their capacity of 15 travelers, how many work trip vehicles leave from origin to destination in Example 8.4 during the peak hour?

8.5. Consider the conditions described in Example 8.4. If an energy crisis doubles the cost of the auto modes (i.e., drive alone and shared rides) and bus costs are not affected, how many workers will take each mode?

8.6. It is known that 4000 automobile trips are generated in a large residential area from noon to 1:00 P.M. on Saturdays for shopping purposes. Four major shopping centers have the following characteristics:

Shopping Center Number	Distance from Residential Area (km)	Commercial Floor Space (in thousands of m²)
1	4	20
2	9	15
3	8	30
4	14	60

If a logit model is estimated with coefficients -0.283 for distance and 0.172 for commercial space (in thousands of m²), how many shopping trips will be made to each of the four shopping centers?

8.7. Consider the shopping trip situation described in Problem 8.6. Suppose that shopping center 3 goes out of business and shopping center 2 is expanded to 50,000 m² of commercial space. What would be the new distribution of the 4000 Saturday afternoon shopping trips?

8.8. If shopping center 3 is closed (see Problem 8.7), how much commercial floor space is needed in shopping centers 1 and 2 to ensure that each of them have the same probability of being selected as shopping center 4?

8.9. Consider the situation described in Example 8.6. If the construction of a new freeway lowers auto and transit travel times to shopping center 2 by 20%, determine the new distribution of shopping trips by destination and mode.

8.10. Consider the conditions descibed in Example 8.6. Heavily congested highways have caused travel times to increase to shopping center 2 by 4 min for both auto and transit modes (travel times to shopping center 1 are not affected). In order to attract as many total trips (auto and transit) as it did before the congestion, how much commercial floor space must be added to shopping center 2 (given that the total number of departing shopping trips remains at 900)?

8.11. A total of 700 auto-mode social/recreational trips are made from an origin (residential area) during the peak hour. A logit model estimation is made and three factors were found to influence the destination choice: (1) population at the destination in thousands (coefficient = 0.2), (2) distance from origin to destination in miles (coefficient = −0.15), and (3) square feet of amusement floor space (e.g., movie theaters, video game centers, etc.) in thousands (coefficient = 0.9). Four possible destinations have the following characteristics:

	Population (in thousands)	Distance from Origin (in km)	Amusement Space (in thousands of m^2)
Destination 1	15.5	12	0.5
Destination 2	6.0	8	1.0
Destination 3	0.8	3	0.8
Destination 4	5.0	11	1.5

Determine the distribution of trips among possible destinations.

8.12. Consider the situation described in Problem 8.11. If a new 1500 m^2 arcade center is built at destination 3, determine the distribution of the 700 peak-hour social/recreational trips.

8.13. Note that with the situation described in Example 8.7, 26.6% ([110 + 23]/ 500) of all social/recreational trips are destined for destination 1. If the total number of trips remains constant, how much additional amusement floor space would have to be added to destination 1 to have it capture 40.0% of the total social/recreational trips?

8.14. Consider the situation described in Problem 8.11. Destination 2 currently attracts 146 of the 700 social/recreational trips. If the total number of trips remains constant, determine the amount of amusement floor space that must be added to destination 2 to attract a total of 250 social/recreational trips.

8.15. Two routes connect an origin and a destination and the flow is 15,000 veh/h. Route 1 has a performance function $t_1 = 4 + 3x_1$, and route 2 has a function of $t_2 = b + 6x_2$, with the x's expressed in thousands of vehicles per hour and the t's in minutes.
  (a) If the user equilibrium flow on route 1 is 9780 veh/h, determine the free-flow speed on route 2 (i.e., b) and equilibrium travel times.
  (b) If population declines reduce the number of travelers at the origin and the total origin-destination flow is reduced to 7000 veh/h, determine user equilibrium travel times and flows.

8.16. An origin–destination pair is connected by a route with a performance function $t_1 = 8 + x_1$, and another with a function $t_2 = 1 + 2x_2$ (x's in thousands of vehicles per hour and t's in minutes). If the total origin–destination flow is 4000 veh/h, determine user equilibrium and system optimal route travel times, total travel time (in vehicle-minutes), and route flows.

8.17. Because of the great increase in vehicle-hours caused by the reconstruction project in Example 8.11, the state transportation department decides to regulate the flow of traffic on the two routes (until reconstruction is complete) to achieve a system optimal solution. How many vehicle-hours will be saved during each peak-hour period if this strategy is implemented and travelers are not permitted to achieve a user equilibrium solution?

8.18. For Example 8.10, what reduction in peak-hour traffic demand is needed to ensure an equality of total vehicle travel time (in vehicle-minutes) assuming a system optimal solution before and during reconstruction?

8.19. Two routes connect an origin and a destination. Routes 1 and 2 have performance functions $t_1 = 2 + x_1$ and $t_2 = 1 + x_2$, where the t's are in minutes and the x's are in thousands of vehicles per hour. The travel time on the routes is known to be in user equilibrium. If an observation on route 1 finds that the gaps between 40% of the vehicles is less than 5 sec, estimate the volume and average travel times on the two routes. (*Hint:* Assume a Poisson distribution of vehicle arrivals as discussed in Chapter 5.)

8.20. Three routes connect an origin and a destination with performance functions $t_1 = 8 + 0.5x_1$, $t_2 = 1 + 2x_2$, and $t_3 = 3 + 0.75x_3$, with the x's expressed in thousands of vehicles per hour and the t's expressed in minutes. If the peak-hour traffic demand is 3000 vehicles, determine user equilibrium traffic flows.

8.21. Two routes connect a suburban area and a city with route travel times (in minutes) given by the expressions $t_1 = 6 + 8(x_1/c_1)$ and $t_2 = 10 + 3(x_2/c_2)$, where the x's are expressed in thousands of vehicles per hour and the c's are the route capacities in thousands of vehicles per hour. Initially, the capacities of routes 1 and 2 are 4000 and 2000 veh/h, respectively. A reconstruction project on route 1 reduces the capacity to 3000 veh/h, but total traffic demand is unaffected. Observational studies note a 35.28 sec increase in average travel time on route 1 and a 68.5% increase in flow on route 2 after reconstruction begins. User equilibrium conditions exist before and during reconstruction. If both routes are always used, determine equilibrium flows and travel times before and after reconstruction begins.

8.22. Three routes connect an origin and destination with performance functions $t_1 = 2 + 0.5x_1$, $t_2 = 1 + x_2$, and $t_3 = 4 + 0.2x_3$ (t's in minutes and x's in thousands of vehicles per hour). Determine user equilibrium flows if the total origin-to-destination demand is (a) 10,000 veh/h and (b) 5000 veh/h.

8.23. For the routes described in Problem 8.22, what is the minimum origin-to-destination traffic demand (in vehicles per hour) that will ensure that all routes are used (assuming user equilibrium conditions).

8.24. An urban freeway has five lanes, four of which are unrestricted (open to all vehicular traffic), and one restricted lane that can be used only by vehicles with two or more occupants. The performance functions, between origin and destination, are $t_r = 4 + 2x_r$ for the restricted lane and $t_u = 4 + 0.5x_u$ for the unrestricted lanes, combined (t's in minutes and x's in thousands of vehicles per hour). During the peak hour, 2000 vehicles with one occupant and 2000 vehicles with two occupants depart for the destination. Determine the distribution of traffic among lanes such that total person-hours is minimized, and compare the savings in person-hours relative to a user equilibrium distribution of traffic among lanes.

8.25. A multilane highway has two northbound lanes. Each lane has a capacity of 1200 vehicles per hour. Currently, northbound traffic consists of 2500 vehicles with 1 occupant, 500 vehicles with 2 occupants, 300 vehicles with 3 occupants, and 20 buses with 50 occupants each. The highway's performance function is $t = t_0[1 + 1.15(x/c)^{6.87}]$, with t in minutes, $t_0 = 15$ minutes, and x and c are volumes and capacities in vehicles per hour. An additional lane is being added (with 1200-veh/h capacity). What would the total person-hours of travel be if the lane was (a) open to all traffic, (b) open to vehicles with 2 or more occupants only, and (c) open to vehicles with 3 or more occupants only? (Assume all qualified higher-occupancy vehicles use only the new lane, no unqualified vehicles use the new lane, and there is no mode shift.)

8.26. Consider the new lane addition in Problem 8.25. First, suppose 500 one-occupant-vehicle travelers take 10 buses (50 on each bus) and the new lane is open to vehicles with two or more occupants. What would the total person-hours be? Second, referring back to part (a) of Problem 8.25, what is the minimum mode shift from one-occupant vehicles to bus (with 50 persons each) needed to make the highway with the new lane (which is restricted to vehicles with two or more occupants) have the person-hours of travel time as low as having all three lanes (the two existing lanes and the new lane) open to all traffic? (Set up the equation and solve to the nearest 100 one-occupant vehicles).

8.27. Two routes connect an origin and destination with performance functions $t_1 = 5 + 3x_1$ and $t_2 = 7 + x_2$, with t's in minutes and x's in thousands of vehicles per hour. Total origin-destination demand is 7000 vehicles in the peak hour. What are user equilibrium and system optimal route flows and total travel times?

8.28. Consider the conditions in Problem 8.27. What is the value of the derivative of the user equilibrium math program evaluated at the system optimal solution with respect to x_1 (with x_1 equal to the system optimal solution)?

8.29. Two routes connect an origin and destination. Their performance functions are $t_1 = 3 + 1.5(x_1/c_1)^2$ and $t_2 = 5 + 4(x_2/c_2)$, with t's in minutes and x's and c's being route flows and capacities, respectively. The origin-destination demand is 6000 vehicles per hour, and c_1 and c_2 are equal to 2000 and 1500 vehicles per hour respectively. Proposed capacity improvements will increase c_2 by 1000 vehicles per hour. It is known that the routes are currently in user equilibrium, and it is estimated that each 1-minute reduction in route travel time will attract an additional 500 vehicles per hour (from latent travel demand and mode shifts). What will the user equilibrium flows and total hourly origin-destination demand be after the capacity improvement?

8.30. Two routes connect an origin–destination pair with performance functions $t_1 = 5 + 4x_1$ and $t_2 = 7 + 2x_2$, with t's in minutes and x's in thousands of vehicles per hour. Assuming both routes are used, can user equilibrium and system optimal solutions be equal at some feasible value of total origin–destination demand (q)? (Prove your answer.)

8.31. Three routes connect an origin–destination pair with performance functions $t_1 = 5 + 1.5x_1$, $t_2 = 12 + 3x_2$, and $t_3 = 2 + 0.2x_3^2$, with t's in minutes and x's in thousands of vehicles per hour. Determine user equilibrium flows if $q = 4000$ veh/h.

8.32. Two routes connect an origin–destination pair with performance functions $t_1 = 6 + 4x_1$ and $t_2 = 2 + 0.5x_2^2$ (with t's in minutes and x's in thousands of vehicles per hour). The origin–destination demand is 4000 veh/h at a travel time of 2 minutes, but for each additional minute added beyond these 2 minutes, 100 fewer vehicles depart. Determine user equilibrium route flows and total vehicle travel time.

8.33. A freeway has six lanes, four of which are unrestricted (open to all vehicle traffic), and two of which are restricted lanes that can be used only by vehicles with two or more occupants. The performance function for the highway is $t = 12 + (2/NL)x$, with t in minutes, NL being the number of lanes, and x in thousands of vehicles. During the peak hour, 3000 vehicles with one occupant and 4000 with two occupants depart for the destination. Determine the distribution of traffic between restricted and unrestricted lanes such that total person-hours are minimized.

8.34. Two routes connect an origin–destination pair with performance functions $t_1 = 6 + (x_1/2)^2$ and $t_2 = 6 + (x_2/2)^2$, with t's in minutes and x's in thousands of vehicles per hour. It is known that at user equilibrium, 75% of the origin–destination demand takes route 1. What percentage would take route 1 if a system optimal solution were achieved, and how much travel time would be saved?

8.35. Two routes connect an origin–destination pair with performance functions $t_1 = 5 + 3.5x_1$ and $t_2 = 1 + 0.5x_2^2$, with t's in minutes and x's in thousands of vehicles per hour. It is known that at $x_2 = 3$, the difference between the first derivatives of system optimal and user equilibrium math programs, evaluated with respect to x_2, is 7 $[dy(x)_{SO}/dx_2 - dy(x)_{UE}/dx_2 = 7]$. Determine the difference in total vehicle travel times (in vehicle-minutes) between user equilibrium and system optimal solutions.

Appendix

Geometric Design of Highways: U.S. Customary Units

A.1   INTRODUCTION

This appendix parallels Chapter 3, but all of the material is presented in U.S. customary units. It is important that U.S. highway engineers understand how to work with both metric and U.S. customary units in the area of geometric design. This is because much geometric work deals with realignment of existing highways whose as-built geometric plans are likely to be drawn in U.S. customary units. Thus engineers must understand the old plans and be able to design the revisions in metric units.

Current guidelines of highway design in U.S. customary units are presented in detail in "A Policy on Geometric Design of Highways and Streets 1990," published by the American Association of State Highway and Transportation Officials (AASHTO 1990). As discussed in Chapter 3, the 1994 version of this publication has the same design guidelines, but they are presented in metric units (AASHTO 1994). The following sections of this appendix parallel those in Chapter 3 exactly, and the reader is encouraged to refer back to Chapter 3 occasionally to gain a better understanding of the subtle differences in metric and U.S. customary unit design.

A.2   PRINCIPLES OF HIGHWAY ALIGNMENT

The alignment of a highway is a three-dimensional problem measured in x, y, and z dimensions. This is illustrated, from a driver's perspective, in Fig. A.1. However, in highway design practice, three-dimensional design computations are cumbersome and, what is perhaps more important, the actual implementation

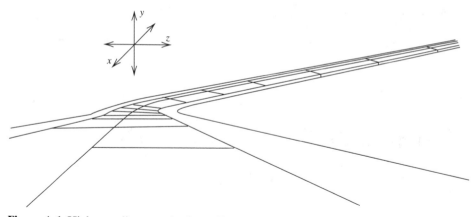

Figure A.1 Highway alignment in three-dimensions.

and construction of a design based on three-dimensional coordinates has historically been prohibitively difficult. As a consequence, the three-dimensional highway alignment problem is reduced to two two-dimensional alignment problems,
as illustrated in Fig. A.2. One of the alignment problems in this figure corresponds
roughly to x and z coordinates and is referred to as horizontal alignment. The
other corresponds to highway length (measured along some constant elevation)
and y coordinates (i.e., elevation) and is known as vertical alignment. Referring
to Fig. A.2, note that the horizontal alignment of a highway is referred to as the
plan view, which is roughly equivalent to the perspective of an aerial photo of
the highway. The vertical alignment is represented in a *profile view,* which gives
the elevations of all points measured along the length of the highway (again,
with length measured along a constant elevation reference).

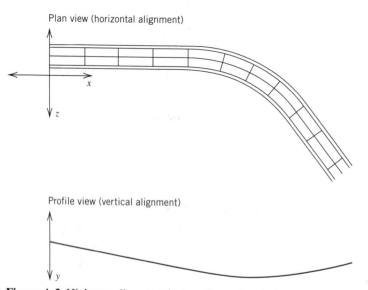

Figure A.2 Highway alignment in two-dimensional views.

Aside from considering the alignment problem as a pair of two-dimensional problems, one further simplification is made. That is, instead of using x and z coordinates, highway positioning and length are defined as the distance along the highway (usually measured along the centerline of the highway, on a horizontal, constant elevation plane) from a specified point. This distance is measured in terms of stations, with each station consisting of 100 ft of highway alignment distance. The notation for stationing distance is such that a point on a highway 1258.5 ft from a specified point of origin is said to be at station $12 + 58.5$ (i.e., 12 stations and 58.5 ft) with the point of origin being at station $0 + 00$. This stationing concept, when combined with the highway's alignment direction given in the plan view (horizontal alignment) and the elevations corresponding to stations given in the profile view (vertical alignment), gives a unique identification of all highway points in a manner that is virtually equivalent to using true x, y, and z coordinates.

A.3 VERTICAL ALIGNMENT

Vertical alignment specifies the elevations of points along a roadway. The elevations of these roadway points are usually determined by the need to provide proper drainage (from rainfall runoff) and an acceptable level of driver safety. A primary concern in vertical alignment is establishing the transition of roadway elevations between two grades. This transition is achieved by means of a vertical curve.

Vertical curves can be broadly classified into crest vertical curves and sag vertical curves as illustrated in Fig. A.3 (AASHTO 1990). In this figure, G_1 is

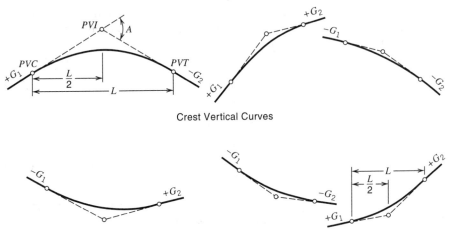

Figure A.3 Types of vertical curves. (Reproduced by permission from American Association of State Highway and Transportation Officials, "A Policy on Geometric Design of Highways and Streets," Washington, D.C., 1990.).

the initial roadway grade (also referred to as the initial tangent grade, viewing Fig. A.3 from left to right), G_2 is the final roadway (tangent) grade, A is the absolute value of the difference in grades (initial minus final, usually expressed in percent), L is curve length, PVC is the point of the vertical curve (the initial point of the curve), PVI is the point of vertical intersection (intersection of initial and final grades), and PVT is the point of vertical tangent, which is the final point of the vertical curve (i.e., the point where the curve returns to the final grade or, equivalently, final tangent). In practice, the vast majority of vertical curves are arranged so that half of the curve length is positioned before the PVI and half after (as illustrated in Fig. A.3). Curves that satisfy this criterion are said to be *equal tangent vertical curves.*

In terms of referencing points on a vertical curve, it is important to note that the profile views presented in Fig. A.3 correspond to all highway points even if a horizontal curve occurs concurrently with a vertical curve (as in Figs. A.1 and A.2). Thus each roadway point is uniquely defined by stationing (which is measured along a horizontal plane) and elevation. This will be made clearer through forthcoming examples.

A.3.1 Vertical Curve Fundamentals

In connecting roadway grades (tangents) with an appropriate vertical curve, a mathematical relationship defining roadway elevations at all points (or, equivalently, stations) along the vertical curve is needed. A parabolic function has been found suitable in this regard because, among other things, it provides a constant rate of change of slope and implies equal curve tangents. The general form of the parabolic equation, as applied to vertical curves, is

$$y = ax^2 + bx + c \tag{A.1}$$

where y is the roadway elevation x stations (or feet) from the beginning of the vertical curve (i.e., the PVC). By definition, c is the elevation of the PVC, because $x = 0$ corresponds to the PVC. In defining a and b, note that the first derivative of Eq. A.1 gives the slope and is

$$\frac{dy}{dx} = 2ax + b \tag{A.2}$$

At the PVC, $x = 0$, so, using Eq. A.2, we have

$$b = \frac{dy}{dx} = G_1 \tag{A.3}$$

where G_1 is the initial slope as previously defined. Also note that the second derivative of Eq. A.1 is the rate of change of slope and is

$$\frac{d^2y}{dx^2} = 2a \tag{A.4}$$

However, the average rate of change of slope, by observation of Fig. A.3, can also be written as

$$\frac{d^2y}{dx^2} = \frac{G_2 - G_1}{L} \qquad \text{(A.5)}$$

Equating Eqs. A.4 and A.5 gives

$$a = \frac{G_2 - G_1}{2L} \qquad \text{(A.6)}$$

These equations define all of the terms in the parabolic vertical curve equation (Eq. A.1). In these equations, it is important to note the units of G_1, G_2, x, and L. If G_1 and G_2 are expressed in percent (i.e., ft/station), x and L must be in stations. If G_1 and G_2 are expressed in ft/ft, x and L must be in feet. The following example gives a typical application of this equation.

EXAMPLE A.1

A 600-ft equal tangent sag vertical curve has the *PVC* at station 170 + 00 and elevation 1000 ft. The initial grade is −3.5 percent and the final grade is 0.5%. Determine the elevation and stationing of the *PVI, PVT,* and the lowest point on the curve.

SOLUTION

Because the curve is equal tangent, the *PVI* will be 300 ft or three stations (measured in a horizontal, profile plane) from the *PVC,* and the *PVT* will be 600 ft or six stations from the *PVC.* Therefore, the stationing of the *PVI* and *PVT* are 173 + 00 and 176 + 00, respectively. For the elevations of the *PVI* and *PVT,* it is known that a −3.5% grade can be equivalently written as −3.5 ft/station (i.e., a 3.5-ft drop per 100 ft of horizontal distance). Because the *PVI* is three stations from the *PVC,* which is known to be at elevation 1000 ft, the elevation of the *PVI* is

$$1000 - 3.5 \text{ ft/station} \times (3 \text{ stations}) = 989.5 \text{ ft}$$

Similarly, with the *PVI* at elevation 989.5 ft, the elevation of the *PVT* is

$$989.5 + 0.5 \text{ ft/station} \times (3 \text{ stations}) = 991.0 \text{ ft}$$

It is clear from the values of the initial and final grades that the lowest point on the vertical curve will occur when the first derivative of the parabolic function (Eq. A.1) is zero because the initial and final grades are opposite in sign. When initial and final grades are not opposite in sign, the low (or high) point on the curve will not be where the first derivative is zero because the slope along the curve will never be zero. For example, a sag curve with an initial grade of -2.0% and a final grade of -1.0% will have its lowest elevation at the *PVT*, and the first derivative of Eq. A.1 will not be zero at any point along the curve. However, in our example problem the derivative will be equal to zero at some point, so the low point will occur when

$$\frac{dy}{dx} = 2ax + b = 0$$

From Eq. A.3 we have

$$b = G_1 = -3.5$$

with G_1 in percent. From Eq. A.6 (with L in stations and G_1 and G_2 in percent),

$$a = \frac{0.5 - (-3.5)}{2(6)} = 0.33333$$

Substituting for a and b gives

$$\frac{dy}{dx} = 2(0.33333)x + (-3.5) = 0$$

$$x = 5.25 \text{ stations}$$

This gives the stationing of the low point at $175 + 25$ (i.e., $5 + 25$ stations from the *PVC*). For the elevation of the lowest point on the vertical curve, the values of a, b, c (elevation of the *PVC*), and x are substituted into Eq. A.1, giving

$$y = 0.33333(5.25)^2 + (-3.5)(5.25) + 1000$$

$$= 990.81 \text{ ft}$$

Note that these equations can also be solved with grades expressed as the decimal equivalent of percent (e.g., 0.02 ft/ft for 2%, which has units of ft/station) if x is expressed in feet instead of stations. Some care should be taken not to mix units. A dimensional analysis of Eq. A.1 must ensure that each right-side element of the equation has resulting units of feet.

Another interesting vertical curve problem that is sometimes encountered is one in which the curve must be designed so that the elevation of a specific location is met. An example might be to have the roadway connect with another (at the same elevation) or to have the roadway at some specified elevation pass under another roadway. This type of problem is referred to as a curve-through-a-point problem and is demonstrated by the following example.

EXAMPLE A.2

An equal tangent vertical curve is to be constructed between grades of -2.0% (initial) and 1.0% (final). The *PVI* is at sation $110 + 00$ and elevation 420 ft. Due to a street crossing the roadway, the elevation of the roadway at station $112 + 00$ must be at 424.5 ft. Design the curve.

SOLUTION

The design problem is one of determining the length of the curve required to ensure that station $112 + 00$ is at elevation 424.5 ft. To begin, we use Eq. A.1:

$$y = ax^2 + bx + c$$

From Eq. A.3,

$$b = G_1 = -2.0$$

and, from Eq. A.6,

$$a = \frac{G_2 - G_1}{2L}$$

Substituting $G_1 = -2.0$ and $G_2 = 1.0$, we have

$$a = \frac{G_2 - G_1}{2L} = \frac{1.0 - (-2.0)}{2L} = \frac{1.5}{L}$$

Now note that c (the elevation of the *PVC*) in Eq. A.1 will be equal to the elevation of the *PVI* plus $G_1 \times 0.5L$ (this is just using the slope of the initial grade to determine the elevation difference between the *PVI* and *PVC*). With G_1 in percent (i.e., ft/station) and the curve length L in stations, we have

$$c = 420 + 2.0(0.5L) = 420 + L$$

Finally, the value of x to be used in Eq. A.1 will be $0.5L + 2$ because the point of interest (station $112 + 00$) is 2 stations from the *PVI* (which is at station $110 + 00$). Substituting $b = -2.0$; the expressions for a, c, and x; and $y = 424.5$ ft (the given elevation) into Eq. A.1 gives

$$424.5 = (1.5/L)(0.5L + 2)^2 + (-2.0)(0.5L + 2) + (420 + L)$$
$$4.5 = 0.375L + 3 + 6/L - 4$$
$$0 = -0.375\,L^2 + 5.5L - 6$$

Solving this quadratic equation gives $L = 13.466$ stations (i.e., the curve must be 1346.6 ft long). Using this value of L,

elevation of $PVC = c = 420 + L = 420 + 13.466 = 433.47$ ft

stationing of $PVC = 110 + 00 - (13 + 46.6)/2 = 103 + 26.7$

elevation of $PVT =$ elevation of $PVI + (0.5L)G_2 = 420 + [0.5(13.466)](1.0)$

$$= 426.73 \text{ ft}$$

stationing of $PVT = 110 + 00 + (13 + 46.6)/2 = 116 + 73.3$

and

$$x = 0.5L + 2.0 = 6.733 + 2.0 = 8.733 \text{ stations from the } PVC$$

To check the elevation of the curve at station $112 + 00$, we apply Eq. A.1 with $x = 8.733$:

$$y = ax^2 + bx + c$$

$$y = \left(\frac{3}{2(13.466)}\right)(8.733)^2 + (-2.0)(8.733) + 433.47$$

$$y = 424.5 \text{ ft}$$

Therefore, all calculations are correct.

Some additional properties of vertical curves can now be formalized. For example, offsets, which are vertical distances from the initial tangent to the curve as illustrated in Fig. A.4, are extremely important in vertical curve design and construction. In Fig. A.4, Y is the offset at any distance, x, from the PVC; Y_m is the midcurve offset; and Y_f is the offset at the end of the vertical curve. From the properties of an equal tangent parabola, it can be readily shown that

$$Y = \frac{A}{200L}x^2 \tag{A.7}$$

where Y is the offset in feet, A is the absolute value of the difference in grades ($|G_1 - G_2|$ expressed in percent), L is the length of the vertical curve in feet, and x is the distance from the PVC in feet. It follows from Fig. A.4 that

$$Y_m = \frac{AL}{800} \tag{A.8}$$

and

$$Y_f = \frac{AL}{200} \tag{A.9}$$

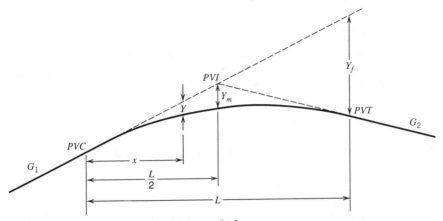

Figure A.4 Offsets for equal tangent vertical curves.

Another useful vertical curve property is one that simplifies the computation of the high and low points of crest and sag vertical curves, respectively (given that the high or low point does not occur at the curve ends, PVC or PVT). Recall that in Example A.1, the first derivative was used to determine the location of the low point. The alternative to this is to use a K-value defined (with L in feet and A in percent) as,

$$K = \frac{L}{A} \qquad (A.10)$$

The K-value can be used directly to compute the high/low points for crest/sag vertical curves by

$$x_{hl} = K \times |G_1| \qquad (A.11)$$

where x_{hl} is the distance from the PVC to the high/low point, and G_1 is the initial grade in percent. In words, the K-value is the horizontal distance, in feet, required to effect a 1% change in the slope of the vertical curve. Aside from high/low point computations, it will be shown in Sections A.3.3 and A.3.4 that the K-value has many important applications in the design of vertical curves.

EXAMPLE A.3

A vertical curve crosses a 4-ft diameter pipe at right angles. The pipe is located at station 110 + 85 and its centerline is at elevation 1091.60 ft. The PVI of the vertical curve is at station 110 + 00 and elevation 1098.4 ft. The vertical curve is equal tangent, 600 ft long, and connects an initial grade of +1.20 percent and a final grade of −1.08 percent. Using offsets, determine the depth, below the surface of the curve, to the top of the pipe, and determine the station of the highest point on the curve.

SOLUTION

The *PVC* is at station 107 + 00 (110 + 00 − 3 + 00, which is half of the curve length), so the pipe is 385 ft (110 + 85 − 107 + 00) from the beginning of the curve (*PVC*). The elevation of the *PVC* will be the elevation of the *PVI* minus the drop in grade over one-half the curve length:

$$1098.4 - (3 \text{ stations} \times 1.2 \text{ ft/station}) = 1094.8 \text{ ft}$$

Using this, the elevation of the initial tangent above the pipe is

$$1094.8 + (3.85 \text{ stations} \times 1.2 \text{ ft/station}) = 1099.42 \text{ ft}$$

Using Eq. A.7 to determine the offset above the pipe at $x = 385$ ft (the distance of the pipe from the *PVC*), we have

$$Y = \frac{A}{200L} x^2$$

$$Y = \frac{|1.2 - (-1.08)|}{200(600)} (385)^2 = 2.82 \text{ ft}$$

Thus the elevation of the curve above the pipe is 1096.6 ft (1099.42 − 2.82). The elevation of the top of the pipe is 1093.60 ft (elevation of the centerline plus one half of the pipe's diameter), so the pipe is 3 ft below the surface of the curve (1096.6 − 1093.6).

To determine the location of the highest point on the curve, we find K from Eq. A.10 as

$$K = \frac{600}{|1.2 - (-1.08)|} = 263.16$$

and the distance from the *PVC* to the highest point is (from Eq. A.11)

$$x = K \times |G_1| = 263.16 \times 1.2 = 315.79 \text{ ft}$$

This gives the station of the highest point at 110 + 15.79 (107 + 00 plus 3 + 15.79). Note that this example could also be solved by applying Eq. A.1, setting Eq. A.2 equal to zero (for determining the location of the highest point on the curve), and following the procedure used in Example A.1.

A.3.2   Minimum and Desirable Stopping-Sight Distances

Construction of a vertical curve is generally a costly operation requiring the movement of significant amounts of earthen material. Thus one of the primary challenges facing highway designers is to minimize construction costs (usually by making the vertical curve as short as possible) while still providing an adequate level of safety. An appropriate level of safety is usually defined as the level of safety that provides drivers with sufficient sight distance to allow them to safely stop their vehicles to avoid collisions with objects obstructing their forward motion. Providing adequate roadway drainage is sometimes an important concern as well, but is not discussed in terms of vertical curves in this book (see AASHTO 1990). Referring back to the vehicle braking performance concepts discussed in Chapter 2, we can compute this necessary stopping sight distance (SSD) simply as the summation of vehicle stopping-sight distance (Eq. 2.48) and the distance traveled during perception/reaction time (Eq. 2.50). That is,

$$\text{SSD} = \frac{V_1^2}{2g(f \pm G)} + V_1 t_p \tag{A.12}$$

where SSD is the stopping-sight distance, V_1 is the initial vehicle speed, g is the gravitational constant, f is the coefficient of braking friction, G is the grade in ft/ft, and t_p is the perception/reaction time.

Given Eq. A.12, the question now becomes one of selecting appropriate values for the computation of SSD. Values of the coefficient of braking friction term, f, are selected to be representative of poor driver skills, low braking efficiencies, and wet pavements, thus providing conservation or near-worst-case stopping-sight distances (see Table 2.6 for such examples of f). Also, a standard driver perception/reaction time of 2.5 seconds is typically used and is a conservative estimate, as individual reaction times are generally less than this. In terms of the assumed initial vehicle speeds used in Eq. A.12, two values are worthy of note. One is the design speed of the highway, which is defined as the maximum safe speed that a highway can be negotiated assuming near-worst-case conditions (wet-weather conditions). The second value is the average vehicle running speed, which is obtained from actual vehicle speed observations, under low traffic-volume conditions, and usually ranges from about 83% to 100% of a highway's design speed, depending on the actual design speed of the highway. At lower highway design speeds, the average running speed is equal to or only slightly less than the highway's design speed. At higher design speeds, the average running speed is significantly less than the highway's design speed. The average running speed reflects drivers' general tendency to drive below the design speed of the highway even under low-volume, uncongested traffic conditions. The application of Eq. A.12 (assuming $G = 0$) produces the stopping-sight distances presented in Table A.1 for average running speeds (the first number in the "assumed speed for conditions" column) and design speeds (the second number in the "assumed speed for conditions" column). Because average running speeds are equal to or lower than the corresponding design speeds (depending on the design speed value), the stopping-sight distance required generally will be lower for average running speeds. Stopping-sight distances derived from average running speeds are referred to as *minimum* SSDs, whereas stopping-sight distances derived from design speeds are referred to as *desirable* SSDs. Whenever possible, desirable SSDs should be used. Many states require special exceptions for design-

Table A.1 Stopping-Sight Distances (minimum and desirable for wet pavements)

Design Speed (mph)	Assumed Speed for Condition (mph)	Brake Reaction Time (sec)	Brake Reaction Distance (ft)	Coefficient of Friction f	Braking Distance on Level (ft)	Stopping-Sight Distance Computed (ft)	Rounded for Design (ft)
20	20–20	2.5	73.3– 73.3	0.40	33.3– 33.3	106.7–106.7	125–125
25	24–25	2.5	88.0– 91.7	0.38	50.5– 54.8	138.5–146.5	150–150
30	28–30	2.5	102.7–110.0	0.35	74.7– 85.7	177.3–195.7	200–200
35	32–35	2.5	117.3–128.3	0.34	100.4–120.1	217.7–248.4	225–250
40	36–40	2.5	132.0–146.7	0.32	135.0–166.7	267.0–313.3	275–325
45	40–45	2.5	146.7–165.0	0.31	172.0–217.7	318.7–382.7	325–400
50	44–50	2.5	161.3–183.3	0.30	215.1–277.8	376.4–461.1	400–475
55	48–55	2.5	176.0–201.7	0.30	256.0–336.1	432.0–537.8	450–550
60	52–60	2.5	190.7–220.0	0.29	310.8–413.8	501.5–633.8	525–650
65	55–65	2.5	201.7–238.3	0.29	347.7–485.6	549.4–724.0	550–725
70	58–70	2.5	212.7–256.7	0.28	400.5–583.3	613.1–840.0	625–850

Source: American Association of State Highway and Transportation Officials, "A Policy on Geometric Design of Highways and Streets," Washington, DC, 1990.

ers to use stopping-sight distances based on anything less than the design speed of the highway.

A.3.3   Stopping-Sight Distance and Crest Vertical Curve Design

In providing sufficient SSD on a vertical curve, the length of curve (L in Fig. A.3) is the critical concern. Longer lengths of curve provide more SSD, all else being equal, but are more costly to construct. Shorter curve lengths are relatively inexpensive to construct but may not provide adequate SSD. What is needed, then, is an expression for minimum curve length given a required SSD. In developing such an expression, crest and sag vertical curves are considered separately.

For the case of a crest vertical curve, consider the diagram presented in Fig. A.5. In this figure, S is the sight distance, L is the length of the curve, H_1 is the

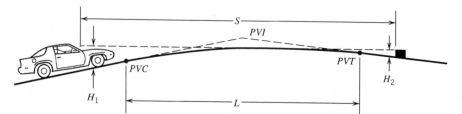

Figure A.5 Stopping-sight distance considerations for crest vertical curves.

driver's eye height, H_2 is the height of a roadway object, and other terms are as previously defined. Using the properties of a parabola for an equal tangent curve, it can be shown that the minimum length of curve, L_m, for a required sight distance, S, is

$$L_m = 2S - \frac{200(\sqrt{H_1} + \sqrt{H_2})^2}{A} \qquad \text{for } S > L \qquad \textbf{(A.13)}$$

$$L_m = \frac{AS^2}{200(\sqrt{H_1} + \sqrt{H_2})^2} \qquad \text{for } S < L \qquad \textbf{(A.14)}$$

For the sight distance required to provide adequate SSD, current AASHTO design guidelines (1990) use a driver eye height, H_1, of 3.5 ft, and a roadway object height, H_2, of 0.5 ft (i.e., the height of an object to be avoided by stopping before a collision). In applying Eqs. A.13 and A.14 to determining the minimum length of curve required to provide adequate SSD, we set the sight distance, S, equal to the stopping-sight distance, SSD. Substituting AASHTO guidelines for H_1 and H_2 and letting $S = $ SSD into Eqs. A.13 and A.14 gives

$$L_m = 2 \, \text{SSD} - \frac{1329}{A} \qquad \text{for SSD} > L \qquad \textbf{(A.15)}$$

$$L_m = \frac{A \, \text{SSD}^2}{1329} \qquad \text{for SSD} < L \qquad \textbf{(A.16)}$$

EXAMPLE A.4

A highway is being designed to AASHTO guidelines with a 70-mph design speed. At one section, an equal tangent vertical curve must be designed to connect grades of +1.0% and −2.0%. Determine the length of curve required assuming provisions are to be made for minimum SSD and desirable SSD.

SOLUTION

The first concern is to determine the SSD to use. If we ignore the effects of grades (i.e., $G = 0$), the SSDs can be read directly from Table A.1. In this case, the assumed average running speed (which is used to determine minimum SSD) is 58 mph, and the corresponding SSD is 625 ft. If we assume that $L > $ SSD (an assumption that is typically made), Eq. A.16 gives

$$L_m = \frac{A \, \text{SSD}^2}{1329} = \frac{3(625)^2}{1329} = \underline{\underline{881.77}}$$

Because $881.77 > 625$, the assumption that $L > $ SSD was correct.

For desirable SSD, the assumed speed is the design speed (70 mph), and the corresponding SSD is 850 ft (from Table A.1). Again assuming that $L > $ SSD, Eq. A.16 gives

$$L_m = \frac{3(850)^2}{1329} = \underline{\underline{1630.93}}$$

Because 1630.93 > 850, the assumption that $L > $ SSD was again correct. Note that providing desirable SSD requires a significantly longer curve than providing SSD for the average running speed (minimum SSD).

The assumption that $G = 0$, made at the beginning of Example A.4, is not strictly correct. If $G \neq 0$, we cannot use the SSD values in Table A.1 and instead must apply Eq. A.12 with the appropriate G value. In this problem, if we use the initial grade in Eq. A.12 (+1.0%) we will underestimate the stopping-sight distance because the vertical curve has a slope as steeply positive as this only at the *PVC*. If we use the final grade in Eq. A.12 (−2.0%) we will overestimate the stopping-sight distance because the vertical curve has a slope as steeply negative as this only at the *PVT*. If we knew where the vehicle began to brake, we could use the first derivative of the parabolic curve function (from Eq. A.2) to give G in Eq. A.12 and set up the equation to solve for SSD exactly. In practice, policies vary as to how this grade problem is handled. Fortunately, because sight distance tends to be greater on downgrades (which require longer stopping distances) than on upgrades, a self-correction for the effect of grades is generally provided. As a consequence, some design agencies ignore the effect of grades completely, while others assume G is equal to zero for grades less than 3% and use simple adjustments to the SSD, depending on the initial and final grades, if grades exceed 3% (see AASHTO 1990). For the remainder of this appendix, we will ignore the effect of grades (i.e., assume that $G = 0\%$ in Eq. A.12). However, it must be pointed out that the use of SSD-grade corrections is very easy and straightforward and all of the equations presented herein still apply.

The use of Eqs. A.15 and A.16 can be simplified if the initial assumption that $L > $ SSD is made, in which case Eq. A.16 is always used. The advantage of this assumption is that the relationship between A and L_m is linear, and Eq. A.10 can be used to give

$$L_m = KA \qquad \text{(A.17)}$$

where

$$K = \frac{\text{SSD}^2}{1329} \qquad \text{(A.18)}$$

Again, K is the horizontal distance, in feet, required to effect a 1% change in the slope (as in Eq. A.10). With known SSD for a given design speed (assuming $G = 0$), K-values can be computed for crest vertical curves as shown in Table A.2. Thus the minimum curve length can be obtained (as shown in Eq. A.17) simply by multiplying A by the K-value read from Table A.2.

Table A.2 Design Controls for Crest Vertical Curves Based on Minimum and Desirable Stopping-Sight Distances

Design Speed (mph)	Assumed Speed for Condition (mph)	Coefficient of Friction f	Stopping-Sight Distance, Rounded for Design (ft)	Rate of Vertical Curvature, K (length [ft] per percent of A)	
				Computed	Rounded for Design
20	20–20	0.40	125–125	8.6– 8.6	10– 10
25	24–25	0.38	150–150	14.4– 16.1	20– 20
30	28–30	0.35	200–200	23.7– 28.8	30– 30
35	32–35	0.34	225–250	35.7– 46.4	40– 50
40	36–40	0.32	275–325	53.6– 73.9	60– 80
45	40–45	0.31	325–400	76.4–110.2	80–120
50	44–50	0.30	400–475	106.6–160.0	110–160
55	48–55	0.30	450–550	140.4–217.6	150–220
60	52–60	0.29	525–650	180.2–302.2	190–310
65	55–65	0.29	550–725	227.1–394.3	230–400
70	58–70	0.28	625–850	282.8–530.9	290–540

Source: American Association of State Highway and Transportation Officials, "A Policy on Geometric Design of Highways and Street," Washington, DC, 1990.

Some discussion about the assumption that $L >$ SSD is warranted. Basically this assumption is made because there are two complications that could arise when SSD $> L$. First, if SSD $> L$, the relationship between A and L_m is not linear, so K-values cannot be used in the $L = KA$ formula (Eq. A.10). Second, at low values of A, it is possible to get negative minimum curve lengths (see Eq. A.15). As a result of these complications, the assumption that $L >$ SSD is almost always made in practice, and thus Eqs. A.17 and A.18 and the K-values presented in Table A.2 are valid. It is important to note that the assumption that $L >$ SSD (upon which Eqs. A.17 and A.18 are based) is usually a safe one because, in many cases, L is greater than SSD. When it is not (i.e., SSD $> L$), the use of the $L >$ SSD formula (Eq. A.16 instead of Eq. A.17) gives longer curve lengths and thus the error is on the conservative, safe side.

A final point relates to the smallest allowable length of curve. Very short vertical curves can be difficult to construct and may not be worth the effort. As a result, it is common practice to set minimum curve length limits that range from 100 ft to 300 ft depending on individual jurisdictional guidelines. A common alternative to these limits is to set the minimum curve length limit (in feet) at three times the design speed, in mph (AASHTO 1990).

EXAMPLE A.5

Solve Example A.4 using the K-values in Table A.2.

SOLUTION

From Example A.4, $A = 3$. For minimum SSD, $K = 290$ (from Table A.2) at a 70-mph design speed (assumed running speed is 58 mph). Therefore, application of Eq. A.17 gives

$$L_m = KA = 290(3) = \underline{\underline{870 \text{ ft}}}$$

which is just slightly different than the 881.77 ft obtained in Example A.4. This difference is due to round-off error. In Example A.4, SSD was rounded from the actual 613.1 ft to 625 ft. In this example, the 613.1-ft SSD is used in Eq. A.18 to determine K, but K is then rounded to 290 from 282.8. This rounding explains the difference in answers, but it is important to note that such small differences are not significant in design.

For desirable SSD, $K = 540$ (from Table A.2) at a 70-mph design speed, giving

$$L_m = 540(3) = \underline{\underline{1620 \text{ ft}}}$$

which is close to the 1630.93 obtained in Example A.4 (again, this difference can be attributed to round-off error in the K-value).

EXAMPLE A.6

If the grades in Example A.4 intersect at station 100 + 00, determine the stationing of the *PVC*, *PVT*, and curve high points for both minimum and desirable SSD curve lengths.

SOLUTION

Using the curve lengths from Example A.5, $L = 870$ ft for minimum SSD. Because the curve is equal tangent (as are virtually all curves used in practice), one-half of the curve will occur before the *PVI* and one-half after, so that

$$PVC \text{ is at } 100 + 00 - L/2 = 100 + 00 - 4 + 35 = 95 + 65$$
$$PVT \text{ is at } 100 + 00 + L/2 = 100 + 00 + 4 + 35 = 104 + 35$$

For the stationing of the high point, Eq. A.11 is used:

$$x = K \times |G_1| = 290(1) = 290 \text{ ft}$$

or at station $95 + 65 + 2 + 90 = 98 + 55$

Similarly, for desirable SSD, the *PVC* can be shown to be at station 91 + 90, *PVT* at station 108 + 10, and the high point at station 97 + 30.

A.3.4 Stopping-Sight Distance and Sag Vertical Curve Design

Sag vertical curve design differs from crest vertical curve design in the sense that sight distance is governed by nighttime conditions because, in daylight, sight distance on a sag vertical curve is unrestricted. Thus the critical concern for sag vertical curve design is the length of roadway illuminated by a vehicle's headlights, which is a function of the height of the headlight above the roadway, H, and the inclined angle of the headlight beam, relative to the horizontal plane of the vehicle, β. The sag vertical curve sight distance design problem is illustrated in Fig. A.6. By using the properties of a parabola for an equal tangent vertical curve (as was done for the crest vertical curve case), it can be shown that the minimum length of curve, L_m, for a required sight distance, S, is

$$L_m = 2S - \frac{200(H + S \tan \beta)}{A} \qquad \text{for } S > L \qquad \textbf{(A.19)}$$

$$L_m = \frac{AS^2}{200(H + S \tan \beta)} \qquad \text{for } S < L \qquad \textbf{(A.20)}$$

For the sight distance required to provide adequate SSD, current AASHTO design guidelines (1990) use a headlight height of 2.0 ft and an upward angle of 1 degree. Substituting these design guidelines and $S = $ SSD (as was done in the crest vertical curve case) into Eqs. A.19 and A.20 gives

$$L_m = 2\,\text{SSD} - \frac{400 + 3.5\,\text{SSD}}{A} \qquad \text{for SSD} > L \qquad \textbf{(A.21)}$$

$$L_m = \frac{A\,\text{SSD}^2}{400 + 3.5\,\text{SSD}} \qquad \text{for SSD} < L \qquad \textbf{(A.22)}$$

As was the case for crest vertical curves, K-values can be computed by assuming $L > $ SSD, which gives us the linear relationship between L_m and A as shown in Eq. A.22. Thus for sag vertical curves (with $L_m = KA$),

$$K = \frac{\text{SSD}^2}{400 + 3.5\,\text{SSD}} \qquad \text{for SSD} < L \qquad \textbf{(A.23)}$$

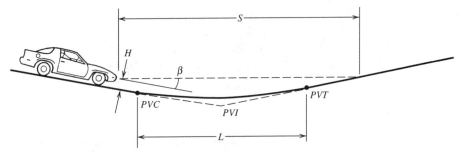

Figure A.6 Stopping-sight distance considerations for sag vertical curves.

Table A.3 Design Controls for Sag Vertical Curves Based on Minimum and Desirable Stopping-Sight Distances

Design Speed (mph)	Assumed Speed for Condition (mph)	Coefficient of Friction f	Stopping-Sight Distance, Rounded for Design (ft)	Rate of Vertical Curvature, K (length [ft] per percent of A)	
				Computed	Rounded for Design
20	20–20	0.40	125–125	14.7– 14.7	20– 20
25	24–25	0.38	150–150	21.7– 23.5	30– 30
30	28–30	0.35	200–200	30.8– 35.3	40– 40
35	32–35	0.34	225–250	40.8– 48.6	50– 50
40	36–40	0.32	275–325	53.4– 65.6	60– 70
45	40–45	0.31	325–400	67.0– 84.2	70– 90
50	44–50	0.30	400–475	82.5–105.6	90–110
55	48–55	0.30	450–550	97.6–126.7	100–130
60	52–60	0.29	525–650	116.7–153.4	120–160
65	55–65	0.29	550–725	129.9–178.6	130–180
70	58–70	0.28	625–850	147.7–211.3	150–220

Source: American Association of State Highway and Transportation Officials, "A Policy on Geometric Design of Highways and Streets," Washington, DC, 1990.

The K-values corresponding to minimum and desirable SSDs are presented in Table A.3. As was the case for crest vertical curves, some caution should be exercised in using this table because the assumption that $G = 0$ (for determining SSD) is used. Also, assuming that $L >$ SSD is a safe, conservative assumption (as was the case for crest vertical curves), and the smallest allowable curve lengths for sag curves are the same as those for crest curves (see discussion in Section A.3.3).

EXAMPLE A.7

An engineering mistake has resulted in the need to connect an already con-structed tunnel and bridge with sag and crest vertical curves. The profile view of the tunnel and bridge is given in Fig. A.7. Devise a vertical alignment to connect the tunnel and bridge by determining the highest possible common design speed (for desirable SSD) for the sag and crest (equal tangent) vertical curves needed. Compute the stationing and elevations of *PVC, PVI,* and *PVT* curve points.

SOLUTION

From left to right (see Fig. A.7), a sag vertical curve (with subscripting *s*) and a crest vertical curve (with subscripting *c*) are needed to connect the tunnel and

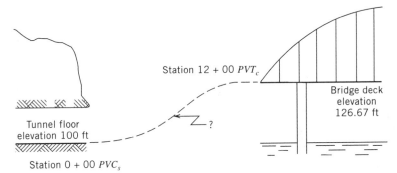

Station 12 + 00 PVT_c

Bridge deck elevation 126.67 ft

Tunnel floor elevation 100 ft

?

Station 0 + 00 PVC_s

Figure A.7 Profile view (vertical alignment diagram) for Example A.7.

bridge. From given information, it is known that $G_{1s} = 0\%$ (i.e., the initial slope of the sag vertical curve) and $G_{2c} = 0\%$ (i.e., the final slope of the crest vertical curve). To obtain the highest possible design speed, we will want to use all of the horizontal distance available. This means we will want to connect the curve such that the *PVT* of the sag curve (PVT_s) will be the *PVC* of the crest curve (PVC_c). If this is the case, $G_{2s} = G_{1c}$, and because $G_{1s} = G_{2c} = 0$, $A_s = A_c = A$, the common algebraic difference in the grades.

Because 1200 ft separates the tunnel and bridge,

$$L_s + L_c = 1200$$

Also, the summation of the end-of-curve offset for the sag curve and the beginning-of-curve offset (relative to the final grade) for the crest curve must equal 26.67 ft. Using the equation for the final offset, Eq. A.9, we have

$$\frac{AL_s}{200} + \frac{AL_c}{200} = 26.67$$

Rearranging,

$$\frac{A}{200}(L_s + L_c) = 26.67$$

Because $L_s + L_c = 1200$,

$$\frac{A}{200}(1200) = 26.67$$

Solving for A gives A = 4.444%. The problem now becomes one of finding K-values that allow $L_s + L_c = 1200$. Because $L = KA$ (Eq. A.17), we can write

$$K_s A + K_c A = 1200$$

Substituting $A = 4.444$,

$$K_s + K_c = 270$$

To find the highest possible design speed, Tables A.2 and A.3 are used to arrive at K-values to solve $K_s + K_c = 270$. By looking at Tables A.2 and A.3 it is apparent that the highest possible design speed is 50 mph at which speed $K_c = 160$ and $K_s = 110$ (i.e., the summation of K's is 270).

To arrive at the stationing of curve points, we first determine curve lengths as

$$L_s = K_s A = 110(4.444) = 488.89 \text{ ft}$$

$$L_c = K_c A = 160(4.444) = 711.11 \text{ ft}$$

Because the station of PVC_s is $0 + 00$ (given), it is clear that $PVI_s = 2 + 44.44$, $PVT_s = PVC_c = 4 + 88.89$, $PVI_c = 8 + 44.44$, and $PVT_c = 12 + 00.00$. For elevations, $PVC_s = PVI_s = 100$ ft, and $PVI_c = PVT_c = 126.67$ ft. Finally, the elevation of PVT_s and PVC_c can be computed as

$$100 + \frac{AL_s}{200} = 100 + \frac{4.444(488.89)}{200} = 110.86 \text{ ft}$$

EXAMPLE A.8

Consider the conditions described in Example A.7. Suppose the horizontal separation between the tunnel and bridge was 1200 ft and a design speed of only 30 mph (desirable) was needed. Determine the lengths of curves required to connect the bridge and tunnel while keeping the connecting grade as small as possible.

SOLUTION

It is known that the 1200 ft separating the bridge and tunnel are more than enough to connect a 30-mph alignment because Example A.7 showed that 50 mph was possible. Therefore, to connect the bridge and tunnel and keep the connecting grade as small as possible, we will place a constant grade section between the sag and crest curves (as shown in Fig. A.8). The elevation change will be the final offsets of the sag and crest curves plus the change in elevation resulting from the constant grade section connecting the two curves. Let G_{con} be the grade of the constant grade section. This means that $G_{2s} = G_{1c} = G_{con}$ and, because $G_{1s} = G_{2c} = 0$ (as in Example A.7), $G_{con} = A_s = A_c = A$. The equation that will solve the vertical alignment for this problem will be

$$\frac{AL_s}{200} + \frac{AL_c}{200} + \frac{A(1200 - L_s - L_c)}{100} = 26.67$$

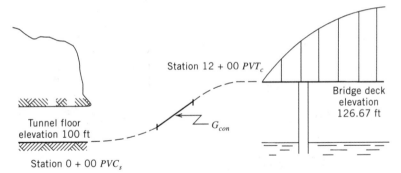

Figure A.8 Profile view (vertical alignment diagram) for Example A.8.

where the third term in this equation accounts for the elevation difference attributable to the constant grade section connecting the sag and crest curves (the 100 in the denominator of this term converts A from percent to ft/ft). Using $L = KA$, we have

$$\frac{A^2 K_s}{200} + \frac{A^2 K_c}{200} + \frac{A(1200 - K_s A - K_c A)}{100} = 26.67$$

From Table A.2, $K_c = 30$, and from Table A.3, $K_s = 40$ (30 mph desirable). Putting these values in the preceding equation gives

$$0.35A^2 + 12A - 0.70\,A^2 = 26.67$$

$$-0.35A^2 + 12A - 26.67 = 0$$

Solving this gives either $A = 2.39$ or $A = 31.0$; $A = 2.39$ is chosen because we want to minimize the grade. For this value of A, the curve lengths are

$$L_s = K_s A = 40(2.39) = \underline{\underline{95.6\ \text{ft}}}$$

$$L_c = K_c A = 30(2.39) = \underline{\underline{71.7\ \text{ft}}}$$

and the length of the constant grade section will be 1032.7 ft. This means that about 24.7 ft of the elevation difference will occur in the constant grade section, with the remainder of the elevation difference attributable to final curve offsets.

Another variation of this type of problem is the case in which the initial and final grades are not equal to zero. This makes the problem a bit more complex, as demonstrated in the following example.

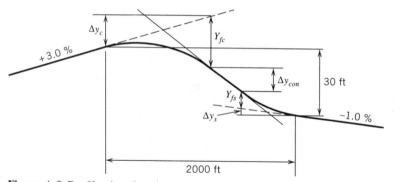

Figure A.9 Profile view (vertical alignment diagram) for Example A.9.

EXAMPLE A.9

Two sections of highway are separated by 2000 ft as shown in Fig. A.9. Determine the curve lengths required for a 40-mph (desirable) vertical alignment to connect these two highway segments while keeping the connecting grade as small as possible.

SOLUTION

Let Y_{fc} and Y_{fs} be the final offsets of the crest and sag curves, respectively. Let G_{con} be the slope of a constant grade section connecting the crest and sag curves. (We will assume that the horizontal distance is sufficient to connect the highway with a 40-mph alignment. If this assumption is incorrect, the following equations will produce an obviously erroneous answer and a lower design speed will have to be chosen.) Finally let Δy_{con} be the change in elevation over the constant grade section, and let Δy_c and Δy_s be the changes in elevation due to the extended curve tangents. The elevation equation is thus (see Fig. A.9)

$$Y_{fc} + Y_{fs} + \Delta y_{con} + \Delta y_s = 30 + \Delta y_c$$

Substituting offset equations and equations for elevation changes (with subscripting c for crest and s for sag),

$$\frac{A_c L_c}{200} + \frac{A_s L_s}{200} + \frac{G_{con}(2000 - L_c - L_s)}{100} + \frac{1.0\,L_s}{100} = 30 + \frac{3.0\,L_c}{100}$$

Using $L = KA$, this equation becomes

$$\frac{A_c^2 K_c}{200} + \frac{A_s^2 K_s}{200} + \frac{G_{con}(2000 - K_c A_c - K_s A_s)}{100} + \frac{1.0\,K_s A_s}{100} = 30 + \frac{3.0\,K_c A_c}{100}$$

From Tables A.2 and A.3, $K_c = 80$ and $K_s = 70$ (at 40 mph desirable). Substituting and defining As, arranging the equation so that G_{con} will be positive, and assuming that G_{con} will be greater than 1%, gives

$$\frac{(3 + G_{con})^2 80}{200} + \frac{(G_{con} - 1)^2 70}{200} + \frac{G_{con}[2000 - 80(3 + G_{con}) - 70(G_{con} - 1)]}{100}$$

$$+ \frac{1.0[70(G_{con} - 1)]}{100} = 30 + \frac{3.0[80(3 + G_{con})]}{100}$$

or

$$-0.75G_{con}^2 + 18.3G_{con} - 33.95 = 0$$

which gives $G_{con} = 2.02$. (The other possible solution is 22.38, which is rejected because we want to minimize the grade.) Using $L = KA$ gives $L_c = \underline{401.6 \text{ ft}}$ (80 × 5.02), and $L_s = \underline{71.4 \text{ ft}}$ (70 × 1.02). Elevations and the locations of curve points can be readily computed with this information.

A.3.5 Passing-Sight Distance and Crest Vertical Curve Design

In addition to stopping-sight distance, in some instances it may be desirable to provide adequate passing-sight distance, which can be an important issue in two-lane highway design (one lane in each direction). Passing-sight distance is a factor only in crest vertical curve design because, for sag curves, the sight distance is unobstructed looking up or down the grade, and, at night, the headlights of oncoming or opposing vehicles will be noticed. In determining the sight distance required to pass on a crest vertical curve, Eqs. A.13 and A.14 will apply, but while the driver's eye height, H_1, will remain 3.5 ft, H_2 will now be set to 4.25 ft, which is the assumed height of an opposing vehicle that could prevent a passing maneuver. Substituting these H-values into Eqs. A.13 and A.14, and letting the sight distance S equal passing-sight distance PSD,

$$L_m = 2 \text{ PSD} - \frac{3093}{A} \qquad \text{for PSD} > L \qquad \textbf{(A.24)}$$

$$L_m = \frac{A \text{ PSD}^2}{3093} \qquad \text{for PSD} < L \qquad \textbf{(A.25)}$$

As was the case for stopping-sight distance, it is typically assumed that the length of curve is greater than the required sight distance (i.e., in this case $L > \text{PSD}$), so

$$K = \frac{\text{PSD}^2}{3093} \qquad \textbf{(A.26)}$$

The passing-sight distance (PSD) used for design is assumed to consist of four distances: (1) the initial maneuver distance (which includes drivers' perception/reaction time and the time it takes to bring the vehicle from its trailing speed

Table A.4 Design Controls for Crest Vertical Curves Based on Passing-Sight Distances

Design Speed (mph)	Minimum Passing Sight Distance, Rounded for Design (ft)	Rate of Vertical Curvature, K Rounded for Design (length [ft] per percent of A)
20	800	210
25	950	300
30	1100	400
35	1300	550
40	1500	730
45	1650	890
50	1800	1050
55	1950	1230
60	2100	1430
65	2300	1720
70	2500	2030

Source: American Association of State Highway and Transportation Officials, "A Policy on Geometric Design of Highways and Streets," Washington, DC, 1990.

to the speed needed to pass), (2) the distance that the passing vehicle traverses while occupying the left lane, (3) the clearance length (a comfortable length between the passing vehicle and the opposing or oncoming vehicle at the end of the pass, assumed to be 110 to 300 ft), and (4) the distance traversed by the opposing vehicle during the passing maneuver. These distances are determined using assumptions regarding the time of the initial maneuver, average vehicle acceleration, and speeds of passing, passed, and opposing vehicles. The summation of these four distances gives a single required passing-sight distance (i.e., there are not *minimums* and *desirables* as was the case with stopping-sight distance). The reader is referred to AASHTO 1990 for a complete description of the assumptions made in determining required passing-sight distances.

The minimum distances needed to pass (PSD) at various design speeds, along with the corresponding K-values as computed from Eq. A.26, are presented in Table A.4. Notice that the K-values in this table are much higher than those required for stopping-sight distance (as shown in Table A.2). As a result, designing a crest curve to provide adequate passing sight distance is often an expensive proposition (due to the length of curve required) and not very common.

EXAMPLE A.10

An equal tangent crest vertical curve is 1575 ft long and connects grades of +2.0% and −1.5%. If the design speed of the roadway is 50 mph, does this curve have adequate passing-sight distance?

SOLUTION

To determine the length of curve required to provide adequate passing-sight distance at a design speed of 50 mph, we use $L = KA$ with $K = 1050$ (as read from Table A.4). This gives

$$L = 1050(3.5) = 3675 \text{ ft}$$

Because the curve is only 1575 ft long, this curve is not long enough to provide adequate passing-sight distance.

A.4 HORIZONTAL ALIGNMENT

The critical aspect of horizontal alignment is the horizontal curve with the focus on design of the directional transition of the roadway in a horizontal plane. Stated differently, a horizontal curve provides a transition between two straight (or tangent) sections of roadway. A key concern in this directional transition is the ability of a vehicle to negotiate a horizontal curve (the provision of adequate drainage is also important, but is not discussed in this book; see AASHTO 1990). As was the case with the straight-line vehicle performance characteristics discussed at length in Chapter 2, highway engineers must design horizontal alignments to accommodate a variety of vehicle cornering capabilities that range from nimble sports cars to ponderous trucks. A theoretical assessment of vehicle cornering at the level of detail given straight-line performance in Chapter 2 is beyond the scope of this book; see Campbell (1978) and Wong (1978). Instead, vehicle cornering performance is viewed only at the practical design-oriented level, with equations simplified in a manner similar to the stopping-distance equation, as discussed in Section 2.9.5.

A.4.1 Vehicle Cornering

Figure A.10 illustrates the forces acting on a vehicle during cornering. In this figure, α is the angle of incline, W is the weight of the vehicle (in pounds), W_n and W_p are the weights normal and parallel to the roadway surface respectively, F_f is the side frictional force (centripetal, in pounds), F_c is the centripetal force (lateral acceleration $\times$ mass, in pounds), F_{cp} is the centripetal force acting parallel to the roadway surface, F_{cn} is the centripetal force acting normal to the roadway surface, and R_v is the radius defined to the vehicle's traveled path (in ft). Some basic horizontal curve relationships can be derived by noting that

$$W_p + F_f = F_{cp} \tag{A.27}$$

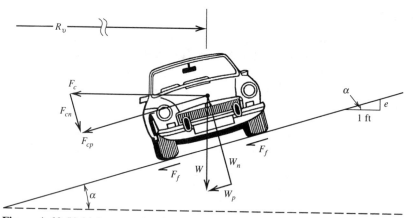

Figure A.10 Vehicle cornering forces.

From basic physics, this equation can be written as [with $F_f = f_s(W_n + F_{cn})$]

$$W \sin \alpha + f_s \left(W \cos \alpha + \frac{WV^2}{gR_v} \sin \alpha \right) = \frac{WV^2}{gR_v} \cos \alpha \qquad \textbf{(A.28)}$$

where f_s is the coefficient of side friction (which is different from the coefficient of friction term, f, used in stopping-distance computations), g is the gravitational constant, and V is the vehicle speed (in feet per second). Dividing both sides of Eq. A.28 by $W \cos \alpha$ gives

$$\tan \alpha + f_s = \frac{V^2}{gR_v}(1 - f_s \tan \alpha) \qquad \textbf{(A.29)}$$

The term $\tan \alpha$ is referred to as the superelevation of the curve and is denoted e. In words, the superelevation is the number of vertical feet of rise per 100 feet of horizontal distance (see Fig. A.10). The term $f_s \tan \alpha$ in Eq. A.29 is conservatively set equal to zero for practical applications due to the small values that f_s and α typically assume (this is equivalent to ignoring the normal component of centripital force). With $e = \tan \alpha$, Eq. A.29 can be arranged so that

$$R_v = \frac{V^2}{g(f_s + e)} \qquad \textbf{(A.30)}$$

EXAMPLE A.11

A roadway is being designed for a speed of 70 mph. At one horizontal curve, the superelevation is 0.08 and the coefficient of side friction is 0.10. Determine the minimum radius of curve (measured to the traveled path) that will provide for safe vehicle operation.

SOLUTION

The application of Eq. A.30 gives (with 1.47 converting mph to ft/s)

$$R_v = \frac{V^2}{g(f_s + e)} = \frac{(70 \times 1.47)^2}{32.3(0.10 + 0.08)} = \underline{\underline{1826.85 \text{ ft}}}$$

This value is the minimum radius, because radii larger than 1826.85 ft will generate centripetal forces lower than those capable of being safely supported by the superelevation and the side frictional force.

In the actual design of a horizontal curve, engineers must select appropriate values of e and f_s. The value selected for superelevation, e, is critical because high rates of superelevation can cause vehicle steering problems on the horizontal curve, and, in cold climates, ice on the roadway can reduce f_s such that vehicles traveling less than the design speed on an excessively superelevated curve could slide inward off the curve by gravitational forces. AASHTO provides general guidelines for the selection of e and f_s for horizontal curve design, as shown in Table A.5. The values presented in this table are grouped by four values of maximum e. The selection of any one of these four maximum e values is dependent on the type of road (e.g., higher maximum e's are permitted on freeways relative to arterials and local roads) and local design practice. Limiting values of f_s are simply a function of design speed.

A.4.2 Horizontal Curve Fundamentals

In connecting straight (tangent) sections of roadway with a horizontal curve, several options are available. The most obvious of these is the simple curve, which is just a standard curve with a single, constant radius. Other options include compound curves, which consist of two or more simple curves in succession, and spiral curves, which are curves with a continuously changing radius. To illustrate the basic principles involved in horizontal curve design, this book will focus on simple curves. For detailed information regarding compound and spiral curves, the reader is referred to standard route-surveying texts (Skelton 1949).

Figure A.11 shows the basic elements of a simple horizontal curve. In this figure, R is the radius (usually measured to the centerline of the road), PC is the point of curve (the beginning point of the horizontal curve), T is the tangent length, PI is the point of tangent intersection, Δ is the central angle of the curve, PT is the point of tangent (the ending point of the horizontal curve), M is the middle ordinate, E is the external distance, and L is the length of curve. Another important term is the degree of curve, which is defined as the angle subtended by a 100-ft arc along the horizontal curve. It is a measure of the sharpness of the curve and is frequently used instead of the radius in construction of the

Table A.5 Limiting Values of e and f_s

Design Speed (mph)	Maximum e	Maximum f_s
20	0.04	0.17
30	0.04	0.16
40	0.04	0.15
50	0.04	0.14
60	0.04	0.12
20	0.06	0.17
30	0.06	0.16
40	0.06	0.15
50	0.06	0.14
60	0.06	0.12
65	0.06	0.11
70	0.06	0.10
20	0.08	0.17
30	0.08	0.16
40	0.08	0.15
50	0.08	0.14
60	0.08	0.12
65	0.08	0.11
70	0.08	0.10
20	0.10	0.17
30	0.10	0.16
40	0.10	0.15
50	0.10	0.14
60	0.10	0.12
65	0.10	0.11
70	0.10	0.10

Note: In recognition of safety considerations, use of maximum $e = 0.04$ should be limited to urban conditions.

Source: American Association of State Highway and Transportation Officials, "A Policy on Geometric Design of Highways and Streets," Washington, DC, 1990.

curve. The degree of curve is directly related to the radius of the horizontal curve by

$$D = \frac{100 \left(\dfrac{180}{\pi} \right)}{R} = \frac{18,000}{\pi R} \qquad \text{(A.31)}$$

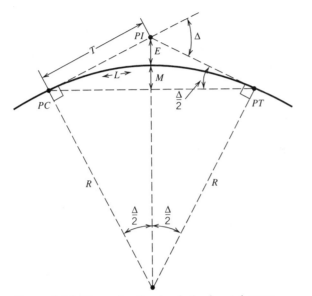

Figure A.11 Elements of a simple horizontal curve.

A geometric and trigonometric analysis of Fig. A.11 reveal the following relationships:

$$T = R \tan \frac{\Delta}{2} \qquad (A.32)$$

$$E = R \left[\frac{1}{\cos(\Delta/2)} - 1 \right] \qquad (A.33)$$

$$M = R \left(1 - \cos \frac{\Delta}{2} \right) \qquad (A.34)$$

$$L = \frac{\pi}{180} R\Delta \qquad (A.35)$$

or, using Eq. A.31,

$$L = \frac{100\Delta}{D} \qquad (A.36)$$

It is important to note that horizontal curve stationing, curve length, and curve radius (R) are typically measured with respect to the centerline of the road (for a two-lane facility). In contrast, the radius determined on the basis of vehicle forces (R_v in Eq. A.30) is measured from the innermost vehicle path, which is assumed to be the midpoint of the innermost vehicle lane. Thus, to be truly correct, a slight correction for lane width is required when equating the R_v of Eq. A.30 with the R in Eqs. A.31 to A.35.

A horizontal curve is designed with a 2000-ft radius. The curve has a tangent length of 400 ft and the *PI* is at station 103 + 00. Determine the stationing of the *PT*.

SOLUTION

Equation A.32 is applied to determine the central angle, Δ.

$$T = R \tan \frac{\Delta}{2}$$

$$400 = 2000 \tan \frac{\Delta}{2}$$

$$\Delta = 22.62°$$

So, from Eq. A.35, the length of the curve is

$$L = \frac{\pi}{180} R\Delta$$

$$L = \frac{3.1416}{180} 2000(22.62) = 789.58 \text{ ft}$$

Given that the tangent is 400 ft,

$$\text{stationing } PC = 103 + 00 - 4 + 00 = 99 + 00$$

Because horizontal curve stationing is measured along the alignment of the road,

$$\text{stationing } PT = \text{stationing } PC + L$$
$$= 99 + 00 + 7 + 89.58 = 106 + 89.58$$

A.4.3 Stopping-Sight Distance and Horizontal Curve Design

As was the case for vertical curve design, adequate stopping-sight distance must be provided in the design of horizontal curves. Sight-distance restrictions on horizontal curves occur when obstructions are present as shown in Fig. A.12. Such obstructions are frequently encountered in highway design due to the cost of right-of-way acquisition and/or the cost of moving earthen materials (e.g., rock outcroppings). When such an obstruction exists, the stopping-sight distance is measured along the horizontal curve from the center of the traveled lane (the assumed location of the driver's eyes). As shown in Fig. A.12, for a specified stopping distance, some distance, M_s (the middle ordinate of a curve that has

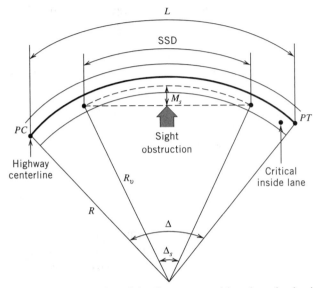

Figure A.12 Stopping-sight distance considerations for horizontal curves.

an arc length equal to the stopping distance), must be visually cleared so that the line of sight provides sufficient stopping-sight distance.

Equations for computing stopping-sight distance (SSD) relationships for horizontal curves can be derived by first determining the central angle, Δ_s, for an arc equal to the required stopping-sight distance. (See Fig. A.12 and note that this is not the central angle, Δ, of the horizontal curve whose arc is equal to L.) Assuming that the length of the horizontal curve exceeds the required SSD (as shown in Fig. 3.12), we have (as with Eq. A.35)

$$\text{SSD} = \frac{\pi}{180} R_v \Delta_s \tag{A.37}$$

where R_v is the radius to the vehicle's traveled path, which is also assumed to be the location of the driver's eyes for sight distance, and again is taken as the radius to the middle of the innermost lane; and Δ_s is the angle subtended by an arc equal to the SSD length. Rearranging terms gives

$$\Delta_s = \frac{180 \, \text{SSD}}{\pi R_v} \tag{A.38}$$

Substituting this into the general equation for the middle ordinate of a simple horizontal curve (Eq. A.34) gives

$$M_s = R_v \left(1 - \cos \frac{90 \, \text{SSD}}{\pi R_v} \right) \tag{A.39}$$

where M_s is the middle ordinate necessary to provide adequate stopping-sight distance, as shown in Fig. A.12. Solving Eq. A.39 for SSD gives

$$\text{SSD} = \frac{\pi R_v}{90} \left[\cos^{-1} \left(\frac{R_v - M_s}{R_v} \right) \right] \qquad \text{(A.40)}$$

Note that Eqs. A.37 to A.40 can also be applied directly to determine sight-distance requirements for passing. If these equations are to be used for passing, distance values given in Table A.4 would apply and SSD in the equations would be replaced by PSD.

EXAMPLE A.13

A horizontal curve on a two-lane highway is designed with a 2000-ft radius, 12-ft lanes, and a 60-mph design speed. Determine the distance that must be cleared from the inside edge of the inside lane to provide sufficient sight distance for desirable and minimum SSD.

SOLUTION

Because the curve radius is usually taken to the centerline of the roadway, $R_v = R - 12/2 = 2000 - 6 = 1994$ ft, which gives the radius to the middle of the inside lane (i.e., the critical driver location). From Table A.1, the desirable SSD is 650 ft, so applying Eq. A.39 gives

$$M_s = R_v \left(1 - \cos \frac{90\,\text{SSD}}{\pi R_v} \right)$$

$$= 1994 \left[1 - \cos \frac{90(650)}{\pi(1994)} \right] = \underline{\underline{26.43\ \text{ft}}}$$

Therefore, 26.43 ft must be cleared, as measured from the center of the inside lane, or 20.43 ft as measured from the inside edge of the inside lane. Similarly, the minimum SSD is 525 ft (Table A.1) and the application of Eq. A.39 gives $M_s = 17.26$ ft. Thus, for minimum SSD, 11.26 ft must be cleared from the inside edge of the inside lane.

EXAMPLE A.14

A two-lane highway (two 12-ft lanes) has a posted speed limit of 50 mph and, on one section, has both horizontal and vertical curves as shown in Fig. A.13. A recent daytime accident (driver traveling eastbound and striking a stationary roadway object) resulted in a fatality and a lawsuit alleging that the 50-mph posted speed limit was an unsafe speed for the curves in question and a major cause of the accident. Evaluate and comment on the roadway design.

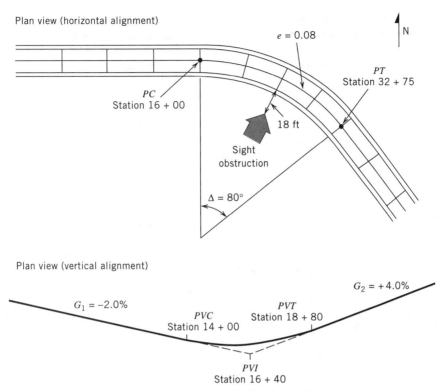

Figure A.13 Horizontal and vertical alignment for Example A.14.

SOLUTION

Begin with an assessment of the horizontal alignment. Two concerns must be considered: the adequacy of the curve radius and superelevation, and the adequacy of the sight distance on the eastbound (inside) lane. For the curve radius, note from Fig. A.13 that

$$L = \text{station of } PC - \text{station of } PT$$

$$L = 32 + 75 - 16 + 00 = 1675 \text{ ft}$$

Rearranging Eq. A.35, we get

$$R = \frac{180}{\pi \Delta} L = \frac{180}{\pi(80)} (1675) = 1198.65 \text{ ft}$$

Using the posted speed limit of 50 mph with $e = 0.08\%$, we find that Eq. A.30 can be rearranged (with the vehicle traveling in the middle of the inside lane, $R_v = R - $ half the lane width or $R_v = 1198.65 - 6 = 1192.65$) to give

$$f_s = \frac{V^2}{gR_v} - e = \frac{(50 \times 1.47)^2}{32.2(1192.65)} - 0.08 = 0.061$$

From Table A.5, the maximum f_s for 50 mph is 0.14. Since 0.061 does not exceed 0.14, the radius and superelevation are sufficient for the 50-mph speed.

For sight distance, the available M_s is 18 ft plus the 6-ft distance to the center of the eastbound (inside) lane, or 24 ft. Application of Eq. A.40 gives

$$SSD = \frac{\pi R_v}{90}\left[\cos^{-1}\left(\frac{R_v - M_s}{R_v}\right)\right]$$

$$= \frac{\pi(1192.65)}{90}\left[\cos^{-1}\left(\frac{1192.65 - 24}{1192.65}\right)\right]$$

$$= 479.3 \text{ ft}$$

From Table A.1, desirable SSD at 50 mph is 475 ft, so the 479.3 ft of SSD provided is sufficient. Turning to the sag vertical curve, the length of curve is

$$L = \text{station of } PVT - \text{station of } PVC$$

$$L = 18 + 80 - 14 + 00 = 480 \text{ ft}$$

Using $A = 6$ (from Fig. A.13) and applying Eq. A.10, we obtain

$$K = \frac{L}{A} = \frac{480}{6} = 80$$

For the 50-mph speed, Table A.3 indicates a necessary K-value of 110 desirable and 90 minimum. Thus the K-value of 80 reveals that the curve is inadequate for the 50-mph speed. However, because the accident occurred in daylight and sight distances on sag vertical curves are governed by nighttime conditions, this design did not contribute to the accident.

NOMENCLATURE FOR APPENDIX A

A	absolute value of the algebraic difference in grades (in percent)
D	degree of curvature
e	rate of superelevation
F_f	frictional side force
F_c	centripetal force
F_{cn}	centripetal force normal to the roadway surface
F_{cp}	centripetal force parallel to the roadway surface
f	coefficient of stopping friction

f_s coefficient of side friction

G grade in ft/ft or percent depending on application

G_1 initial roadway grade in ft/ft or percent

G_2 final roadway grade in ft/ft or percent

g gravitational constant

H height of vehicle headlights

H_1 height of driver's eyes

H_2 height of roadway object for stopping, height of oncoming car for passing

K horizontal distance required to effect a 1% change in slope

L length of curve

L_m minimum length of curve

M middle ordinate

M_s middle ordinate for stopping-sight distance

PC initial point of horizontal curve

PI point of tangent intersection (horizontal curve)

PSD   passing-sight distance

PT final point of horizontal curve

PVC initial point of vertical curve

PVI point of tangent intersection, vertical curve

PVT final point of vertical curve

R radius of curve measured to roadway centerline

R_v radius of curve measured to center of vehicle

S sight distance

SSD   stopping-sight distance

T tangent length

t_p driver perception/reaction time

V vehicle speed

V_1 initial vehicle speed

W vehicle weight

W_n vehicle weight normal to the roadway surface

W_p vehicle weight parallel to the roadway surface

x distance from beginning of the vertical curve

x_{hl} distance from beginning of the vertical curve to high or low point

Y vertical curve offset

Y_f end-of-curve offset, vertical curve

Y_m midcurve offset, vertical curve

α angle of superelevation

β angle of upward headlight beam

Δ central angle

Δ_s central angle subtended by the stopping-sight (SSD) arc

REFERENCES

American Association of State Highway and Transportation Officials. "A Policy on Geometric Design of Highways and Streets." Washington, DC, 1990.

Campbell, C. *The Sports Car: Its Design and Performance.* Cambridge, MA: Robert Bently, Inc., 1978.

Wong, J. H. *Theory of Ground Vehicles.* New York: Wiley, 1978.

Skelton, R. R. *Route Surveys.* New York: McGraw-Hill, 1949.

PROBLEMS

A.1. A 1600-ft-long sag vertical curve (equal tangent) has a *PVC* at station 120 + 00 and elevation 1500 ft. The initial grade is −3.5% and the final grade is +6.5%. Determine the elevation and stationing of the low point, *PVI*, and *PVT*.

A.2. A 500-ft-long equal tangent crest vertical curve connects tangents that intersect at station 340 + 00 and elevation 1322 ft. The initial grade is +4.0% and the final grade is −2.5%. Determine the elevation and stationing of the high point, *PVC,* and *PVT*.

A.3. Consider Example A.3. Solve this problem without using offsets (i.e., use Eq. A.1).

A.4. Again consider Example A.3. Does this curve provide adequate stopping-sight distance (desirable) for a speed of 60 mph?

A.5. An equal tangent sag vertical curve is designed to provide desirable SSD. The *PVC* of the curve is at station 109 + 00 (elevation 950 ft), the *PVI* is at station 110 + 92.5 (elevation 947.11 ft), and the low point is at station 110 + 65. Determine the design speed of the curve.

A.6. An equal tangent vertical curve was designed in 1996 (to 1990 AASHTO guidelines) for desirable SSD at a design speed of 70 mph to connect grades $G_1 = +1.0\%$ and $G_2 = -2.0\%$. The curve is to be redesigned for a 70-mph design speed in the year 2030. Vehicle braking technology has advanced so that coefficients of stopping friction, f's, have increased by 40% relative to the 1990 guidelines given in Table A.1, but, due to the higher percentage of older people in the driving population, design reaction times have increased by 20%. Also, vehicles have become smaller so that the driver's eye height is assumed to be 2.75 ft above the pavement and roadway objects are assumed to be 0.25 ft above the pavement. Compute the difference in design curve lengths for 1996 and 2030 designs.

A.7. A highway reconstruction project is being undertaken to reduce accident rates. The reconstruction involves a major realignment of the highway so that a 60-mph design speed is attained. At one point on the highway, an 800-ft equal tangent crest vertical curve exists. Measurements show that at 3 + 52 stations from the *PVC*, the vertical curve offset is 3 ft. Assess the adequacy of this existing curve in light of the reconstruction design speed of 60 mph, and, if the existing curve is inadequate, compute a satisfactory curve length. (Consider both minimum and desirable SSDs.)

A.8. Two level sections of an east-west highway ($G = 0$) are to be connected. Currently, the two sections of highway are separated by a 4000-ft (horizontal distance), 2% grade. The westernmost section of highway is the higher of the two and is at elevation 100 ft. If the highway has a 60-mph design speed, determine, for the crest and sag vertical curves required, the stationing and elevation of the *PVC*s and *PVT*s given that the *PVC* of the crest curve (on the westernmost level highway section) is at station 0 + 00 and elevation 100 ft. In solving this problem, assume desirable SSDs and that the curve *PVI*s are at the intersection of $G = 0$ and the 2% grade; that is, $A = 2$.

A.9. Consider Problem A.8. Suppose it is necessary to keep the entire alignment within the 4000 ft that currently separate the two level sections. It is determined that the crest and sag curves should be connected (i.e., the *PVT* of the crest and *PVC* of the sag) with a constant grade section that has the lowest grade possible. Again using 60-mph design speed, determine, for the crest and sag vertical curves, the stationing and elevation of the *PVC*s and *PVT*s given that the westernmost level section ends at station 0 + 00 and elevation 100 ft. (Note that A must now be determined and will not be equal to 2.)

A.10. An equal tangent crest vertical curve is designed for 60 mph desirable. The initial grade is +4.0% and the final grade is negative. What is the elevation difference between the *PVC* and the high point of the curve?

A.11. An equal tangent crest curve connects a +1.0% and a −0.5% grade. The *PVC* is at station 54 + 84 and the *PVI* is at station 57 + 44. Is this curve long enough to provide passing-sight distance at a 55-mph design speed?

A.12. Due to accidents at a railroad crossing, an overpass (with a roadway surface 24 ft above the existing road) is to be constructed on an existing level highway. The existing highway has a design speed of 50 mph (desirable SSD). The overpass structure is to be level, centered above the railroad, and 200 ft long. What length of the existing level highway must be reconstructed to provide an appropriate vertical alignment?

A.13. A section of a freeway ramp has a +4.0% grade and ends at station 127 + 00 and elevation 138 ft. It must be connected to another section of the ramp (which has a 0.0% grade) that is at station 162 + 00 and elevation 97 ft. It is determined that the crest and sag curves required to connect the ramp should be connected (i.e., the *PVT* of the crest and *PVC* of the sag) with a constant grade section that has the lowest grade possible. Design a vertical alignment to connect between these two stations using a 40-mph design speed (desirable). Provide the lengths of the curves and constant grade section.

A.14. A 1200-ft equal tangent crest vertical curve is currently designed for 50 mph (desirable). A civil engineering student contends that 60 mph is safe in a van because of the higher driver's eye height. If all other design inputs are standard, how high must the van driver's eye height be for the student's claim to be valid?

A.15. A roadway has a design speed of 50 mph. At station 100 + 00 a +3.0% grade roadway section ends, and at station 130 + 00 a +2.0% grade roadway section begins. The +3.0% grade section of highway (at station 100 + 00) is at a higher elevation than the +2.0% grade section of highway (at station 130 + 00). If a −5.0% constant grade section is used to connect the crest and sag vertical curves that are needed to link the +3.0% and +2.0% grade sections, what is the elevation difference between stations 100 + 00 and 130 + 00 if the roadway is designed for desirable SSD? (The entire alignment, crest and sag curves, and constant grade section must fit between stations 100 + 00 and 130 + 00.)

A.16. You are asked to design a horizontal curve for a two-lane road. The road has 12-ft lanes. Due to expensive excavation, it is determined that a maximum of 34 ft can be cleared from the road's centerline toward the inside lane to provide for stopping-sight distance. Also, local guidelines dictate a maximum superelevation of 0.08 ft/ft. What is the highest possible design speed for this curve?

A.17. A horizontal curve on a single-lane highway has its *PC* at station 124 + 10 and its *PI* at station 131 + 40. The curve has a superelevation of 0.06 and is designed for 70 mph (desirable). What is the station of the *PT*?

A.18. A horizontal curve is being designed through mountainous terrain for a four-lane road with lanes that are 10 ft wide. The central angle (Δ) is known to be 40 degrees, the tangent distance is 510 ft, and the stationing of the tangent intersection (*PI*) is 2700 + 00. Under specified conditions and vehicle speed, the roadway surface is determined to have a coefficient of side friction of 0.08, and the curve's superelevation is 0.09 ft/ft. What is the stationing of the *PC* and *PT*, and what is the safe vehicle speed?

A.19. A new interstate highway is being built with a design speed of 70 mph. For one of the horizontal curves, the radius (measured to the innermost vehicle path) is tentatively planned as 900 ft. What rate of superelevation is required for this curve at 70-mph design speed?

A.20. A developer is having a single-lane raceway constructed with a 100-mph design speed. A curve on the raceway has a radius of 1000 ft, a central angle of 30 degrees, and *PI* stationing at 1125 + 10. If the design coefficient of side friction is 0.20, determine the superelevation required at the design speed. (Do not ignore the normal component of the centripetal force.) Also, compute the degree of curve, length of curve, and stationing of the *PC* and *PT*.

A.21. A horizontal curve is being designed for a new two-lane highway (12-ft lanes). The *PI* is at station 250 + 50, design speed is 65 mph, and a maximum superelevation of 0.08 ft/ft is to be used. If the central angle of the curve is 35 degrees, design a curve for the highway by computing the radius and stationing of the *PC* and *PT*.

A.22. You are asked to design a horizontal curve with a 40-degree central angle ($\Delta = 40$) for a two-lane road with 10-ft lanes. The design speed is 70 mph and superelevation is limited to 0.06 ft/ft. Give the radius, degree of curvature, and length of curve that you would recommend.

A.23. A freeway exit ramp has a single 12-ft lane and consists entirely of a horizontal curve with a central angle of 90 degrees and a length of 628 ft. If the distance cleared from the centerline for sight distance is 19.4 ft, what design speed was used? (Assume desirable SSD.)

A.24. For the horizontal curve in Problem A.21, what distance must be cleared from the inside edge of the inside lane to provide adequate sight distance for minimum and desirable SSDs?

Appendix **B**

Unit Conversions

LENGTH

1 millimeter (mm)	=	10^{-1} cm
		10^{-3} m
		10^{-6} km
		0.03937 in.
		3.281×10^{-3} ft
		6.214×10^{-7} mi
1 centimeter (cm)	=	10 mm
		10^{-2} m
		10^{-5} km
		0.3937 in.
		3.281×10^{-2} ft
		6.214×10^{-6} mi
1 meter (m)	=	10^3 mm
		100 cm
		10^{-3} km
		39.37 in.
		3.281 ft
		6.214×10^{-4} mi
1 kilometer (km)	=	10^6 mm
		10^5 cm
		1000 m
		3.937×10^4 in.
		3281 ft
		0.6214 mi
1 inch (in.)	=	25.40 mm
		2.540 cm
		2.540×10^{-2} m
		2.540×10^{-5} km
		8.333×10^{-2} ft
		1.578×10^{-5} mi

1 foot (ft)	=	304.8 mm
		30.48 cm
		0.3048 m
		3.048×10^{-4} km
		12 in.
		1.894×10^{-4} mi
1 mile (mi)	=	1.609×10^{6} mm
		1.609×10^{5} cm
		1609 m
		1.609 km
		6.336×10^{4} in.
		5280 ft

AREA

1 square centimeter (cm²)	=	10^{-4} m²
		1.0×10^{-8} ha
		0.1550 in.²
		1.076×10^{-3} ft²
		2.471×10^{-8} acres
1 square meter (m²)	=	10^{4} cm²
		1.0×10^{-4} ha
		1550 in.²
		10.76 ft²
		2.471×10^{-4} acres
1 hectare (ha)	=	1.0×10^{8} cm²
		1.0×10^{4} m²
		1.56×10^{7} in.²
		1.076×10^{5} ft²
		2.471 acres
1 square inch (in.²)	=	6.452 cm²
		6.452×10^{-4} m²
		6.944×10^{-3} ft²
		6.4×10^{-8} ha
		1.59×10^{-7} acres
1 square foot (ft²)	=	929.0 cm²
		9.290×10^{-2} m²
		144 in.²
		9.290×10^{-6} ha
		2.296×10^{-6} acres
1 acre	=	4046.7×10^{4} cm²
		4046.7 m²
		6.27×10^{6} in.²
		43,560 ft²
		0.4046 ha

MASS

1 gram (gm)	=	0.001 kg
		6.852×10^{-5} slug
1 kilogram (kg)	=	1000 gm
		6.852×10^{-2} slug
1 slug	=	1.459×10^4 gm
		14.59 kg

DENSITY

1 slug per cubic foot (slug/ft^3)	=	515.4 kg/m^3
1 kilogram per cubic meter (kg/m^3)	=	1.940×10^{-3} slug/ft^3

SPEED

1 meter per second (m/s)	=	3.6 km/h
		3.281 ft/s
		2.237 mph
1 kilometer per hour (km/h)	=	0.2778 m/s
		0.9113 ft/s
		0.6214 mph
1 foot per second (ft/s)	=	1.097 km/h
		0.3048 m/s
		0.6818 mph
1 mile per hour (mph)	=	1.609 km/h
		0.4470 m/s
		1.467 ft/s

FORCE

1 newton (N)	=	0.2248 lb
		2.248×10^{-4} kip
1 pound (lb)	=	4.448 N
		0.001 kip
1 kip	=	4.448×10^3 N
		1000 lb

PRESSURE

1 kilopascal (kPa)	=	0.145 psi
		0.1 N/cm^2
1 newton per square centi-	=	10 kPa
meter (N/cm^2)		1.45 psi
1 pound per square inch (psi)	=	6.895 kPa
		0.6895 N/cm^2

WORK

1 foot-pound (ft-lb)	=	1.356 Nm
1 newton-meter (Nm)	=	0.7376 ft-lb

POWER

1 kilowatt (kW)	=	737.6 ft-lb/s
		1.341 hp
1 foot-pound per second	=	1.356×10^{-3} kW
(ft-lb/s)		1.818×10^{-3} hp
1 horsepower (hp)	=	550 ft-lb/s
		0.7457 kW

Index